Commercialization of Nanotechnologies–A Case Study Approach

Dermot Brabazon · Eva Pellicer
Fatima Zivic · Jordi Sort
Maria Dolors Baró · Nenad Grujovic
Kwang-Leong Choy
Editors

Commercialization of Nanotechnologies–A Case Study Approach

WIMB

Co-funded by the Tempus Programme of the European Union

Springer

Editors
Dermot Brabazon
School of Mechanical and Manufacturing Engineering
Dublin City University
Dublin
Ireland

Eva Pellicer
Physics Department
Universitat Autònoma de Barcelona
Bellaterra
Spain

Fatima Zivic
Faculty of Engineering
University of Kragujevac
Kragujevac
Serbia

Jordi Sort
ICREA
Physics Department
Universitat Autònoma de Barcelona
Bellaterra
Spain

Maria Dolors Baró
Physics Department
Autonomous University of Barcelona
Bellaterra
Spain

Nenad Grujovic
Faculty of Engineering
University of Kragujevac
Kragujevac
Serbia

Kwang-Leong Choy
Institute for Materials Discovery
University College London
London
UK

ISBN 978-3-319-56978-9 ISBN 978-3-319-56979-6 (eBook)
DOI 10.1007/978-3-319-56979-6

Library of Congress Control Number: 2017938128

Printed on acid-free paper

This Springer imprint is published by Springer Nature
The registered company is Springer International Publishing AG
The registered company address is: Gewerbestrasse 11, 6330 Cham, Switzerland

Preface

This book is the result of contributions from senior researchers in the various areas of nanotechnology which presents a review of the current state of development. Authors possess extensive experience in both theoretical and practical aspects in relevant subjects and have presented insights into both material development and aspects of practical end applications. Nanotechnology as one of the main drivers of products and services still lacks more focused and clear guidance on commercialization. The focus of this book is to indicate challenges and possible opportunities, fostered by discoveries in nanotechnology, towards real market products and services. Numerous novelties have emerged pertaining to both theoretical foundations to further build upon knowledge, and practical applications especially in sectors of energy, advanced manufacturing, biomedicine, environment sustainability and preservation. Along with great enthusiasm for these innovations, issues related to possible dangers of new nanomaterials and their long-term effects have started to be perceived, strongly indicating the need for a more comprehensive approach. Science has made large steps by taking advantage of the discoveries in the nanoworld, but at the same time this new field is at its infancy, facing different controversies especially regarding practical applications in real products.

Areas of nanotechnology commercialization are very diverse and it is not possible to include all of these in one book. However, some areas can be distinguished due to their prominent already proven results. The study of the multi-scale phenomena of the micro- and macroscale effects simultaneously with nanostructures and their mutual influence is a must, in aiming to develop real products. Together with all the parameters and variables important for the proper functioning, green products and environmental preservation are one of the obligatory constraints to consider, especially in nanotechnologies and nanomaterials. Possible toxicity and detrimental environmental effects therefore need to be well characterised and evaluated.

This book covers diverse areas in which nanoscience and nanotechnology have led to significant technological advances and practical applications, with special emphasis on novel types of nanomaterials and their applicability into a new generation of nano- and microdevices. Different nanomaterials are reviewed with a focus on several practical application areas and their commercial utilization. Production technologies of nanomaterials are presented as one of the challenges today—efficient commercial scale production of nanomaterials to satisfy the needs for large quantities. Both the fabrication and storage of nanoparticles impose great challenges and the costs, in general, with scarcely available technologies, which are still too costly and slow for mass production.

The energy sector has shown perhaps the most prominent recent advancements in nanotechnology commercialization, if energy production, storage and power devices are considered altogether. Ferroelectric and piezoelectric nanomaterials as well as nanomaterials for sustainable energy production and storage have already led to different practical applications, and these are reviewed along with basic theoretical foundations. The integration of nanostructured thermoelectric materials in micropower generators based on nanostructured materials is presented. Examples of several developed thin-film thermogenerators are shown, as basic elements for portable devices in microelectronics. A review of existing solutions and developing challenges is given regarding sustainable energy production, photovoltaics, solar cells, hydrogen economy and improved classes of batteries as contributions to green products and circular economy. Novel, highly promising areas in nanotechnology are also shown, such as voltage-driven nanospintronics, from theoretical and future application aspects. Recent advances in information processing nanotechnologies are also described, as well as friction characterisation at the nanolevel.

Health-related products represent another area which has taken great advantage of the nanotechnology revolution, especially in cosmetics, drug delivery systems and improved medical implants. Cosmetic products have incorporated different nanoparticles and gained highly improved effects, such as skin revitalization, and some insights into this area are collected in this book. Novel drug delivery systems have been increasingly developed and modelling of such structures is very important for proper system design and maintenance in the long run, as well as for the determination of influential properties. Methodologies in modelling of self-healing materials with nanocontainers are reviewed as one example in designing these systems that incorporate nanoparticles. Contributions and possibilities shown by nanotechnology in ophthalmology are also reviewed.

Material properties change when going from macro-bulk characteristics down to micro- and nanoscales. Nanoparticles synthesized so far have changed traditional materials in such extent that this area alone shows great further opportunities and completely justifies enormous investments into fabrication, processing, storage, transport and utilization of various classes of new nanomaterials, starting from its basic form—nanoparticles. The health sector is one of the major drivers of economies today with gigantic profits and return on investments. Moreover, the commercialization of research results form nanotechnologies is very strong in this area, with some of the largest investments ever in the history of our world. Several proven nanomaterials have been reviewed in this book pertaining to biomedicine. The use of silver nanoparticles and related nanomaterials in the cosmetic industry, and the potential for iron-based nanomaterials in biomedicine are reviewed, also with recognised challenges and possible threats of non-controlled use of nanomaterials.

Potential applications are great and commercially realised products can exhibit large benefits, but long-term effects still need to studied and considered, especially when it comes to medical products and environmental sustainability. Innovation in practical commercial products, ranging from consumer goods, medical implants, power devices to large civil constructions and heavy loaded structures, in nanotechnologies is confronted with demand for new sustainable green products. We believe that this book will provide young researchers and professionals with valuable insights into opportunities offered by these new technologies, from aspects of different expertise: medical, engineering, physics, chemistry and material science, while emphasizing safety as an important input, to motivate them to pursue further research and practical applications in this broad and multifaceted field of science.

This work is the result of joint efforts of different companies, and academic and research institutions participating in the large international project, WIMB Tempus, 543898-TEMPUS-1-2013-1-ES-TEMPUS-JPHES, "Development of Sustainable Interrelations between Education, Research and Innovation at WBC Universities in Nanotechnologies and Advanced Materials where Innovation Means Business", co-funded by the Tempus Programme of the European Union.

Dublin, Ireland — Dermot Brabazon
Bellaterra, Spain — Eva Pellicer
Kragujevac, Serbia — Fatima Zivic
Bellaterra, Spain — Jordi Sort
Bellaterra, Spain — Maria Dolors Baró
Kragujevac, Serbia — Nenad Grujovic
London, UK — Kwang-Leong Choy
January 2017

Contents

Introduction—The Current Status and Momentum in Nanotechnology Commercialisation 1
Fatima Zivic, Nenad Grujovic, Inam Ul Ahad and Dermot Brabazon

Review of Production Routes of Nanomaterials 15
Shei Sia Su and Isaac Chang

Nanomaterials for Sustainable Energy Production and Storage: Present Day Applications and Possible Developments 31
Igor Hut, Lidija Matija, Milica Peric, Petar Nikolovski and Svetlana Pelemis

Integration of Nanostructured Thermoelectric Materials in Micro Power Generators 73
D. Dávila, A. Tarancón and L. Fonseca

Towards Voltage-Driven Nano-Spintronics: A Review 99
Jin Zhang, Eva Pellicer and Jordi Sort

Ferroelectric and Piezoelectric Nanomaterials—Basic Properties, Characterization and Applications 117
Vesna Paunović, Zoran Prijić and Dragan Antić

Friction at Nanoscale—Self-assembled Monolayers 153
Jelena Manojlović

Modeling of Self-healing Materials with Nanocontainers Using Discrete and Continuum Methods 177
Nenad Filipovic, Dejan Petrovic, Milica Nikolic, Svetlana Jovanovic, Aleksandar Jovanovic and Nebojsa Zdravkovic

Nanomaterials for Skin Care 205
Suzana Miljković, Marija Tomić, Igor Hut and Svetlana Pelemis

Characteristics and Applications of Silver Nanoparticles 227
Fatima Zivic, Nenad Grujovic, Slobodan Mitrovic, Inam Ul Ahad and Dermot Brabazon

Nanotechnology in Ophthalmology 275
Nenad Petrovic, Mirjana Janicijevic Petrovic, Suncica Sreckovic, Svetlana Jovanovic, Dusan Todorovic and Tatjana Sarenac Vulovic

Ever-Expanding Application Potentials for Iron-Based Nanomaterials: Catalyses and Biomedicine 299
Suzana Gotovac Atlagić and Verica Pavlić

Introduction—The Current Status and Momentum in Nanotechnology Commercialisation

Fatima Zivic, Nenad Grujovic, Inam Ul Ahad and Dermot Brabazon

Everybody is talking about nanoscience and nanotechnology. Nowadays 'nano' is very popular in science and technology, as well as in everyday news in journals, magazines and newspapers. If you ask non-scientific people "What is nano?", the majority would probably respond "something small and new"… We can use some of the definitions, starting from Feynman's famous sentence, "There's Plenty of Room at the Bottom" or some officially adopted such as "Nanotechnology is science, engineering, and technology conducted at the nanoscale, which is about 1 to 100 nm.", USA National Nanotechnology Initiative. Today, the term nano has become more than this simple definition. It is probably also becoming a kind of philosophy, as once in ancient Greece, philosophy was the science comprising mathematics, biology and other fields of natural science. Today, again, one discipline is uniting a number of scientific areas into one—nanotechnology. The distinguished feature of nanoscience is its multidisciplinarity at its core and its boosting of cooperation between different scientific fields. Nanoscience is continue to evolve only through the joint efforts from many areas—physics, chemistry, engineering, medicine, biology, mathematics, to name a few. Furthermore, if a nanometer is basic unit in nanoscience, then quantum physics and quantum theories become the foundation. There is no other scientific area that caused so many

F. Zivic (✉) · N. Grujovic
Faculty of Engineering, University of Kragujevac, Kragujevac, Serbia
e-mail: zivic@kg.ac.rs

N. Grujovic
e-mail: gruja@kg.ac.rs

I.U. Ahad · D. Brabazon
Advanced Processing Technology Research Centre, Dublin City University, Dublin 9, Ireland
e-mail: inamul.ahad@dcu.ie

D. Brabazon
e-mail: dermot.brabazon@dcu.ie

D. Brabazon et al. (eds.), *Commercialization of Nanotechnologies–A Case Study Approach*, DOI 10.1007/978-3-319-56979-6_1

controversies, complete acceptances and complete denials, as the appearance of quantum theories in scientific circles. Quantum mechanics is considered revolutionary because it introduced completely new approaches to the view of the world. In 1965, in one of his famous lectures, Richard Feynman said: "I think I can safely say that nobody understands quantum mechanics". Quantum mechanics is one of the foundation concepts for nanoscience. Schrödinger's cat thought experiment underlines the quantum mechanics paradox. Can we now say that we understand nanoscience? The things that in reality were considered absolutely impossible now exist in quantum world. It was a theory for many years, finally to start appearing in a real world with nanotechnologies and nanoparticles. Many unresolved and unexplained phenomena still exist in quantum and nano world, especially if scaled up to the consequences in a real macro world, but it obviously works, whether we fully understand it or not! Perhaps the best description of this new nano world is that by Feynman: "The 'paradox' is only a conflict between reality and your feeling of what reality 'ought to be'".

Material science has become one broad field of rapid discoveries. New materials have started to appear regularly in scientific literature. The pace of novelty is so fast that the experimentalists have difficulty keeping up with all the tests that common traditional practice should involve for proper introduction of the new materials as a real market product. Nanoscience has already shown results which can significantly improve the world in different ways, but at the same time it opened up so many questions and we will all see how the future will unfold.

If the object is observed as energy, then the object disappears, yet in a real world it exists strongly. Even the smallest changes at nano scale can result in dramatic changes in the macro properties. For example, the simple consequence of small changes in the crystal lattice of iron can result in iron becoming a non-magnetic material. For centuries, iron was the synonym for magnetic material. Traditional materials that were not conductive were changed to conduct electricity. Invisible materials were discovered and applied, previously existing only in fairy tales, like invisibility cloaks. Combinations between different material classes became possible, even between those traditionally incompatible to each other, and standard classification of materials to metals, polymers, ceramics, and natural materials started to dim. Combinations between non-organic and organic materials became possible. When materials are combined into new composites, such as metal—ceramic, or metal—polymer composites, or the structure is changed if compared to traditional materials, like in glassy metals or metamaterials, it produces wide possibilities for resulting products improvement or designing of completely new products. High capacity storage systems small in size, fast computers, sustainable energy production, fuel cells, new flexible aircraft parts with significantly increased load bearing and crack resistance, new lighter construction materials with highly improved resistance to earthquakes, new electronic components with superb qualities, biodegradable plastics, textiles that can withhold ultra-high temperatures, light materials that can endure high impact, food containers that enable significantly longer storage of food simultaneously providing protection from heat, light or decay and many other novelties have appeared in the recent years.

Medicine is one of the fields that can benefit greatly from nanotechnology. Micro needles controlled by magnetic field can be used for painless and blood-less surgeries. New highly improved pharmaceutical and cosmetic products are almost regularly appearing. Drug delivery systems are developed to fight against incurable diseases. Implants have been significantly improved towards human-like tissues to be fully accepted by the body and have already improved quality of life of millions of patients around the globe. New optical systems with high resolution enabled significantly improved medical diagnosis and more precise treatments with less side-effects for the patient. They also enabled further studies to understand and cure some incurable diseases, such as in neuroscience. High resolution microscopy and spectroscopy, together with very recent high resolution in situ microscopy, enabled to study the nature in a way that scientists can understand its essential foundations, thus supporting development of yet another area—biomimetics. Biomimetics has transferred structures, units and elements existing in nature onto solving the human problems, such as self-healing property. Surfaces that have extremely high hydrophobic nature are used for self-cleaning glass. The gecko effect—ultra high adhesion by designing structural properties is used for several applications.

Smart materials that can interact with their environment are studied for medical applications but also aiming for many other areas. Intelligent components that integrate different functionalities provide novel concepts in the development of structures. Changes of physical properties (shape, color, viscosity) as the response to external stimulus (temperature, stress, magnetic or electrical field) or changes in structure (crystal lattice orientation) due to external loading are only some of the examples that have been proven. Such materials can repair themself, suppress vibration, or inform the user of internal damage. Bioactive surfaces promote self-regeneration and tissue healing.

New material structures often can be applied in several other areas along with the ones that initiated research, unlike traditional materials that were designed mainly for one or several similar applications. Anti-bacterial surfaces are applied both in medicine and consumer goods, such as textile or bathroom accessories. Silver nanoparticles and ceramic nanoparticles can be used in medical implants, but also in water filtration systems, or cosmetics, or consumer goods. Self-repairing films in medical devices and self-repairing concrete in construction use similar principle. This diversity of applications further promotes collaboration between scientific disciplines.

Nanotechnology has marked our century with numerous astonishing discoveries in almost all branches of science, technologies, and products. The nanotechnology market has emerged and has initiated enormous investments, justified by highly promising scientific results. Numerous national and international, global funding programs are focused on nanotechnologies, from fundamental research aspects, to applied research and towards commercialization. Not a single scientific area has been left untouched by the nano world. Even the social sciences have begun to study the effects of nanotechnologies on the urban world. Ethics became one of the rising questions, especially important because medicine is one of the fields that is strongly effected by the nanotechnology developments in different ways.

The distinctive multidisciplinarity of nanoscience can be seen as its advantage, but sometimes it also becomes its disadvantage. Intellectual property rights were always important but in nanoscience it can be a rather difficult issue, probably due to the involvement of many actors in the discovery. The legal aspects have also emerged to deal with the specifics of commercialization that are unique to nanotechnology.

In general, the basic units of nanoscience are nanoparticles. Nanomaterials can be classified into 2D—nanolayers, 1D—nanopillars, nanowires and nanotubes and 0D—nanoparticles. The history of nanoparticles started with discovery of the C60—fullerene. A carbon molecule can be in a form of hollow sphere, tube, ellipsoid or other shape. Buckminsterfullerenes or 'buckyballs' are the spherical fullerenes discovered in 1985. After the discovery of the first nanoparticles of carbon, a range of other nanoparticles have been synthesized and fabricated and this is still an area of cutting-edge research. Graphene is a carbon nanomaterial with astonishing properties, including the strength of 130 GPa, and a conductivity capability almost as good as superconductors. Perhaps the most utilized carbon nanoparticles are carbon nanotubes (CNTs) because these are the longest studied so far and their addition to the traditional materials showed different improvements in properties, dominantly increase of strength with rather small amounts of CNTs within the structure. Different nanoparticles are studied theoretically and experimentally, but in general the cost of production is high and mass production is still unavailable. The future will show which types of other nanoparticles will be fabricated and the effects they can provide. There is still lack of fully established procedures in production, storage and transportation for many nanoparticles and many nanoparticles are experimentally still out of reach. For example, pure ceramic nanoparticles have been successfully produced only in recent months. In theory, nanomaterials can be based on any existing chemical element, but in reality, only a limited range of nanomaterials have been experimentally fabricated and investigated.

There is a very large collection of literature pertaining to different nanoparticles and related research, as well as various effects that can be obtained by their use. Very good overview of already commercialized nanomaterials is given in nanomaterial consumer products inventory by Vance et al. (2015) with dedicated website: http://www.nanotechproject.org/cpi/browse/nanomaterials/. Some nanoparticles that have been produced and used in commercial products are shown in Table 1, but this list is not exhaustive, as new products are being discovered on regular bases. Furthermore, each nanoparticle can be combined with other nanoparticles to form different hybrid materials. Carbon based nanomaterials represent the largest group of investigated nanomaterials so far, but silver nanoparticles can be noted as already heavily applied in numerous real world applications, including anti-bacterial effects.

There are numerous news reports about the effects promised by nanoparticles utilization, appearing almost every day. For example, seamlessly integrated wearable technology that can inform the user about their health condition, or deliver drugs, such as high tech bra that fights breast cancer, or shoes that can inform about the risk of injury, http://www.dailymail.co.uk/health/article-2987245/The-hi-tech-bra-helps-beat-breast-cancer-clothes-treat-prevent-illness.html.

Table 1 Commercially used nanoparticles (http://www.nanotechproject.org)

Nanoparticles	Applications
Aluminum oxide	Alumina powders used in medical implants, cosmetics, electronics, sport and fitness devices—bicycles, automotive accessories—polishing creams
Calcium	Added in health supplements, food and beverages, cosmetics, dental mouthwash; Research aimed at materials for medical implants
Carbon based (including carbon nanotubes, fullerenes, graphene, graphite)	Numerous different applications: added to improve mechanical properties (e.g. sporting goods: bicycles—tyres, frames; handlebars in fitness devices or tennis and badminton racquets; golf clubs and balls; bowling balls; baseball bats); food and beverage containers; textile and clothing—underwear; sports shoes; cleaning agents; consumer goods (e.g. toothbrushes, safety masks and vests); water filtration systems; air conditioners; aircraft elements; display thin coatings; electronics and computers (RAM memory)
Ceramics	Consumer goods (hairdryers, hair straightener, cooking pans); coatings (cutting tools, durable heavy loaded surfaces); water filters and purifiers; automotive paint finishes; automotive accessories—batteries; coatings in construction elements—flooring; added in civil engineering materials in general aiming to obtain wear resistant, nice looking surfaces; home furnishings; photo paper; cleaning agents
Clay	Beer bottle plastics; cosmetics; outdoor paints; food and beverage containers
Cobalt	Electronics—Metal Nano Dot (MND) memory; food supplements
Copper	Water filtration systems; food supplements; cosmetics; Electronics—Intel® Core™ Duo Processor
Gold	Cosmetics; food supplements; automotive engine oil; consumer goods (toothpaste); clothing; filtration; cleaning agents; air conditioners
Iron	Food supplements; kitchen appliances; food and beverage storage; sporting goods—Sandvik Nanoflex® Alloy
Lead	Cosmetics—Sunscreen
Magnesium	Food supplements; sporting goods (tennis and badminton racquets); cosmetics
Nanocellulose	Food supplements; clothing
Platinum	Cosmetics; food supplements

(continued)

Table 1 (continued)

Nanoparticles	Applications
Polymer	Textiles and clothing; Hybrid Membrane Technology (HMT); ecology coatings; coatings in construction materials—water repellents
Quantum dots	Sensors; Research aimed at solar cells, computing and quantum calculations—quantum computers; optical applications such as high resolution cell imaging or light emitting diodes
Silicon (including Silicon dioxide)	Electronics and computers—Flash memory; 45 nm processors; AMD® Athlon™ 64 Processors; IBM Hard Disk Drive (High Capacity—GMR head); IBM® PowerPC® 970FX/970MP Processors; Intel® Processors; Intel® StrataFlash® Cellular Memory; Sanyo® Organic Electroluminscent (OEL) Displays; Air filtration systems; cosmetics; Siemens® Hearing Instruments; food supplements; sporting goods—wet suits; auto sealants; anti fog products; cleaning agents; protective paints—anti-graffiti paint; flooring Treatment (absorbent surfaces); nano-coatings for: glass, ceramic, microporous surfaces, sand-blasted surfaces, stones, textiles
Silver	Added to numerous different materials to obtain anti-bacterial effects: textiles (wound dressings, cotton sheets, socks, towels), refrigerators, door handles, water taps, food containers, cosmetic products, consumer goods—toothpaste, toothbrushes, hair brushes; filtration systems—air purifiers, vacuum cleaners; food supplements; and many more products
Titanium (including Titanium dioxide)	Sporting goods: tennis and badminton racquets, surf boards, fishing rods; hairdryers; cleaning agents; paints; glues and sealants; filtration systems—air purifiers; cosmetics—protection sunscreens, lip balms, Dove deodorant; Head and Shoulders 2 in 1 Shampoo, Pantene shampoo and conditioner; kitchen appliances—refrigerators; cleaning agents; Self-cleaning glass; Photocatalyst Environment Cleaner
Tungsten disulfide	Lubricants in automotive industry
Zeolite	Food and beverage supplements
Zinc oxide	Protection sunscreens; cosmetics; health supplements; paints; self-cleaning coatings
Zirconia	Research aimed at materials for medical implants

This technology is based on micro-encapsulation of drugs into the fabrics and is developed by Master students.

Another example is the enhancement of the touch sensing by nanomembranes, https://www.newscientist.com/article/dn22162-fingertip-tingle-enhances-a-surgeons-sense-of-touch.

The liquid solution made from ultra-fine conductive nanoparticles that can change the gloves to touchscreen friendly is another good example, https://www.thegrommet.com/nanotips.

There are also numerous new ideas, conceived in promised nano-products, such as idea of a kitchen that can dynamically flex itself as per wishes of the user—appearing and disappearing kitchen fixtures by using nanobots, http://dornob.com/living-kitchen-wall-uses-multi-touch-nanotechnology/.

Another interesting application of nanotechnology in biomedical engineering is highly promising recently commercialized biodegradable generators that can produce energy to power implantable medical devices for some time and then degrade and vanish from the body, http://www.hexapolis.com/2016/03/09/biodegradable-nanogenerators-could-soon-power-medical-implants/.

Some nanomaterials really shows astonishing properties, such as newly developed TiO_2 based nanomaterial that can produce energy, generate hydrogen and purify water, http://inhabitat.com/scientists-develop-wonder-nanomaterial-that-can-produce-energy-clean-water-and-hydrogen/#ixzz2aeIG00dr&i.

This list of news can go on and on, and it clearly shows that commercialization of nanomaterials and in nanoscience, is progressing at fast pace and we shall see what the future will bring. The renowned sources that provide lectures, instructional materials, activities and labs and reference material are listed in Annex 1. However this list is indeed not exhaustive.

One aspect of nanoscience that must not be overlooked is the toxicity that can be induced by nanoparticles. Nano today has also raised very serious questions on the health risks and environmental sustainability. Along with increasing waste and environment pollution in the world, the appearance of nanoparticles with proven toxic effects adds up to environment sustainability concerns. There are many literature sources with investigations pertaining to the toxic aspects of nanoparticles. Based on numerous analyses, particles with nano sizes can be toxic with different pathways of exposure, depending on the organism in question. The majority of studies are related to the consequences of nanoparticle on human health, though there are also significant studies on potential risks to plants and animals, such as cell mutations, soil and water contamination. However, this research area is still in infancy and further research is needed. In general, there is a lack of numerous standards related to nanoparticles in the food sector, medical devices (particularly implantable devices), novel drugs and drug delivery systems which use some form of nano/micro structures, and filtration systems and waste management of nanoparticles released into the soil and water and their further behaviour. It is obvious that detailed studies related to nanotoxicology are necessary but the extent of this is also conditioned by the level of development in the field of nanoparticles fabrication, storage, transport and use.

Annex 1

Useful links with information about nanoscience and nanotechnology

A. Lectures and Instructional Material

1. Trynano-Nanomaterials
   Explore nanomaterials
   http://www.trynano.org/
2. Trynano-Nanotechnology Applications
   Nanotechnology Applications
   http://www.trynano.org/
3. Micro/Nano Processing Technology
   A course to introduce the theory and technology of micro/nano fabrication.
   http://ocw.mit.edu/
4. Multi-Scale System Design
   Multi-scale systems (MuSS) consist of components from two or more length scales (nano, micro, meso, or macro-scales)
   http://ocw.mit.edu/
5. Introduction to Nanoelectronics
   This course covers the "bottom up" approach to electronic devices
   http://ocw.mit.edu/
6. Nano-tribology and Macro-tribology
   This course considers the relationship between nano-tribology and macro-tribology
   http://ocw.mit.edu/
7. Design and Fabrication of Microelectromechanical Devices
   An introduction to microsystem design
   http://ocw.mit.edu/
8. Integrated Microelectronic Devices
   This course examines the physics of microelectronic semiconductor devices for silicon integrated circuit applications
   http://ocw.mit.edu/
9. Nanomechanics of Materials and Biomaterials
   Latest scientific developments and discoveries in the field of nanomechanics
   http://ocw.mit.edu/
10. The Power of Graphene
    Lesson focuses on graphene and its electrical properties and applications.
    http://tryengineering.org/
11. Try Your Hand at Nano
    Lesson focuses on two simple activities younger students can do to gain an appreciation of nanotechnology
    http://tryengineering.org/
12. Be A Scanning Probe Microscope

Lesson focuses on how engineers have developed and use special tools that can observe the landscape of materials when they are working at the nano scale.
http://tryengineering.org/

13. Fizzy Nano Challenge
Students learn about nanotechnology and how engineers can harness the differences in how materials behave when small to solutions for challenges in many industries.
http://tryengineering.org/
14. Sugar Crystal Challenge
Students explore surface area, nanostructures, and work in teams and participate in hands-on activities.
http://tryengineering.org/
15. What is a Nanometer?
Lesson focuses on how to measure at the nano scale and provides students with an understanding of how small a nanometer really is.
http://tryengineering.org/
16. Nano Waterproofing
Lesson focuses on how nanotechnology has impacted the design and engineering of many everyday items, from paint to fabrics.
http://tryengineering.org/
17. Exploring at the Nanoscale
Lesson focuses on how nanotechnology has impacted our society and how engineers have learned to explore the world at the nanoscale.
http://tryengineering.org/
18. NanoSense Size Matters: Introduction to Nanoscience
Lecture notes promote the learning of science concepts that account for nanoscale phenomena
http://nanosense.sri.com/
19. NanoSense Clear Sunscreen: How Light Interacts with Matter
Lecture notes promote the learning of science concepts that account for nanoscale phenomena
http://nanosense.sri.com/
20. NanoSense—Clean Energy: Converting Light to Energy
Lecture notes promote the learning of science concepts that account for nanoscale phenomena
http://nanosense.sri.com/
21. NanoSense—Fine Filters: Filtering Solutions for Clean Water
Lecture notes promote the learning of science concepts that account for nanoscale phenomena
http://nanosense.sri.com/
22. NanoDays
The NanoDays digital kit contains hands-on activities and programs to engage a public audience in nanoscale science, technology, and engineering.
http://nisenet.org

23. NISENet
    Nisenet nanotechnology educational video
    http://nisenet.org
24. NISENET—Programs and Activities
    Nisenet nanotechnology educational activities
    http://nisenet.org
25. NISENET—Media
    Nisenet nanotechnology educational activities
    http://nisenet.org
26. What is Nano
    Nanotechnology educational video
    http://whatisnano.org/
27. Scale of the Universe
    Video lessons attempt to comprehend the scale of the universe
    http://www.khanacademy.org
28. Nanotechnology
    Uses of nanotechnology, and descriptions of nanomaterials and lesson plans for teachers and students
    http://www.understandingnano.com/

B. Activities and Labs

1. Nanohub
   Online hub for nanotechnology resources and activities
   http://www.nanohub.org
2. Nanohub Courses
   Nanotechnology resources and courses
   http://www.nanohub.org
3. Bone Regrowth
   Scientists using nanotechnology to help regenerate nerves and bones
   http://pbskids.org/dragonflytv
4. Hockey Sticks
   Do nanotubes really make a better hockey stick? http://pbskids.org/dragonflytv
5. What's Nano
   Science fair project idea for your elementary or middle school science fair
   http://pbskids.org/dragonflytv
6. Where's Nano
   Science fair project idea for your elementary or middle school science fair
   http://pbskids.org/dragonflytv
7. Nanotechnology—101
   What is nanotechnology and how it works
   http://www.nano.gov/
8. Nanotechnology and You
   Nanotechnology benefits and applications
   http://www.nano.gov/

9. Macro Concerns in a Nano World
   QUEST looks further into nanotechnology, as this rapidly expanding field begins to play a larger part in our lives.
   http://kqed.org
10. Nanotechnology Takes Off
    Discover the nanotech boom where researchers are working to unlock the potential of nanoscience to battle global warming, and disease.
    http://kqed.org
11. Science on the SPOT: Color By Nano
    How Kate uses the phenomenon known as "structural color" to transform nanotechnology into creativity.
    http://kqed.org
12. Solar City: The Future of Nanosolar
    Nanosolar is creating paper-thin solar panels harnessing nanotechnology, a product that could revolutionize solar power.
    http://kqed.org
13. The World's Most Powerful Microscope
    Electron microscope ability to make images to a resolution half the width of a hydrogen atom made it the most powerful microscope in the world.
    http://kqed.org
14. Atomic Force Microscopy
    Videos explaining basic techniques of microscopy
    http://umassk12.net
15. Nanofilm: Oleic Acid Langmuir Film
    Videos explaining basic techniques of nano fabrication
    http://umassk12.net
16. Spin Coating
    Videos explaining basic techniques of nano fabrication
    http://umassk12.net
17. Photolithography
    Videos explaining basic techniques of nano fabrication
    http://umassk12.net
18. Electrochemical Deposition
    Videos explaining basic techniques of nano fabrication
    http://umassk12.net
19. Magnetism
    Videos explaining basic techniques of a section on concepts of magnetism
    http://umassk12.net
20. What's so BIG About Nano-technology?
    Nano related activities
    http://www.sciencebuzz.org
21. NNIN Education and Training
    NNIN's education programs address the explosive growth of nanotechnology and its growing need for a skilled workforce and informed public by offering education and training to individuals (school-aged students to Higher educations).

http://www.nnin.org

22. Nanotechnology Poster
    A series of posters with accompanying educator and learner guides for Public engagement with Nanotechnology
    http://www.jivemedia.co.za
23. Nanozone
    Nano related activities
    http://www.nanozone.org
24. A Hands-On Introduction to Nanoscience and Technology
    A lab-based hands-on introduction to nanoscience and nanotechnology for early undergraduates
    http://www.virlab.virginia.edu/

C. Reference Material

1. Secret Worlds: The Universe Within
   Video Milky Way to Nano scale illustration
   http://micro.magnet.fsu.edu/primer/java/electronmicroscopy/magnify1/index.
2. Tejal Desai
   Tejal Desai is a biomedical engineer who designs tiny, nano-sized capsules to transport medicine in the body to the exact spot it is needed
   http://pbskids.org/dragonflytv
3. UVA Virtual Lab
   The UVA Virtual Lab is based at the University of Virginia. It employs emerging software visualization tools to explain technologies affecting our daily lives
   http://www.virlab.virginia.edu/
4. Graphene the 2D Material That Could Change Everything
   Graphene the 2D material that could change everything infograph
   http://www.visualcapitalist.com/
5. Nano-to-Macro Transport Processes
   Video overview of the course and the research in the field of nanoscience and technology
   http://ocw.mit.edu/
6. Submicrometer and Nanometer Technology
   A course that surveys techniques to fabricate and analyze submicron and nanometer structures, with applications
   http://ocw.mit.edu/
7. Nanotechnologies: Principles, Applications, Implications and Hands-On Activities. A compendium for educators
   A publication by the European Commission outlines nanoscience and nanotechnology. Educational material that will help inform, motivate and inspire young people about nanotechnologies and their applications
   http://ec.europa.eu/research/industrial_technologies/e-library.cfm

8. Nanotechnology: The Invisible Giant Tackling Europe's Future Challenges
   A publication by the European Commission outlines nanoscience and nanotechnology
   http://ec.europa.eu/research/industrial_technologies/e-library.cfm
9. Nanometrics—A Technometric and Socio-Economic Analysis System to Support the Development of the European Nanotechnology Strategy Options
   A publication by the European Commission outlines nanoscience and nanotechnology
   http://ec.europa.eu/research/industrial_technologies/e-library.cfm
10. Nanometrics—A Technometric and Socio-Economic Analysis System to Support the Development of the European Nanotechnology Strategy Options—Case Study
    A publication by the European Commission outlines nanoscience and nanotechnology
    http://ec.europa.eu/research/industrial_technologies/e-library.cfm
11. Reaching Out to the Future
    A publication by the European Commission outlines nanoscience and nanotechnology
    http://ec.europa.eu/research/industrial_technologies/e-library.cfm
12. Photovoltaics and Nanotechnology: From Innovation to Industry
    A publication by the European Commission outlines nanoscience and nanotechnology
    http://ec.europa.eu/research/industrial_technologies/e-library.cfm
13. Successful European Nanotechnology Research
    A publication by the European Commission outlines nanoscience and nanotechnology
    http://ec.europa.eu/research/industrial_technologies/e-library.cfm
14. Communicating Nanotechnology
    A publication by the European Commission outlines nanoscience and nanotechnology
    http://ec.europa.eu/research/industrial_technologies/e-library.cfm
15. Knowledge, Attitudes and Opinions on Nanotechnology Across European Youth
    A publication by the European Commission outlines nanoscience and nanotechnology
    http://ec.europa.eu/research/industrial_technologies/e-library.cfm
16. Report on Legal, Ethical Issues Related to Nanotechnologies
    A publication by the European Commission outlines nanoscience and nanotechnology
    http://ec.europa.eu/research/industrial_technologies/e-library.cfm
17. Nanomedicine the Future of Medicine
    How Nanomedicine Works and applications
    http://www.associates-degree-in-nursing.org/

18. Fabricating a Tiny High-Tech Future
How advanced manufacturing drives the tiny, high-tech future
http://www.jabil.com/blog/fabricating-a-tiny-high-tech-future-infographic.
19. Big Things from a Tiny World
Brochure that explains nanotechnology and its potential in a format that appeals to general audiences
http://www.nano.gov/
20. Nano and Energy: Powerful Things from a Tiny World
Brochure provides an overview of nanotechnology's potential applications in the energy sector, including information on batteries, catalysts, solar cells, and green fuels
http://www.nano.gov/
21. An Introduction to Nanotechnology
An Introduction to Nanotechnology and its applications
http://www.understandingnano.com/
22. Nanotechnology Applications
Nanotechnology applications
http://www.understandingnano.com/
23. Nanoreisen
Nano-journeys to micro- and nano-cosmos
http://www.nanoreisen.de/
24. Nanokids
NanoKids(TM) is an education and outreach program for intermediate-level (middle-school) students intended to increase their knowledge of nanotechnology and emerging research and technology in this field,
http://www.nanokids.rice.edu/
25. NanoTech: Insights into a Nano-Sized World
Students learn about nanotechnology, its extreme smallness, and its vast and growing applications
http://www.worldteachengineering.org.

Reference

Vance ME, Kuiken T, Vejerano EP et al (2015) Nanotechnology in the real world: redeveloping the nanomaterial consumer products inventory. Beilstein J Nanotechnol 6:1769–1780

Review of Production Routes of Nanomaterials

Shei Sia Su and Isaac Chang

Abstract Manufacturing and commercialization of nanomaterials are often hampered due to lack of appropriate large scale production of these nanomaterials. In this review article, we discuss different types of commercial scale production, including vapour, liquid and solid phase synthesis. The choice of production methods will influence the nanomaterials' properties and functionalities, scalability and production costs. This review emphasized on the fundamental process of each production method, including merits and drawbacks.

1 Introduction

Nanomaterials can be classified as materials with the size less than ~100 nm diameter and exhibit unique properties due to their nanoscale dimensions. With the development of new techniques and tools to synthesis and manipulate the size, shape and structure of nanomaterials, nanomaterials have been utilised in numerous technological applications, including catalytic systems, hydrogen storage, chemical sensors, biomedical applications, electronic, etc.

Although there is an extraordinary growth of using nanomaterials in the consumer end-products, many unprecedented materials innovations are still remain on the laboratory bench. One of the main challenges faced in commercialising the application of these nanomaterials is lack of appropriate large scale production of these nanoparticles. This chapter will discuss different types of commercial scale

S.S. Su
School of Metallurgy and Materials, College of Engineering and Physical Sciences, University of Birmingham, Edgbaston, Birmingham B15 2TT, UK
e-mail: sheisiasu@gmail.com

I. Chang (✉)
Brunel Centre for Advanced Solidification Technology (BCAST), Institute of Materials and Manufacturing, Brunel University London, Kingston Lane, Uxbridge UB8 3PH, UK
e-mail: Isaac.Chang@brunel.ac.uk

D. Brabazon et al. (eds.), *Commercialization of Nanotechnologies–A Case Study Approach*, DOI 10.1007/978-3-319-56979-6_2

production, with the focus on the inorganic nanoparticles and carbon-related nanoparticles.

There are two main approaches to fabricate nanomaterials, namely top down and bottom up approaches. Figure 1 showed the schematic illustration of top down and bottom up approaches. The top down approach involved decreasing the size of the large uniform piece of material in order to create the required nanostructures from it. This method is very much used in the electronic industry, whereby lithography and etching techniques are applied to create the unique electronic circuitry. Mechanical milling is another common top down approach to generate nanoparticles. However, it poses some difficulties to generate uniform nanoparticles and is more likely to introduce internal stress, structural defects and contaminations. The bottom up approach, on the other hand, arranges the basic building blocks such as atoms or molecules into large nanostructures using chemical or physical forces.

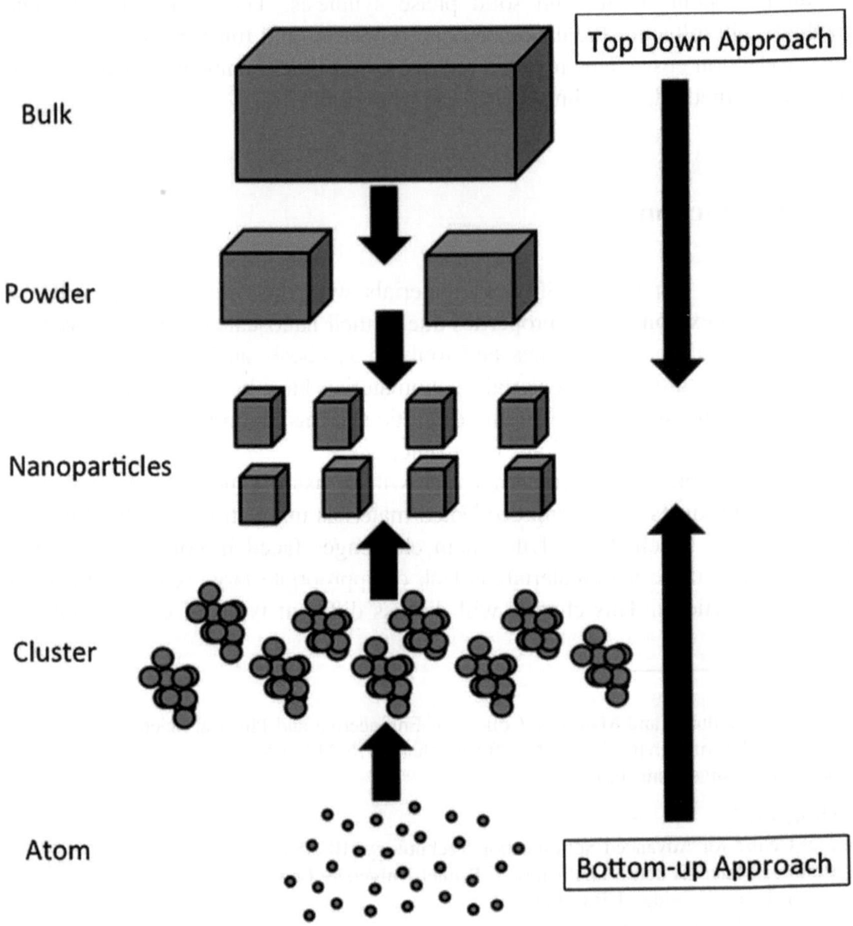

Fig. 1 Top down and bottom up approaches of nanoparticles production

The growth and self-assembly of the building blocks allows designing and controlling nanoparticles with more precise structures, shape, size and chemical composition.

This chapter will discuss the pros and cons of different production methods of creating nanomaterials. Scientists and engineers have to weigh between the needs for quantity or quality of the nanoparticles before they make their decision on choosing the right production methods to produce the nanoparticle with the desired properties.

2 Vapour Phase Synthesis

The works on vapour phase synthesis have evolved considerably since the early 1960s. Since then, there has been numerous of reports and references detailing the works on direct gas to particles conversion synthesis methods. Formation of nanoparticles through the vapour phase synthesis generally undergoes three main processes. It first started with nucleation of clusters from the supersaturated vapours, either homogenously or heterogeneously. Homogeneous nucleation occurs when the vapour molecules condense to form nuclei without the presence of foreign particles or ions. Heterogeneous nucleation, on the other hand, occurs on the foreign nuclei, ions or surfaces. The nanocrystal nuclei condense from the constituent vapour later on undergoes growth due to collision and coalescence. Finally, the primary particles coagulate and form aggregates as the temperature decreases.

A number of different vapour phase methods of synthesis have been developed to increase the production of nanoparticles. This section will discuss the vapour synthesis methods including flame synthesis, chemical vapour condensation, arc discharge and laser ablation in more details.

2.1 Flame Synthesis

Flame synthesis is an inexpensive and widely used method for commercial production of nanoparticles. The most common examples of nanoparticles produced using the flame synthesis are titania, fumed silica and carbon blacks (Swihart 2003; Stamatakis et al. 1991), with the annual production rate reaching several million metrics tons. These nanoparticles are used extensively as pigments, catalysts, flowing aids and in telecommunication.

In flame synthesis, synthesis of particles is carried out within the flame, produced by the combustion reactions. Flame heat is employed to initiate the chemical reactions and to form flames of dust clouds comprising the corresponding metals. The volatile precursors are oxidized within the flame to form metal-oxide monomers, which later nucleate and coalescence to form aggregated nanoparticles. Unlike the wet chemistry synthesis route, the desired crystal phases are formed directly without the need for a calcination step due to the high temperature of the flame. Figure 2 shows a typical flame synthesis unit, which consists of a precursor unit, a burner incorporated with the gas delivery systems, and a filter unit for particle collection.

The main fuels used in flame synthesis are methane, hydrogen or ethylene and air or oxygen as oxidants. These gases sometimes are diluted with inert gases such as argon, helium or nitrogen. An example of the flame synthesis is synthesis of iron oxide nanoparticles from iron pentacarbonyl precursor using $CH_4/O_2/Ar$ flame, as reported by Buyukhatipoglu and Morss Clyne (2010). It was found that change of flame configuration from diffusion flame configuration to inverse diffusion flame configuration reduces the size range of the nanoparticles from 50–60 to 6–12 nm.

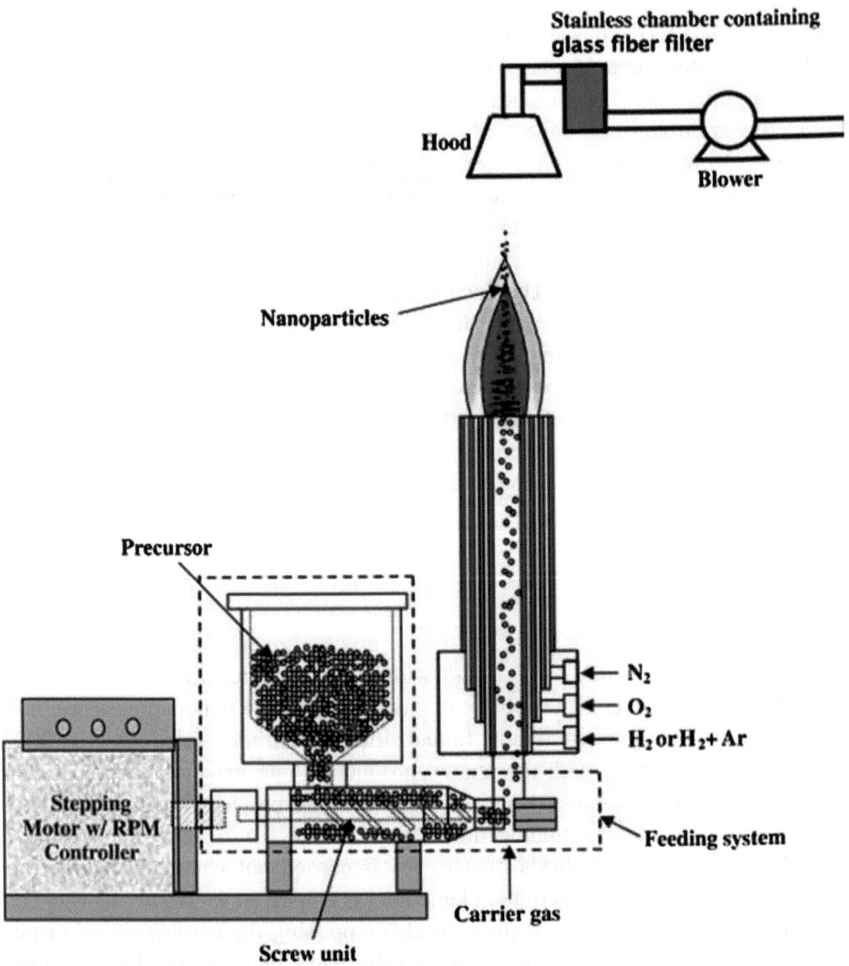

Fig. 2 Typical set up for flame synthesis (Yang et al. 2010). Reprinted from Powder Technology, 2010. 197(3), Yang, S. et al., A flame metal combustion method for production of nanoparticles, pp. 170–176. Copyright 2010, with permission from Elsevier

The addition of N_2 gas cooling further lowers the flame temperature and leads to the formation of magnetite Fe_3O_4 instead of hematite α-Fe_2O_3 (Buyukhatipoglu and Morss Clyne 2010). The main parameters influencing the flame synthesis of iron oxide are the flame temperature and particle residence time. The characteristics of the flame-made nanoparticles are also controlled by the composition of the precursors, burner configuration, turbulence, precursor injection and particle collection location, supplementary laser irradiation and external electric fields (Buyukhatipoglu and Morss Clyne 2010; Yang et al. 2010).

Apart from the simplicity of experimental set up, flame synthesis involves only a 'one-step' process. A wide array of precursors can be employed using the flame synthesis. Thus, a broad range of the nanosized powders can be synthesized using this method. Johannessen et al. also reported other complex products including composite metal oxide (e.g. $ZnAl_2O_4$, $MgAl_2O_4$ spinel) and supported noble catalysts are synthesized using the flame synthesis (Johannessen et al. 2004). The synthesized products require no subsequent post-processing and less process waste is obtained as flame synthesis is a solvent free process. In spite of simplicity and efficiency of flame synthesis of nanoparticles, agglomeration of nanoparticles remains a significant challenge for this synthesis method.

2.2 Laser Ablation Method

Since the discovery of laser decades ago, laser has been used extensively in various applications including nanomaterials synthesis, surface cleaning and surface nanopatterning (Zhong et al. 2012). Synthesis of nanoparticles using laser ablation involves vaporization of source materials or precursor through laser irradiation. A wide variety of materials including high melting point elements or compounds can be produced at low temperature, including metals such as silver (Tsuji et al. 2008), silicon (Khang and Lee 2010), ceramics, oxide composites and nitrides (Kononenko et al. 1997). Figure 3 displays an illustration of laser ablation systems connected with particles analyser.

The most common source laser source for ablation is neodymium-doped yttrium aluminium garnet (Nd:YAG). During laser ablation, the laser beam is irradiated onto the surface of the target materials at specific incident angle, either 90 or 45 degree. The energy of the laser beam can be transferred to the target materials through different pathways depending on the laser power, wavelength and pulse duration. Photothermal ablation will take place if low laser power (100–500 MW/cm^2) and longer pulse length (>10 ns) are used (Wang et al. 2002). During the photothermal ablation, the free or bound electrons on the target materials will absorb the energy. The excited electrons transfer the energy through the collision interaction with other electrons and lattice phonons, resulting in heat transfer and increase of the temperature of the material, which melts the target material (Zhong et al. 2012). The melted material continues to absorb the energy from the laser beam and subsequently forms the plasma plumes on the surface. The atoms on

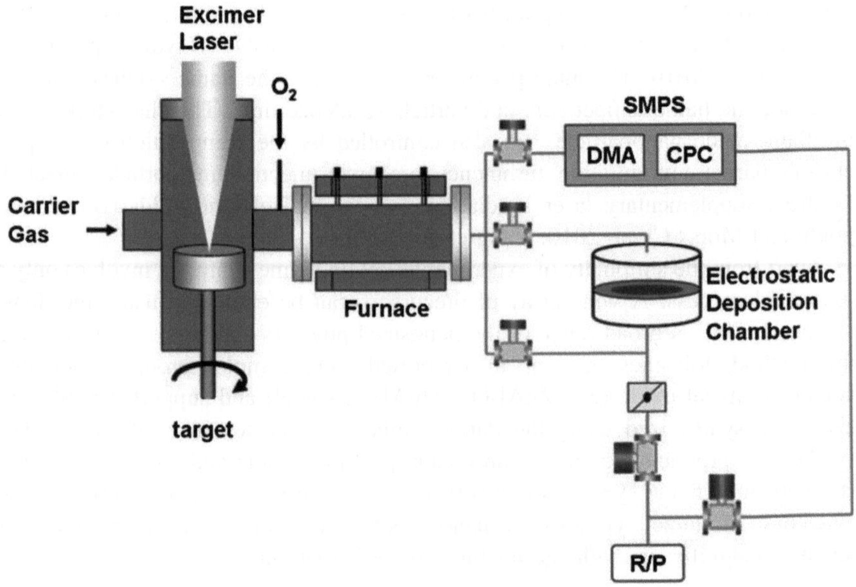

Fig. 3 Schematic illustration of laser ablation systems connected with particles analyser (Khang and Lee 2010). Reprinted from Journal of Nanoparticle Research, 2010. 12(4), Khang, Y. and Lee, J., Synthesis of Si nanoparticles with narrow size distribution by pulsed laser ablation, pp. 1349–1354. Copyright 2009, with permission from Springer

the heated surface are finally ejected and vaporized due to the rapid volume expansion.

Plasma assisted ablation is the second pathway, in which temporary plasma is produced above the target material when high peak power in the laser pulse is used. Generally, high laser power (>500 MW/cm^2) is applied to provide high-energy ions, which bombard the surface of the target materials. The atoms of the target materials are then heated and ejected from the surface. Due to the high energy of the laser ablation, both photothermal and plasma assisted ablation will leave crates on the surface of the target material. At the same time, the liquid droplets of the melted materials solidified and create a rim. Photochemical or photolytic ablation, on the other hand, occurs when laser photon energy similar to the bonding energy of the molecules is absorbed and lead to dissociation of the bond. This later causes the ejection of the material. Unlike photothermal and plasma assisted ablation, only the target material irradiated by the laser beam is ablated and the remaining area remains unaffected for the photochemical ablation Wang et al. (2002).

Laser ablation technique is applicable to synthesis a wide range of materials with the complex chemical structure and stoichiometry composition similar to the target. The properties of nanoparticles synthesized using this method can be easily controlled by altering the laser pulse energy, pulse fluence, repetition rate, pulse duration and laser spot size. Due to the monochromatic characteristics of laser, selective excitation of molecules will occur by applying the appropriate laser wavelength. Apart from having the ablation carries out in vacuum and gas, there are also numerous studies reported to have laser ablation in the liquid medium (Tsuji et al. 2008; Neddersen et al. 1993).

In the last decade, Pulsed Laser Ablation in Liquid (PLAL) has emerged as an alternative synthesis technique to generate, excite, fragment, and conjugate a large variety of ultrapure nanostructures for large scale production (Intartaglia et al. 2014). This technique has overcome most of the shortcomings of traditional wet chemical synthesis, presenting a simple, straightforward and versatile technique to produce ligand-free nanoparticles which are a prerequisite for specific biological or electrical applications (Barcikowski and Compagnini 2013). In addition, the liquid phase can be tailored in order to produce viscosities suitable for ink jet printing or other conventional controlled fluid surface coating techniques (Bagga et al. 2015) Ablation in liquid medium offers a number of advantages including higher cooling rate which tends to form smaller size nanoparticles, allows surface modification, and increase stability of nanoparticles. While the liquid can be as benign as deionised water, other solvent species used could affect the chemical composition of the nanoaprticles synthesized in a positive or negative manner depending on application.

2.3 Arc Discharge Method

Fabrication of nanoparticles using arc discharge method has gained much interests due to its simplicity in apparatus set up and ability to scale up the production rate of nanoparticles (Ashkarran et al. 2009). Different types of nanoparticles have been reported to synthesize using this method, ranging from metal to refractory metals. One of the most widely reported nanoparticles synthesized using this method is carbon nanotubes (CNTs) using the direct current (DC) arc discharge. For the production of single walled carbon nanotubes (SWNTs), catalyst metals for instance Ni, Y, Fe, Co or Mo are required, whereas it is not required for the multi-walled carbon nanotubes (MWNTs).

Arc discharge involves very simple setups with two electrodes acting as cathode and anode respectively. These electrodes are subjected to a potential and subsequently cause an electrical breakdown. Arc discharge generally requires low voltage (few to several tens of V) and high current density (10^5–10^{11} A/m^2) (Xu 2007). Synthesis of nanoparticles using arc discharge method can be either in pulsed (Su et al. 2014) or continuous mode. To produce MWNTs and SWNTs, high purity graphite are used as electrodes and arc discharge can be carried out in helium (Waldorff et al. 2004) or hydrogen gas (Zhao et al. 1997). Figure 4 showed a schematic illustration of the experimental set up for the arc discharge.

During the arc discharge, electrons are emitted either through heating or field emission due to large electric field generated. Plasma is thus formed due to the ionization and this exerts large amount of heat, which subsequently evaporate the

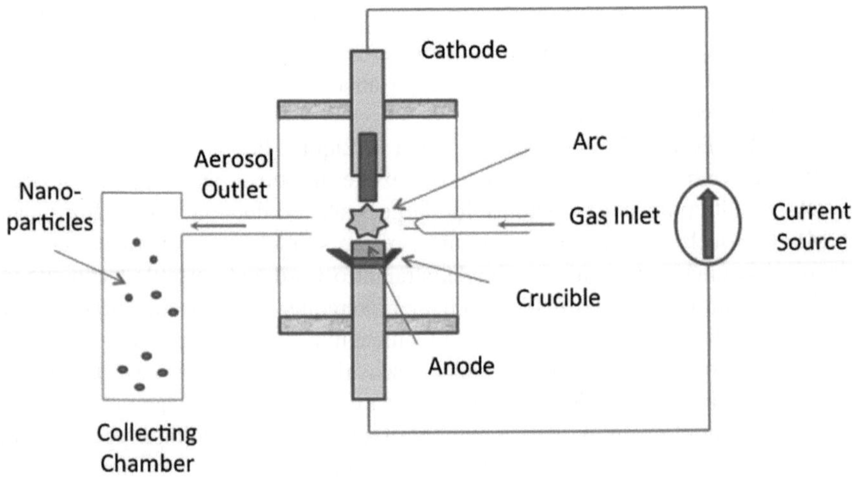

Fig. 4 Schematic illustration of the experimental set up for arc discharge in gas chamber

metal of interest into metal vapor. The plasma also enhances the kinetics of the chemical reactions, dissociating atoms and radicals. The metal vapor later nucleates and forms nanoparticles in the surrounding cooling medium (Ashkarran et al. 2008). The metal vapour can either be quenched by mixing with cold gas or by the surrounding liquid medium.

Su et al. has reported synthesis of ZnO nanoparticles with the size varies from 10 to 25 nm using the pulsed arc discharge in deionized water (Su et al. 2014). Compare with the arc discharge in vacuum, arc discharge in deionized water is more economical, does not involve the use of organic solvent and chemicals and so has potential for the environmentally friendly and biocompatible production of nanoparticles. At the same time, it also avoids the use of toxic, explosive and corrosive gases and reduces the use of costly vacuum equipment. In most cases, nanoparticles synthesized using the arc discharge method are subjected to high cooling rates. Due to the non-uniform cooling, the uniformity of the nanoparticles produced using this method often deteriorates. Thus, control of particle nucleation and growth continuously require much attention.

3 Liquid Phase Synthesis

Liquid phase synthesis or wet chemical method is another important nanoparticles synthesis approaches widely used by the industries. A typical liquid phase synthesis involves mixing solution of different ions with well-defined quantities and subjected it to controlled heat, temperature and pressure to promote the formation of insoluble nanoparticles, which precipitates out of the solution. Unlike the vapour phase synthesis, liquid phase synthesis is more energy efficient and generally carried out at low temperatures and below the atmospheric pressure. It does not require high temperature, expensive energy source or high pressure. Besides, liquid phase synthesis also enable more precise control of the stoichiometry composition of the end products and the nanoparticles synthesized can be stabilized or functionalized during the synthesis process.

Although liquid phase synthesis offers many advantages, it still faces some challenges for their industrial applicability and scalability. Very often, the chemicals used in the liquid phase synthesis are toxic, explosive and corrosion, which pose a significant risk to occupational health and environments. Furthermore, nanoparticles synthesized require further processing and purification. The purity of the nanoparticles synthesis using liquid phase synthesis is usually less than those synthesized using the vapour phase synthesis.

3.1 Chemical Precipitation

Chemical precipitation is one of the most widely used liquid phase synthesis method to synthesize complex oxides. The main advantages of this method are due to its simplicity and ability to yield final products of near perfect stoichiometry without high temperature treatment. Figure 5 summarized an overall process involves in the chemical precipitation.

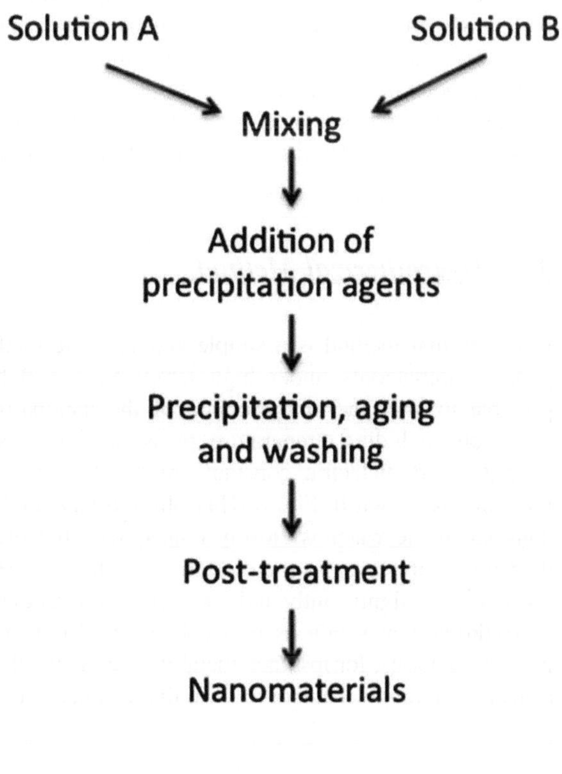

Fig. 5 A summary of the procedures for chemical precipitation of nanoparticles (Zhong et al. 2012)

It first starts with dissolving the starting materials, usually metal oxides, metal salts or their hydrates to form aqueous solution. The solution later on mixed with another solution with dissolved precipitation agent either directly or by drop-by-drop method. Meanwhile, co-precipitation will take place when the reactant products dissolve in the solvent and precipitation agent is added to solution to precipitate the reaction products. The precipitates later undergo aging process to obtain larger particles and these particles are collected through filtration or centrifugation. Further washing with distilled water, ethanol or other solvents is required to clean the impurities and to obtain high purity nanoparticles. Post treatment, including annealing, sintering or calcination is carried out to obtain nanoparticles with desired crystal structures and morphology.

Chemical precipitation involves simultaneous occurrence of nucleation, growth, coarsening and/or agglomeration. However, the fundamental mechanism of chemical precipitation is not fully understood due to the difficulties in isolating each process for independent studies (Cushing et al. 2004). The critical step to precipitation is nucleation, which is governed by the degree of supersaturation. A large number of small crystallites are initially formed and subsequently grow through secondary processes, for instance Ostwald ripening and aggregation to form more stable particles. Control of the nucleation and growth processes is necessary, as it will affect the characteristic of the nanoparticles synthesized. The main parameters influencing the as-synthesized nanoparticles are surface charge or zeta potential, reactant concentration, reaction temperature, pH, degree of mixing and recrystallization of the precipitates. By controlling these parameters, the particles size, particle size distribution and stoichiometry of the nanoparticles can be manipulated. The drawbacks of this method are it requires the use of a lot of chemicals and may generate potentially hazardous waste products. Besides, most of the nanoparticles synthesized using chemical precipitation requires stabilizers and a number of post-processing treatments are required to improve the purity of the end products.

3.2 *Hydrothermal Method*

Hydrothermal method is a simple and versatile method for the synthesis of inorganic nanomaterials under high temperature and high-pressure condition. It is possible to alter the characteristic of the nanoparticles by adjusting processing conditions including temperature, pressure or precursor concentration. The density of water and dielectric constant are highly dependent on the temperature and pressure, as shown in Fig. 6 (Hayashi and Hakuta 2010). A drop in the dielectric constant of the water is closely related with the increase of the temperature and decrease in pressure. As the dielectric constant of the water decreases, the reaction rate enhances significantly and ionic species starts to precipitate to form solid phase.

Hydrothermal synthesis is usually carried out in a closed vessel and the precursors solutions, for instance metal nitrate, metal chloride or acetates undergo two steps formation mechanism to form the nanoparticles, as shown in Eqs. (1) and (2).

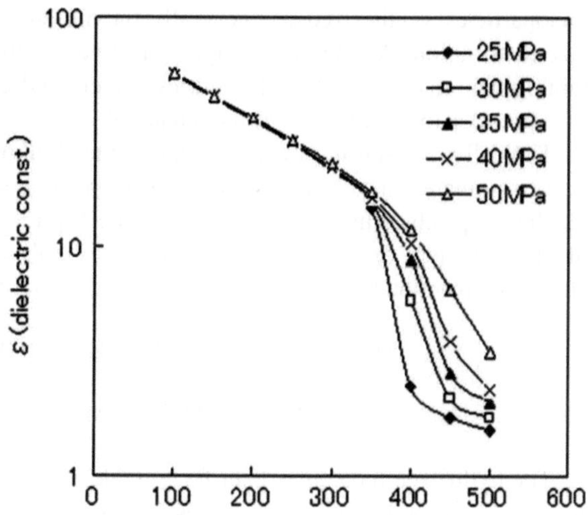

Fig. 6 Dielectric constant of water at various temperatures and pressures (Hayashi and Hakuta 2010). Reprinted from Materials, 2010. 3(7), Hayashi, H. and Hakuta, Y., Hydrothermal Synthesis of Metal Oxide Nanoparticles in Supercritical Water, pp. 3794–3817. CC BY-NC-SA 3.0 licence

M denotes the metal and X the anions. The hydrated metal ions are initially hydrolysed to form metal hydroxide. The metal hydroxides later undergo dehydration to form metal oxide (Adschiri et al. 1992).

$$\textbf{Hydrolysis:}\quad MX_x(aq) + xH_2O\,(l) \rightarrow M\,(OH)_x(s) + xHX\,(aq) \tag{1}$$

$$\textbf{Dehydration:}\quad M\,(OH)_x(s) \rightarrow MO_{x/2}(s) + x/2H_2O\,(l) \tag{2}$$

The hydrothermal synthesis method has been used extensively to fabricate TiO_2, which is an important photocatalyst for the decomposition of toxic chemicals. Mi et al. has reported a fast and one step synthesis of rutile TiO_2 nanoparticles using isopropanol-water as solvent at temperature of 300 °C and pressure of 25 MPa (Mi et al. 2012). TiO_2 with the average size of 22 nm was synthesized at the supercritical condition, which allow precipitation of nanoparticles instantaneously as the temperature of the metal salt brought very rapidly above the critical point. Hydrothermal method is easily scalable with the development of the continuous flow hydrothermal reactors. Furthermore, it can be used to synthesize complex inorganic nanoparticles by altering the synthesis parameter. However, the disadvantage of this method is there is a need to find proper chemical reactions and processing conditions for different nanoparticles.

3.3 *Sol-Gel Method*

Similar with the chemical precipitation and hydrothermal method, sol gel synthesis takes place in an inexpensive reactors, allow precise stoichiometry control and the nanoparticles synthesized are generally having smaller particles size and narrow size distribution. As the name implies, it involves the formation of a sol-gel containing nanomaterials which later dries, sinters and calcinates at specific temperature in order to obtain desired nanomaterial.

Typically, the starting materials can be metal alkoxides or metal chlorides used to form solvated metal precursor (sol). The precursors are hydrolysed with water or alcohol to produce hydroxide, following the Eq. (3) below. Apart from water and alcohol, an acid or base can also help to hydrolyse the precursor. Equation (4) shows the reaction takes place between alkoxide and the acid (Tavakoli et al. 2007)

$$M\,(OR)_n + xH_2O \leftrightarrow M\,(OR)_{n-x}(OH)_x + xROH \tag{3}$$

$$-M{-}OR + AOH \rightarrow M{-}O{-}A + ROH \tag{4}$$

The hydroxide molecules later form oxide- or alcohol bridged network (the gel) through polycondensation or polyesterifcation reactions. At this stage, gelation occurs and a dense gel consisting of three-dimensional skeleton with interconnected pores is obtained. The properties of the final products are largely influenced by the rates of hydrolysis and condensation. Slower and more controlled hydrolysis will yield smaller particles. The rate of hydrolysis is dependent on the electronegativity of metal atoms and its coordination number. More electronegative metal tends to have faster hydrolysis rate. Meanwhile, the hydrolysis rate reduces as the coordination number of the metal increase due to the steric hindrance (Srivastava 2013).

After the solution condensation to a gel, it will follow by aging of the gel, during which polycondensation continues until the gel form a solid mass. At the same time, the solvent is removed from the gel pores and accompanied by contraction of gel network. The gel is subsequently calcined at high temperature. This results in densification and decomposition of the gel, whereby the pores of the gel network collapsed and remaining organic species are volatilized (Cushing et al. 2004). The thermal treatment is crucial to obtain nanoparticles with desired crystal structure, crystallinity and size.

Sol gel method has been used to synthesize numerous of metal oxide nanoparticles and nanocomposites. Niznansky has reported synthesis of Fe_2O_3/SiO_2 nanocomposites by one-step sol-gel method. Commercial precursors (TEOS and iron (III) nitrate) were dissolved in the methanol solution. After gelation and drying, the transparent nanocomposites with the size 2–10 nm was obtained, depending on the Fe concentration and heating temperature (Niznansky et al. 1997). Despite the advantages of nanoparticles synthesis using sol-gel method, there are also some drawbacks on using this method including high cost of raw materials, presence of residual carbon, health hazard of organic compounds and possible cation segregation during thermal decomposition (Wang et al. 2002).

4 Solid Phase Synthesis

Compare with vapour phase and liquid phase synthesis, solid phase synthesis is the oldest techniques to synthesize nanomaterials. Solid phase synthesis involves physically pounding coarse powders into finer powders. Most of the solid phase synthesis is carried out using planetary or rotating ball mills. The advantages of these techniques are they are simple and require low cost equipment. However, large particle size, broad size distribution and contaminations often pose a major challenge for this type of synthesis.

4.1 Mechanical Milling

Mechanical milling refers to mechanical disintegration of the solid metals into smaller particles after subjected to the high-energy collision from the balls. This is a common process used for the production of hard metals, oxide powder, intermetallics, ceramics and composites. Mechanical milling is largely applied for particle size reduction, solid state alloying, mixing or blending and particle shape changes.

Grinding process is carried out, by rotating the jar filled with hard balls and powdered metals to be milled at high speed. Collision between the balls and the metals later crushed it into powders with finer size. Apart from causing fractures, milling operation also deforms and cold-welds the impacted particles. As a result of cold-working during the mechanical milling, the average grain size can be reduced from the range of 50–100 μm to the range of 2–20 nm. Milling can be performed in either wet or dry conditions. To reduce oxidation on the freshly created particle surfaces, mechanical milling is preferably conducted in protective atmosphere.

Mechanical milling is relatively simple operation with the ease of scalability and ability to form nanostructured powders with different structures (nanocrystalline, amorphous, quasicrystalline) and atomic bonding (metal/metal, metal/semiconductor, metal/ceramic, etc.) (Wang et al. 2002). However, the major drawback of mechanical milling is non-uniformity of particles sizes. Although there has been significant improvement in the milling instrument and quality of grinding media, nanoparticles produced using mechanical milling generally has large particles size distribution. Furthermore, mechanical milling also have high tendency to introduce contamination from the milling media and atmosphere. Milling products are very sensitive to the grinding conditions. Very often, surface-active grinding aids are used to reduce surface energy of particles and enable shorter grinding time. To further reduce the particle size, longer milling time is required (Schatt and Wieters 1997). This subsequently introduced more impurities into the powder particles. It is always problematic to remove these impurities and grinding aids.

5 Summary

Vapour phase synthesis generally is a clean process and able to yield nanoparticles with high purity. However, the nanoparticles synthesized using this method are mainly polycrystalline in nature and it also suffers from the agglomeration problem which can annihilates the desired nanoparticles properties. Liquid phase synthesis or wet chemical method, on the other hand, allows precise control of the stoichiometry composition and it is relatively easy to scale-up as it does not require high temperature and high-pressure condition. However, purity of the nanoparticles synthesized remains a major issue as stabilizers or capping agents needs to be added to prevent agglomeration. Similar with the liquid phase synthesis, solid phase synthesis also faces the problem on contaminations. However, it can be used to form alloy compounds at low temperature and is a relative simple and efficient method for scale up production.

All the vapour, liquid and solid synthesis methods have its own merits and drawbacks. Fabrication and processing of nanoparticles often faces the challenges to ensure nanoparticles have desired size, uniform size distribution, morphology, crystallinity, chemical composition and microstructures. Furthermore, the issues of Ostwald ripening and agglomeration of nanoparticles need to be overcome. Although there has been continuous research and development to meet these challenges, scalability and production costs of the production methods need to take into consideration. Hence, the choice of the production methods is very much dependent on the need and application of the nanoparticles.

References

Adschiri T, Kanazawa K, Arai K (1992) Rapid and continuous hydrothermal crystallization of metal oxide particles in supercritical water. J Am Ceram Soc 75(4):1019–1022. doi:10.1111/j.1151-2916.1992.tb04179.x

Ashkarran AA, Iraji Zad A, Ahadian MM et al (2008) Synthesis and photocatalytic activity of WO_3 nanoparticles prepared by the arc discharge method in deionized water. Nanotechnology 19(19):195709

Ashkarran AA, Iraji Zad A, Mahdavi SM et al (2009) ZnO nanoparticles prepared by electrical arc discharge method in water. J Mater Chem Phys 118(1):6–8. doi:10.1016/j.matchemphys.2009.07.002

Bagga K, McCann R, Wang M, Stalcup A, Vazquez M, Brabazon D (2015) Laser assisted synthesis of carbon nanoparticle with controlled viscosities for printing applications. J Colloid Interface Sci 447:263–268

Barcikowski S, Compagnini G (2013) Advanced nanoparticle generation and exitation by lasers in liquids. Phys Chem Chem Phys 15:3022–3026

Buyukhatipoglu K, Morss Clyne A (2010) Controlled flame synthesis of αFe_2O_3 and Fe_3O_4 nanoparticles: effect of flame configuration, flame temperature, and additive loading. J Nanopart Res 12(4):1495–1508. doi:10.1007/s11051-009-9724-9

Cushing BL, Kolesnichenko VL, Charles J, O'Connor CJ (2004) Recent advances in the liquid-phase syntheses of inorganic nanoparticles. Chem Rev 104(9):3893–3946. doi:10.1021/cr030027b

Hayashi H, Hakuta Y (2010) Hydrothermal synthesis of metal oxide nanoparticles in supercritical water. Materials 3(7):3794
Intartaglia R, Bagga K, Brandi F (2014) Study on the productivity of silicon nanoparticles by picosecond laser ablation in water: towards gram per hour yield. Opt Express 22(3):3117–3127
Johannessen T, Jensen JR, Mosleh M et al (2004) Flame synthesis of nanoparticles. Chem Eng Res Des 82(11):1444–1452. doi:10.1205/cerd.82.11.1444.52025
Khang Y, Lee J (2010) Synthesis of Si nanoparticles with narrow size distribution by pulsed laser ablation. J Nanopart Res 12(4):1349–1354. doi:10.1007/s11051-009-9669-z
Kononenko TV, Garnov SV, Klimentov SM et al (1997) Laser ablation of metals and ceramics in picosecond–nanosecond pulsewidth in the presence of different ambient atmospheres. Appl Surf Sci 109–110:48–51. doi:10.1016/S0169-4332(96)00905-1
Mi JL, Clausen C, Bremholm M et al (2012) Rapid hydrothermal preparation of rutile TiO_2 nanoparticles by simultaneous transformation of primary brookite and anatase: an in situ synchrotron PXRD study. Cryst Growth Des 12(12):6092–6097. doi:10.1021/cg301230w
Neddersen J, Chumanov G, Cotton TM (1993) Laser ablation of metals: a new method for preparing SERS active colloids. Appl Spectrosc 47(12):1959–1964. doi:10.1366/0003702934066460
Niznansky D, Viart N, Rehspringer JL (1997) Nanocomposites Fe_2O_3/SiO_2-preparation by sol-gel method and physical properties. J Sol-Gel Sci Technol 8(1–3):615–618. doi:10.1007/BF02436910
Schatt W, Wieters KP (1997) Powder metallurgy: processing and materials. European Powder Metallurgy Association, Shrewsbury
Srivastava AK (2013) Oxide nanostructures: growth, microstructure and properties. Taylor & Fraincis Group, Florida
Stamatakis P, Natalie CA, Palmer BR et al (1991) Research needs in aerosol processing. Aerosol Sci Technol 14(3):316–321. doi:10.1080/02786829108959493
Su SS, Chang ITH, Kuo WCH et al (2014) Pulsed electrical discharge synthesis of red photoluminescence zinc oxide nanoparticles. J Nanopart Res 16(9):1–12. doi:10.1007/s11051-014-2611-z
Swihart MT (2003) Vapor-phase synthesis of nanoparticles. Curr Opin Colloid Interface Sci 8(1): 127–133. doi:10.1016/S1359-0294(03)00007-4
Tavakoli A, Sohrabi M, Kargari A (2007) A review of methods for synthesis of nanostructured metals with emphasis on iron compounds. Chem Pap 61(3):151–170. doi:10.2478/s11696-007-0014-7
Tsuji T, Thang DH, Okazaki Y et al (2008) Preparation of silver nanoparticles by laser ablation in polyvinylpyrrolidone solutions. Appl Surf Sci 254(16):5224–5230. doi:10.1016/j.apsusc.2008.02.048
Waldorff EI, Waas AM, Friedmann PP (2004) Characterization of carbon nanotubes produced by arc discharge: effect of the background pressure. J Appl Phys 95(5):2749–2754. doi:10.1063/1.1642737
Wang Z, Liu Y, Zhang Z (2002) Handbook of Nanophase and Nanostructured Materials-Synthesis. In: Wang Z, Liu Y, Zhang Z (eds) Handbook of Nanophase and Nanostructured Materials-Synthesis. Kluwer Academic Plenum Publishers, USA
Xu YH (2007) Direct synthesis of multifunctional heterostructured magnetic nanoparticles in gas phase. Doctor of Philosophy, Faculty of the graduate school of the University of Minnesota, University of Minnesota
Yang S, Jang YH, Kim CH et al (2010) A flame metal combustion method for production of nanoparticles. Powder Technol 197(3):170–176. doi:10.1016/j.powtec.2009.09.011
Zhao X, Ohkohchi M, Wang M et al (1997) Preparation of high-grade carbon nanotubes by hydrogen arc discharge. Carbon 35(6):775–781. doi:10.1016/S0008-6223(97)00033-X
Zhong WH, Li B, Maguire RG et al (2012) Nanoscience and nanomaterials: synthesis, manufacturing and industry impacts. DEStech Publications Inc, USA

Nanomaterials for Sustainable Energy Production and Storage: Present Day Applications and Possible Developments

Igor Hut, Lidija Matija, Milica Peric, Petar Nikolovski and Svetlana Pelemis

Abstract Currently used technologies in energy industry are oftentimes potentially harmful for the environment and/or not sustainable in the long run (e.g. fossil and mineral based fuels, nuclear energy). Therefore the oncoming energy crisis must be solved by developing new sustainable energy supply that has lower or, preferably, negligible impact on the environment. Any new sustainable energy solutions will have to envelope three important sets of technologies: production, storage and transportation. Additionally, in order to determine whether these solutions are truly sustainable, adequate life cycle assessment methodology (LCA) needs to be developed and properly applied. This article gives a general survey of state of the art and current research trends in application of nanostructured materials for sustainable energy production, storage and transportation, as well as review of LCA methodology implemented in this domain. A special attention is given to already existent applications and possible future developments of nanotechnology for solar energy production, hydrogen economy, batteries and electrical energy transmission.

Keywords Sustainable energy production • Photovoltaics • Solar cell • Nanotechnology • Nanostructures • Nanomaterials • Hydrogen economy • Batteries • LCA

I. Hut (✉) · M. Peric · P. Nikolovski
Faculty of Mechanical Engineering, Innovation Center, University of Belgrade, Kraljice Marije 16, 11120 Belgrade, Serbia
e-mail: ihut@mas.bg.ac.rs

L. Matija
Faculty of Mechanical Engineering, University of Belgrade, Kraljice Marije 16, 11120 Belgrade, Serbia

S. Pelemis
Faculty of Technology, University of East Sarajevo, Karakaj bb, 75400 Zvornik, Republic of Srpska

D. Brabazon et al. (eds.), *Commercialization of Nanotechnologies–A Case Study Approach*, DOI 10.1007/978-3-319-56979-6_3

1 Introduction

During ongoing century, humankind has a vitally important problem to solve, namely securing a long term sustainable energy supply. As a matter of fact this is nothing less but the question of survival for our modern society. Accelerating industrial development, economy based on offshore production and transportation, ever rising demands of living comfort paired with constantly growing world population, increased connectivity and mobility, especially in terms of road and air traffic yield steady rise of global energy consumption. Presently, main energy sources for human activity are fossil and mineral fuels, nuclear fusion and hydroelectric potential, of which only the last one is considered to be renewable and clean. Energy production based on fossil and mineral fuels places various stresses on the environment. It is considered to be one of the main culprits behind the global warming, biosphere destruction and ecological devastation. The risks and dangers connected with nuclear energy production are well known and unfortunately experienced in two major accidents with severe and long-lasting consequences (Chernobyl and Fukushima). In fact majority of present-day methods and technologies used for energy production should be considered harmful in terms of their environmental impact. As it is written by Boden et al.: "Since 1751 approximately 356 billion metric tons of carbon have been released to the atmosphere from the consumption of fossil fuels and cement production alone. Half of these fossil-fuel CO_2 emissions have occurred since the mid-1980s" (Boden et al. 2012). In addition to this, availability of fossil-based fuels is in steady decline, leading to a constant price rise of energy. Apart from concerns related to environmental and climate changes, factors such as reliability and growing risk for future social and political unrest are important issues which also call for dedicated efforts to change the present energy system to a sustainable one. It is obvious that solutions to aforementioned problems will not come overnight. Truly sustainable energy systems are, at best, few decades away. However, there are already many different clean power generation technologies developed, that are based on renewable energy sources such as solar, wind, geothermal, ocean thermal and tidal. Unfortunately, all these alternatives to conventional methods of energy production are, more or less, immature and their exploitation is still rather insignificant in global perspective. Major limiting factors are their comparatively high manufacturing cost versus efficiency ratio when compared to conventional technologies and environmental implications caused by their utilization, which are still not completely understood (Serrano et al. 2009). The intention behind this article is to provide an overview of how nanoscience and nanotechnology may contribute to the development of more efficient and more sustainable energy production and storage.

More than half a century has passed since Noble laureate Richard Feynman delivered his, today famous, lecture "There's plenty room at the bottom". In this speech, practically, the conceptual basis of what we today call nanotechnology was established. Feynman theorized about the number of possible developments of a general ability to manipulate matter on an atomic scale (Gribbin and Gribbin 1998).

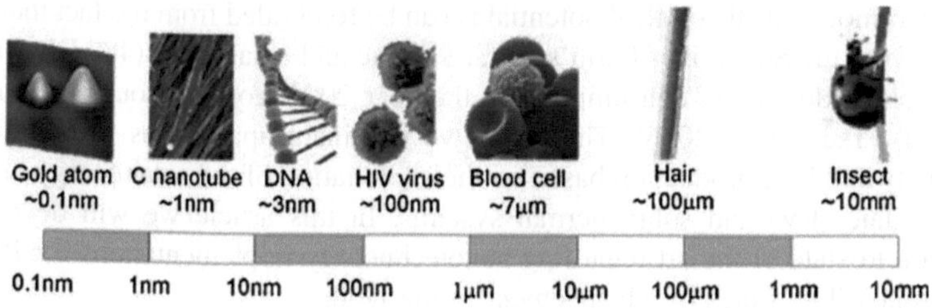

Fig. 1 Length scale and some related examples. Reprinted with permission from (Serrano et al. 2009)

Though at the time the talk went unnoticed, nowadays it is considered to be a seminal event in the field.

Norio Taniguchi a professor of Tokyo University of Science created the term "nanotechnology" in 1974. He defined it as follows: "nanotechnology mainly consists of the process of separation, consolidation, and deformation of materials by one atom or one molecule" (Taniguchi 1974). Today this term has a broader meaning and is typically used to describe materials and phenomena at nanoscale, i.e., on the scale of 1 billionth to several tens of billionths of a meter (see Fig. 1). It specifically implies both the miniaturization and the precise manipulation of atoms and molecules to design and control the properties of the nanomaterials or nanosystems. One of the peculiarities of nanomaterials lies in the fact that many of their physical properties are size-dependent, due to the occurrence of quantum phenomena in the 1–100 nm range (Cao 2004). This is one of the major reasons why nanotechnology could have a significant impact in terms of improvement of systems for energy production and storage. Nanostructured materials exhibit properties that are rather different from those of the same material in the bulk form. This potentially enables the development of devices with features and performances not feasible with employment of conventional materials.

In this article special attention will be given to already existent applications, as well as possible future developments, of nanotechnologies for photovoltaic systems, hydrogen economy, batteries and electrical energy transmission. When thinking about development of mentioned technologies one of the most important aspects that needs to be considered is whether they really support the concept of sustainability, so the ending part of the article will be devoted to the life cycle assessment for nanotechnologies used in systems for sustainable energy production and storage.

2 Nanotechnology for Photovoltaic Systems

Taking into account the fact that sunlight has by far the highest theoretical potential of Earth's renewable energy sources (Tsao et al. 2006), technologies that rely on solar energy utilization seem to be a holy grail of sustainable energy production.

How enormous this theoretical potential is can be foreboded from the fact that more energy from the Sun strikes Earth's surface in one and a half hour (480 EJ) than the total worldwide energy consumption in the year 2001 from all sources combined (430 EJ) (Tsao et al. 2006). There are two dominant approaches for sustainable electricity and heat production based on the exploitation of the solar energy, namely photovoltaic (PV) and solar thermal systems. In this article we will devote our attention to state of the art regarding nanotechnology implementations for PVs, as well as possible routes of advancement in the field.

Production of electricity by photovoltaic solar cells relies on photoelectric effect. First generation of photovoltaic cells still prevails in the global PV market. These are silicon wafer-based solar cells (thick around 150–250 μm), made of crystalline silicon (c-Si) in either of two forms, i.e. multicrystalline silicon (multi-Si), or monocrystalline silicon (mono-Si). The second-generation of PVs are the so called thin film solar cells that feature, as the name implies, thin film layers of semiconductor materials. Film thickness ranges between few nanometers (nm) and tens of micrometers (μm). Usually thin epitaxial deposits of semiconductors on lattice-matched wafers are used. This type of solar cells has the least manufacturing cost, when compared to other technologies. Traditional drawback of thin film PVs is that they exhibit low conversion efficiencies. Nevertheless it has to be said that cell efficiencies of more than 21% have been reported for cadmium telluride (CdTe) and copper indium gallium diselenide (CIGS), in laboratory conditions (Green et al. 2016). This surpasses efficiency of conventional multi-Si solar cells.

A possible approach for reducing or overcoming existing limitations of conventional solar cells is exploitation of nanostructured materials. Such solar cells, which are still in an early stage of ongoing research or have limited commercial availability, belong to what is often referred to as third generation photovoltaics.

Currently there are a large number of research groups working on this type of technologies for energy conversion. Since the performance of mentioned systems is strongly dependent on the characteristics of materials used for their construction, various nanomaterials are being developed for this purpose.

These new materials bring some very promising advantages such as the ability to precisely control the electronic energy band-gap so that it matches different ranges of the sunlight spectrum, enhancement of the effective optical path or significant decrease of the probability of charge recombination.

Research into deployment of nanostructures in solar cells, actually, has a long history. Early applications were primarily aimed towards improving the efficiency of conventional thin film PV devices. Since these beginnings several PV technologies based on different nanostructures and their unique properties have been developed in parallel. We could group nanostructures used in photovoltaic systems into four major classes: nanocomposites and nanostructured materials (especially nanostructured organic cells), quantum wells, nanowires and nanotubes, and nanoparticles and quantum dots (Tsakalakos 2010).

2.1 Nanostructures in Thin Film Solar Cells

Amorphous silicon (a-Si), the non-crystalline form of silicon is used for production of thin-film solar cells for already several decades. Moreover, hydrogenated amorphous silicon (a-Si:H) is the actual material of choice for absorbing material in thin-film solar cells as unhydrogenated a-Si has a very high defect density. So from the beginning of 1970s a-Si:H was explored and implemented in solar cells with steady increase of efficiency to about 14% in 1989 (Street 2005). Further improvement of performance required the introduction of novel materials with a low defect density and favorable carrier transport properties. In the last 25 years, hydrogenated nanocrystalline silicon (nc-Si:H) has been intensively studied as a promising material for the role of low band-gap absorber in the multijunction cells. Hydrogenated nanocrystalline silicon is actually an amalgamate of nanometer-size Si crystallites embedded in an amorphous silicon matrix along with grain boundaries and micro-voids (Jiang et al. 2011). The presence of nano-sized crystallites provides crystalline paths for carrier transport enabling much higher carrier mobility than a pure amorphous system would provide. Along with this, nc-Si:H has one more distinctive advantage when compared to a-Si:H or, for that matter, hydrogenated amorphous silicon germanium (a-SiGe:H) and that is the reduction in light-induced degradation of the solar cell efficiency. Thus the film, i.e. the absorber layer, can be thicker, allowing for generation of high photocurrent density.

2.2 Nanoparticles in Dye-Sensitized Solar Cells

Dye-sensitized, i.e. Grätzel solar cells (Fig. 2) offer another alternative to conventional Si-based solar cells. This type of solar cells is probably the earliest one in which nanostructured materials were intentionally implemented with the aim of performance enhancement (Vlachopoulos et al. 1988). As already mentioned, when compared to their bulk form nanostructured materials feature much larger surface area to volume ratio and this can substantially alter their physical and chemical properties. Moreover such characteristics can be fine-tuned according to characteristic dimensions and shape of the nanostructure. This specific aspect has been intensively explored with the goal of boosting the absorption of dye-sensitized solar cells (DSSCs) and hence improving their conversion efficiency. In this implementation of PV technology light absorption predominantly occurs in dye molecules that are anchored to the surface of nanoparticles of a wide band-gap semiconductor, typically TiO_2 (O'Regan and Gratzel 1991; Hagfeldt and Graetzel 1995). Following excitation with visible light the dye injects electrons from its excited state into the TiO_2 conduction band, leading to a charge separated state, i.e. difference of potentials, thus enabling the production of electrical energy in the form of voltage or current. The huge surface area of nanoporous TiO_2 significantly increases the effective light absorption area of the dye when compared to

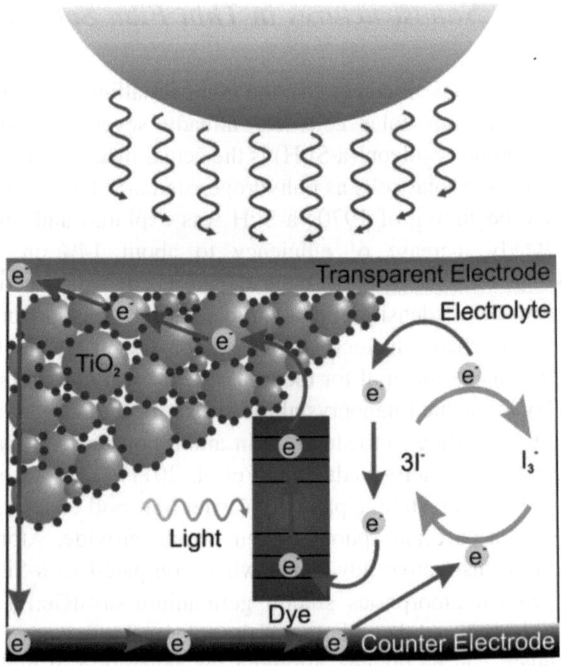

Fig. 2 Schematic illustration of a generic dye-sensitized (Grätzel) solar cell (Jones 2009). The performance of such solar cell is critically dependant on nanoscale components. As shown, photon absorption occurs in a light apsorbing dye. This dye, in return, is attached to a nanostructured semiconductor layer

conventionally used polycrystalline electrodes, which are relatively flat. This in turn improves the cell conversion efficiency to higher than 10% (Grätzel 2004). The particle size of TiO_2 nanocrystals is a decisive factor in the design of DSSCs. For the most part this has to do with the electron transport process in the nanoporous TiO_2 electrode. This process is parameterized by the diffusion length of electrons $L=(D\times\tau)^{\frac{1}{2}}$, where D is the electron diffusion coefficient and τ is the electron lifetime. It has been shown that D increases with the size of TiO_2 nanocrystals, probably due the decreased number of boundaries between particles. On the other hand larger particles result in size reduction of the total cross-sectional area of boundaries, which in turn increases the resistance for the charge transport. Taking into account these two opposite effects the conclusion can be drawn that there is an optimal particle size in terms of the resulting cell performance. As it turns out today's high efficient DSSCs feature TiO_2 nanoparticles with typical diameters in the range of between 10 and 20 nm.

Dye-sensitized nanocrystalline solar cells have also been constructed based on nanoparticles of several other semiconductor materials (Corma et al. 2004; Rodriguez et al. 2007; Seabold et al. 2008). Also approaches exploiting different nanostructures such as nanotubes, photonic crystals or photonic sponges instead of nanoparticles (Rodriguez et al. 2007; Seabold et al. 2008), have been tried, with variable results in terms of their efficiency.

2.3 Nanostructures for Organic Solar Cells

First observations and measurements of photovoltaic effect in organic systems happened in 1958 in a cell based on magnesium phthalocyanine (MgPc) (Kearns and Calvin 1958), and in 1959 in a cell composed of anthracene single crystal set between two electrodes (Kallmann and Pope 1959). The measured open circuit voltages were around 200 mV in both cases and the efficiency was extremely low due to the inefficient exciton dissociation. Since these humble organic solar cells have been very actively developed. Today, organic solar cells are regarded as a possible low cost alternative to silicon solar cells. Moreover, and in general, organic solar cells bring some other advantages, apart from less expensive and simpler fabrication, such as electronic tunability by molecular engineering (e.g. changing the length and functional group of polymers), flexible substrates, portability and they are kinder to our environment. Here, we will discuss hybrid solar cells composed of organic thin films and inorganic nanostructure. This type of solar cell is composed of inexpensive and readily processable organic materials with inorganic semiconductor nanostructures embedded in the film in order to improve the absorption efficacy and enable fine tuning of the absorption spectra trough the bandgap modulation by the quantum confinement effect (QCE) (Huynh et al. 2002).

There are several more or less different strategies deployed in the production of such hybrid solar cells. One approach is to build heterojunction organic solar cell that incorporates bulk nanostructured materials as the electron-accepting layer. The idea behind this application is that such electron acceptors in planar or bilayer heterojunctions should increase the cell efficiency. Quite a few nanostructured materials have been tested for this role. The allotrope form of carbon C_{60}, also known as the buckminsterfullerene, or popularly bucky-ball, is of the strongest known electron acceptors. It can accept up to six electrons per molecule (Ohsawa and Saji 1992). For the detailed information on the development of fullerene derivatives, interested readers can be referred to a thorough review by He and Li (2011).

The first known application of C_{60} in a PV device was the one explored by Saricifci et al. (1993). However these first devices exhibited rather low power conversion efficiencies of only 0.02% (monochromatic illumination at 514.5 nm). This was likely a consequence of narrow interface width between the polymer and fullerene molecules which resulted in a limited interfacial area and low excitation separation. Drees et al. tried to address this issue by using "thermally controlled interdiffusion" (Drees et al. 2002). Conversion efficiency of such devices reached a maximum of 0.30% under monochromatic illumination at 470 nm (Drees et al. 2004).

Other nanostructured materials have been also investigated as the electron acceptors for the use in organic polymer bilayer heterojunctions. TiO_2 has been explored in these terms since 1998 (Savenije et al. 1998). Reported power conversion efficiency was 0.15% under AM1.5 illumination. Thereafter different groups developed devices with TiO_2 deposited by various means and using different materials for the back electrodes. Dye-sensitized solar cells constructed with composite photoanodes made out of TiO_2 nanoparticles/nanobelts (TiO_2 P–B) have shown efficiencies up to 4.81% (Fan et al. 2014).

Apart from TiO_2 other materials such as zinc oxide (ZnO) have been explored for the application in organic polymer bilayer heterojunctions. ZnO nanorods can be readily grown using wet chemistry techniques and have yielded efficiencies comparable to those achieved with nanocrystalline TiO_2 (Peiro et al. 2006).

The bulk heterojunction polymer solar cells were introduced by the works of Haall et al. (1995) and Yu et al. (1995). The idea was to overcome the limitations of bilayer devices and increase the interfacial surface area. In order to obtain a bulk heterojunction, a heterogeneous composite is formed by mixing electron acceptors into the polymer. By blending a p-type (electron donor) and an n-type (electron acceptor) material together in the solution and controlling the morphology of the devices, a high interfacial area throughout the bulk can be achieved resulting in enhanced exciton dissociation and charge transfer (McLeskey and Qiao 2010). Implementation of various inorganic semiconductor nanocrystals including CdSe, $CuInS_2$, CdS, or PbS has been reported for this type of PVs (Alivisatos 1996; Greenham et al. 1996; Huynh et al. 2002; Arici et al. 2003; McDonald et al. 2005). The bulk heterojunction polymer PVs exhibit certain characteristics that are potentially advantageous: implemented inorganic nanostructured material have high absorption coefficients, good photoconductivity and stability; it is relatively easy to fine tune doping characteristics of used nanostructures; since the QCE is size dependent it is relatively simple to change the electronic and optical properties of nanostructures, hence the hybrid solar cells containing nanostructures of different sizes can be used for tandem solar cells.

Recently, high expectations of perovskite solar cells have been evoked. They exhibit significant performance advantages when compared to conventional hybrid PV cells (Hodes and Cahen 2014; Collavini et al. 2015). These solar cells own their name to compound known as perovskite. Most commonly this is a metal halide based material like methylammonium lead trihalide $CH_3NH_3PbX_3$ (where X is a halogen atom such as Cl, Br or I) in the form of nanocrystal. Optical bandgap is determined by halide content and usually varies between 1.5 and 2.3 eV. Formamidinum lead trihalide, $H_2NCHNH_2PbX_3$, has also been considered, with optical bandgaps in a similar range and theoretical capabilities for higher conversion efficiency than in the case of methylammonium lead trihalide (Eperon et al. 2014). As inclusion of Pb in perovskite materials is regarded as problematic from the stance of their life cycle and environmental impact PVs with tin-based perovskite absorbers, such as $CH_3NH_3SnI_3$, have also been constructed, though with less satisfying power conversion efficiencies (Noel et al. 2014; Hao et al. 2014; Sun et al. 2016; Cortecchia et al. 2016).

A structural diagram of the hybrid perovskite solar cell, as constructed by Kim et al. (2012a) is schematically shown in Fig. 3.

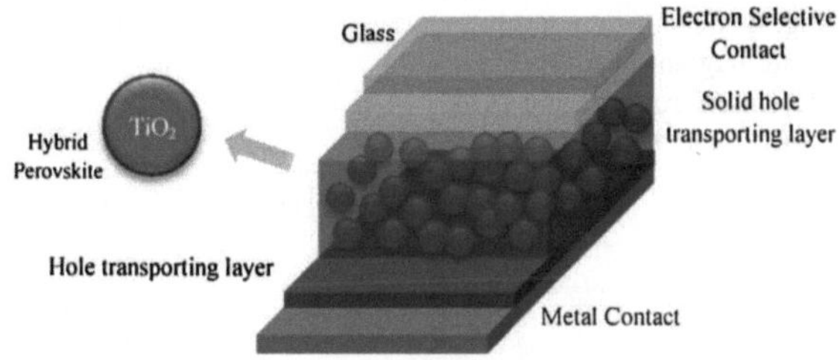

Fig. 3 Schematic diagram of a hybrid perovskite solar cell architecture. Reprinted from (Kandjani et al. 2015) under CC BY 3.0 licence

The first article regarding perovskite solar cell performance reported power conversion efficiency of 3.8% in 2009 (Kojima et al. 2009). Since then, the conversion efficiency of perovskite PV cells has been increasing at an unprecedented rate, first of all, thanks to exploration of novel nanomaterials and device architectures (Kojima et al. 2009; Zhang et al. 2009; Hetsch et al. 2011; Im et al. 2011, 2012; Lee et al. 2012; Qiu et al. 2013; Noh et al. 2013; He et al. 2014; Saliba et al. 2016; Seo et al. 2016; Tsai et al. 2016; Abdelraouf and Allam 2016). Current world record is 21.1% efficiency at normal operating conditions (Saliba et al. 2016).

Though these results seem rather promising there are still few issues to be resolved before perovskite solar cells can be successfully commercialized. First of all environmental and photo-stabilities should be guaranteed, which is still not the case. The structure and function of organic components are easily compromised by penetration of the ambient gas or moisture through the packaging materials. Actually this is a common problem for most types of the organic solar cells. Hence the successful development of encapsulation based technology is a priority. Advances in fabrication methods with good uniformity and reproducibility are also important challenge in regard to commercial mass production. Finally, adequate substitution of Pb with other elements is vital in terms of environmental protection.

2.4 *Quantum Well Solar Cells*

Taking into account only the efficiency the quantum well solar cell (QWSC) is a good candidate for a champion in the world of nanostructured solar cells. The quantum well (QW) is a 1D confinement nanostructure. Quantum wells have been intensively researched in past decades. Moreover they have been put into practical use in plethora of semiconductor devices, most prominent of being light-emitting diodes (LEDs), photodetectors, high-electron-mobility transistors (HEMTs), lasers… What makes QWs adequate for this wide implementation is the fact that their optical and electrical properties can be readily engineered to meet the requirements of the concrete application. This suggests that QWs have significant potential for applications in PV. The band gap of QW nanostructures can be modulated by quantum confinement effect (QCE) so that it matches better the relevant part of the solar spectrum. This in turn provides higher conversion

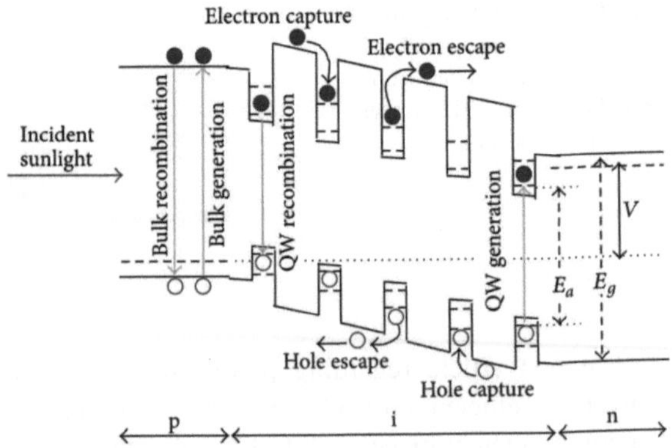

Fig. 4 Schematic energy band diagram of a QWSC. Reprinted from (Zhang et al. 2015) under CC BY 3.0 licence

efficiency. Area of QWs for PV implementations is pioneered by the quantum photovoltaics (QPV) group at Imperial College London (Barnham et al. 2002). In the early 1990s two prominent members of this group, Barnham and Duggan, proposed the utilization of multi-quantum-well (MQW) or superlattice systems as the absorbers in concentrator solar cells (Barnham and Duggan 1990). They envisaged efficiencies with upper limits in excess of 40%. Figure 4 schematically shows a band diagram of a QWSC.

This diagram depicts an intrinsic region, where QWs are located, which is inserted into a p-n junction of a conventional solar cell in order to extend the field-bearing region. What this means is that QWs enable the absorption to occur bellow the band gap E_g of the bulk host material, all the way to the threshold E_a. The photocarrier generation and recombination are happening in both the bulk and QW regions. In order for photocarriers, generated in QWs, to contribute to the total photocurrent, they need to escape from QWs to the bulk cell, so they can be effectively collected at the external electrodes. Two processes are crucial for this to happen, quantum tunneling and thermal escape. Preconditions for these are enough thermal energy and transverse electric field maintained over the i-region. Fortunately room temperatures provide enough thermal energy for high escape efficiency, in fact the probability for photoexcited carriers escape can be close to unity in the presence of a strong enough electrical field (Barnham et al. 1991; Nelson et al. 1993; Barnes et al. 1996; Serdiukova et al. 1999).

Generally, in regard to lattice constant difference between barrier and QW layers, two material systems are used in this type of solar cells. These are lattice matched systems such as AlGaAs/GaAs, InGaP/GaAs, InP/InGaAs, and InGaAsP/InP and lattice mismatched systems including GaAs/InGaAs, InP/InAsP, GaAsP/InGaAs, GaAs/InGaAs, and InGaAs/InGaAsP. Production wise these are quite desirable materials as there are well developed growth technologies, such as metalorganic vapour phase epitaxy (MOVPE) or molecular beam epitaxy (MBE), which can provide monolayer accuracy for their deposition (Zhang et al. 2015).

Exploration of QWSCs began with lattice matched material systems, based mainly on GaAs or InP as the QW materials. These materials enable good solar cell performance by adjusting the absorption threshold to higher energy levels compared to conventional homojunction cells formed from GaAs or InP. Also, there is no strain induced in these material systems, making them easier to work with.

Nevertheless even more desirable option would be to gain lower absorption energy threshold values in comparison to e.g. conventional high efficiency GaAs solar cells. The principal obstacle in this case is that there are no, naturally occurring, lattice matched lower band gap systems to be used for QWs. For instance InGaAs is the preferable choice to be used for QW material in combination with GaAs as the barrier material, but it features larger lattice constant thus the compressive strain is required to grow the InGaAs QW structures. In each layer of a MQW stack this strain adds up. Consequently numerous threading dislocations are induced by the strain relaxation at the top and the bottom of the stack. This leads to larger leakage current, decreased quantum efficiency for short wavelengths, as well as loss in the V_{oc} (Griffin et al. 1996; Mazzer et al. 1996). In order to overcome this issue, i.e. to grow QWs without strain induced dislocations, strain-balancing techniques have been developed. For instance thin $In_xGa_{1-x}As$ QW layers are grown in compressive strain, alternating with $Ga_xA_{s1-x}P$ barriers in tensile strain. Thickness and composition of these layers can be precisely regulated, thus securing that there is no additional strain induced in the GaAs substrate. This technology enables growing more than 65 QWs without dislocations being generated (Ekins-Daukes et al. 2001; Adams et al. 2009).

Quantum well solar cells seem to be a rather promising PV technology in terms of tuning the absorption spectrum and hence achieving high power conversion efficiency. The challenge is to increase the absorption of the MQW stack, which could be solved by growing larger number of QWs, but this is a technological challenge that still needs to be addressed. Light trapping structures are also intensively researched (Ekins-Daukes et al. 2009; Turner et al. 2013; Inoue et al. 2015).

2.5 *Other Nanotechnology Approaches for PV*

There is a plethora of other approaches for potential PV technology enhancement that rely on different nanostructures. Further, we will give a brief overview of those that are currently explored, without going too much into detail, as we are restricted by the format of this survey.

Nanowires are nanostructures with a diameter in the range of tens of nanometers down to just few nanometers, while the ratio of the length to width is greater than 1000. The nanowire geometry brings many potential advantages over conventional planar wafer based or thin-film PV cells (Hochbaum and Yang 2010; Garnett et al. 2011). Thanks to the lack of grain boundaries along the length of a nanowire its use as a direct conduction path for charge carriers enables high charge transport rates and small carrier recombination probabilities to be achieved. These nanostructures also exhibit exceptional light trapping characteristics, which leads to reduced surface reflection. Besides they exhibit other beneficial properties such as: improved band gap tuning, facile strain relaxation and increased defect tolerance (Garnett et al. 2011). These improvements do not necessarily lead to maximum conversion efficiencies above standard limits for conventional PVs, what the real potential merit is the reduction of material quantity and quality needed to approach those limits. This could lead to substantial cost reduction for mass production of nanowire based solar cells. Moreover, production technology which utilizes nanowires allows manufacturing of complex single crystalline semiconductor devices directly on inexpensive substrates, being it aluminum foil, inox steel or conductive glass (Law et al. 2005; Fan et al. 2009; Zhang et al. 2014). Nanowire solar cells offer great opportunities for PV mass production, nevertheless this technology still has challenges that need to be addressed before successful commercial exploitation. Those that need to be stressed are: "surface and interface recombination, surface roughness, mechanical and chemical stability, fine morphology and doping control, nanowire array uniformity, and synthetic scalability" (Garnett et al. 2011).

Application of quantum dots (QD) for solar cells is also intensively studied, especially thanks to their highly tunable, size dependent, properties. QD tandem solar cells have attracted a lot of attention in recent years. Results have shown that these are capable of high conversion efficiencies (Conibeer et al. 2006; Luther et al. 2008; Conibeer 2011; Choi et al. 2011). In these PV cells band-gap engineered nanocrystals are utilized as absorbing material. One of the main advantages of these solar cells is that they are mostly designed to use the common elements and compounds, which are readily available, such as Si, Ge, PbSe and PbS. This brings the reduction in material costs and secures a long term availability of raw materials for mass production. One of the greatest challenges faced by this PV architecture, especially in terms of large scale manufacturing, is reliable production, i.e. controllable synthesis, of nanocrystals, with needed properties. Moreover this needs to be accomplished by utilization of inexpensive fabrication methods. Additionally, we need better understanding of how nanocrystals' electrical properties can be controlled, in order to be able to successfully implement such technology for production of quantum dot tandem solar cells (Conibeer et al. 2011).

Potential use of nanostructures for absorber materials has also been explored in conjunction with intermediate band solar cells (Martí et al. 2000, 2006), hot carrier solar cells (Westland et al. 1988; Nozik et al. 1990; Rosenwaks et al. 1993; Conibeer et al. 2008; König et al. 2010), as well in some more advanced concepts, still in rather early stage of research and development, such as multiple carrier generation, up/down conversion, thermophotonics, and thermophotovoltaics (Catchpole et al. 2002; Švrček et al. 2004; Franceschetti et al. 2006; Chen et al. 2009).

Apart from being used and advanced absorber materials, nanostructures offer promising advantages as alternative light trapping systems for thin film solar cells (Kanamori et al. 2001; Zeng et al. 2008; Park et al. 2009; Gjessing et al. 2010; Yu et al. 2010b). Plasmonic nanoparticles have also been studied for nanophotonic light trapping schemes (Atwater and Polman 2010; Aubry et al. 2010; Akimov and Koh 2011; Catchpole et al. 2011; Tan et al. 2012b).

3 Hydrogen Economy

Hydrogen is a promising fuel for sustainable energy production (Andrews and Shabani 2012). Before the hydrogen can be used as a sustainable energy source, number of problems need to be solved. Broadly speaking, these problems can be divided into three categories: energy production, storage and conversion.

3.1 Hydrogen Production

There are number of ways to produce hydrogen. The usual method is steam reforming, though this is not a sustainable energy solution. Today almost all production is based on raw fossil materials, while hydrogen production from water electrolysis makes up only 4% of total production (Chiarello et al. 2010). Water electrolysis can also be called water splitting. Water splitting refers to the processes of decomposition of water to hydrogen and oxygen. Currently there is also much research focused on the subject of sustainable hydrogen production by utilization of solar energy. When solar energy is used then water splitting is achieved via photocatalytic water electrolysis, also called artificial photosynthesis. Photocatalytic water electrolysis was first demonstrated in 1972 by Fujishima and Honda (1972).

Photocatalytic water electrolysis can be approached in two ways: by an electrode system, or by a particle system (Chen et al. 2012). Both processes are similar and they can be described as follows: (a) a semiconductor absorbs light and electrons are excited from valence band to conduction band, (b) as a result an electron-hole pairs are generated, and (c) electrons are transported to the metal electrode or metal particle (metal acts as the electron trapper or a co-catalyst) (Chen et al. 2012).

TiO_2 is the well-known photocatalyst, and one that has been widely investigated in past several decades (Ahmad et al. 2015). Though it possess excellent characteristics, it can use only UV light for photocatalysis and it suffers from rapid recombination of electron-hole pairs (Serrano et al. 2009). Therefore, nanomaterials are utilized in order to overcome these drawbacks. For example various morphologies of TiO_2 have been investigated, such as nanotubes, nanoparticles, nanofibers, composite nanofibers, and fibers with porous structures (Xu et al. 2013; Hou et al. 2014; Shi et al. 2014). Fabrication of N-doped TiO_2 nanomaterials using hydrothermal method (Chainarong et al. 2011) and heat treatment of sol-gel derived TiO_2 (Petala et al. 2014) were also reported, and their photocatalytic activities investigated. Generally, nanomaterials increase photocatalytic activity because of their high active surface area.

Graphene in combination with TiO_2 has also been considered. Graphene-based 3D aerogel embedded with TiO_2 nanoparticles may have a potential in photoelectrochemical hydrogen production and pollution removal (Han et al. 2014). Another approach to combination of graphene and TiO_2 is nanocomposite of graphene oxide with N-doped TiO_2 (Pei et al. 2013). It was reported that this nanocomposite exhibits high photocatalytic efficiency of 716.0 or 112.0 μmol/gh (Pei et al. 2013).

Undoped and Er-doped PbSe nanoparticles can be synthesized in an environmentally friendly process (Khataee et al. 2015). The study on reusability demonstrated that these synthesized nanoparticles still have good capability after 10 sets of photocatalytic reactions (Khataee et al. 2015).

The effects of ball milling time and proportion of molybdenum phosphide-graphite (MoP) nanosheets on electrocatalytic activity were investigated by Aravind et al. (Jyothirmayee Aravind et al. 2015). It was reported that this material showed superior electrocatalytic performance. Aravind et al. also noted that their study offers new method for development of non-noble electrocatalysts by a simple solid state ball milling.

Investigation into CdO-CdS nanocomposite showed that CdO-CdS has significantly enhanced photocatalytic activity when compared to CdS nanoparticles (Kahane et al. 2013).

SiC nanomaterials with different morphologies (whiskery, wormlike and particulate) have been studied by Hao et al. (2013). It was found that SiC nanoparticles exhibit high hydrogen evolution rate of 83.9 μL/gh, due to high surface specific surface area. Hao et al. found that all samples that were tested generally show good photocatalytic performance without sacrificial agent under visible light irradiation.

In a recent review by Preethi et al. efficiency of different nanomaterials in photocatalytic hydrogen production was examined (Preethi and Kanmani 2013). They concluded that lack of industrial application of this technology can be mainly attributed to two reasons: the low photocatalytic efficiency and lack of agreement on how to quantify this efficiency. They also noted that other challenge is successful scale-up of experiments to an industrially relevant scale.

3.2 *Hydrogen Storage*

Hydrogen may be stored in several forms: as pressurized gas, in liquefied form, or it can be bonded to a solid-state material (Chen et al. 2012). Storage of a hydrogen as a pressurized gas or in the liquid form is mainly inefficient (Serrano et al. 2009). Currently storage of hydrogen on a solid-state material is considered to be the most efficient hydrogen storage method (Serrano et al. 2009). There are two mechanisms by which hydrogen can be bonded to a solid-state material either with strong chemical associations (i.e. chemisorption) or weak dispersive interactions (i.e. physisorption) (Mao et al. 2013). In chemisorption, hydrogen atoms are integrated into the lattice of a solid material, while in physisorption adsorption of hydrogen molecules is achieved through the van der Waals interactions (see Fig. 5) (Chen et al. 2012). However, these two techniques do not come without difficulties. Chemisorption bounds too strongly the hydrogen atoms, and as a consequence temperatures required for hydrogen release must be considerably higher than the

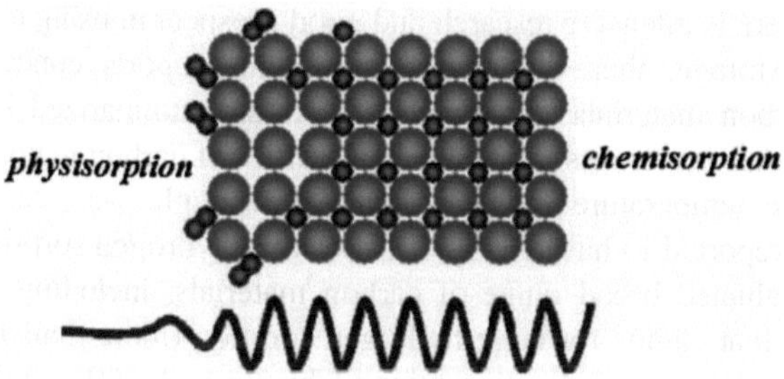

Fig. 5 Schematic illustration of basic mechanisms of solid-state hydrogen storage. By physisorption, hydrogen is attached to the surface of a material either as hydrogen molecules or as hydrogen atoms. By chemisorption, a hydrogen molecule is dissociated into hydrogen atoms, and then the hydrogen atoms are incorporated/distributed into the solid lattice. The potential energy for hydrogen sorption is also illustrated. Reprinted with permission from (Mao and Chen 2007)

room temperature. In physisorption hydrogen is bounded by weak forces, which leads to a low storage capacity at the room temperature.

Materials based on magnesium are potential candidates as solid-state hydrogen storage materials. This is because of their potentially high hydrogen storage capacity. MgH_2 has high hydrogen mass percentage of 7.6 wt% (Mao et al. 2013). Mg is also abundant and has a relatively low price. However, Mg can absorb hydrogen only on temperatures higher than 300 °C. Therefore, desorption of hydrogen from MgH_2 generally cannot occur bellow this temperature.

Using various methods for Mg-based hydrogen storage materials fabrication (e.g. ball milling, thin film synthesis, etc.) Shao et al. found that nanostructure and catalysis can improve the kinetics, but desorption enthalpy stays the same in the range of 5–300 nm (Shao et al. 2012, 2014). In order to solve high enthalpy problem of MgH_2, usual procedure is to make alloys of Mg with other metals, though this leads to reduction of storage capacity (Mao et al. 2013). Shao et al. studied $Mg_{50}Co_{50}$ alloy with body centered cubic structure (BCC) and found that studied sample shows capacity of 2.65 wt% at temperature of 258 K (Shao et al. 2009). They also reported superior hydrogen absorption kinetics for this alloy, though poor thermal conductivity (0.432 W/m/K) (Shao et al. 2014). Same authors also noted that absorption temperature of 258 K could be currently the lowest reported absorption temperature for Mg-based materials. In contrast, group of authors investigated ball milled $Mg_{10}NiNb_2$ with Nb2-1.5Nb, which were ball milled with multi-walled carbon nanotubes (MWCNTs), and found to poses high hydrogen storage capacity (Aminorroaya et al. 2011). Storage capacity of around 6 wt% was reported, though the investigated temperature was 200 °C and 2 MPa of hydrogen pressure.

Recently an interesting approach has been investigated for possible use of Mg alloys from waste material (Pistidda et al. 2014). Alloys AZ91 and Mg-10 wt% Gd, collected in form of shavings, were investigated, and results suggest that production of hydrogen storage from such waste material is possible.

Although there is intensive research and good prospect in using carbon materials for hydrogen storage, there have been conflicting reports concerning possible capacity of carbon materials. A survey of literature summarized in (Schur et al. 2002) showed various reports of storage capabilities of carbon nanomaterials, as an example in the temperature range of 80–300 K single wall carbon nanotubes (SWCNT) are reported to have from 1.1 to 10 wt% hydrogen sorption. In contrast, other study evaluated broad range of carbon materials, including SWCNTs, and results suggest that claims for large hydrogen sorption (more than 1 wt%) at room temperature should be viewed with caution (Tibbetts et al. 2001). Nikitin et al. used atomic hydrogen as hydrogenation agent on SWCNTs, and reported hydrogen storage capacity of 7 wt% at room temperature for SWCNT with 2 nm diameter (Nikitin et al. 2008). Recent study by Barghi et al. (2014) reports that they have calculated chemisorption and physisorption isotherms of hydrogen in MWCNT independently. They have reported maximum measured hydrogen physisorption to be 0.13 wt% and chemisorption 0.058 wt% at the room temperature.

Nanoporous carbon materials are also good candidates for hydrogen storage applications (Terrés et al. 2005; Wang et al. 2009; Yang et al. 2012). Recent study proposed nanoporous material made of carbon nanotubes (CNTs) named Super Diamond (Tylianakis et al. 2014). Theoretical calculations suggest possible value of hydrogen storage to be 8 wt% at room temperature. Thus, it meets the US Department of Energy (DOE) proposed goal for 2015 (EERE 2014), though this material is not yet synthesized. Also recently, cost-effective nanoporous carbon materials were synthesized by nano-casting with agglomerated rice husk silica using two different carbon precursors (sucrose and glycerol) (Attia et al. 2013). When glycerol precursor was used it was found that hydrogen adsorption is 2.41 wt% at 77 K and 7.3 MPa, with the total hydrogen uptake of 5.7 wt% (Attia et al. 2013).

Important factors in porous materials are active surface area and pore size. Cabria et al. simulated the hydrogen storage in nanoporous carbons by altering pore shapes (Cabria et al. 2011). They found that optimal sizes of planar, cylindrical and spherical pores are in the ranges 5.8–7.5, 7.0–8.6 and 8.5–10.7 Å at 300 K, respectively.

In order to investigate significant differences between theoretical and experimental hydrogen storage capacities for carbon materials a group of authors recently conducted a first-principles study (Robledo et al. 2014). They investigated nanoporous reduced graphite oxide and noted that water formation was observed for high O/C ratio which lead to reduction of hydrogen absorption energy. As a result they recommended that content of oxygen should be low for hydrogen storage purposes in reduced graphite oxides (Robledo et al. 2014).

Another approach to possible solution for a meaningful hydrogen storage capacity is to embed nanoparticles in porous carbon (Zhang et al. 2007). It has been found that Pd nanoparticles in 3D activated porous carbon (Pd/PC-850) show a good hydrogen storage capacity (4.6 wt% at 77 K and 20 bar) (Zhu et al. 2014). The Pd/PC-850 has great number of 0.6–0.7 nm micropores with small amount of 2–3.4 nm mesopores (Zhu et al. 2014).

Yet another promising class of materials for hydrogen storage are materials based on borane. Ammonia borane (H_3NBH_3), often abbreviated as AB, is considered to be most promising candidate, because it can contain 19.6 wt% of hydrogen, however it can only release hydrogen at temperatures much higher than the room temperature (Marder 2007). Detailed history and properties of AB has been recently summarized in (Li et al. 2014). Using Ni-Co/r-GO catalyst with AB Chou et al. reported that the gravimetric density of storage capacity can be reached as high as 6.46 wt% (Chou et al. 2014). In experimental study of oxygen-doped boron nitride nanosheet Lei et al. reported storage capacity of 5.7 wt% under 5 MPa and at a room temperature, which is the highest storage capacity reported for boron nitride materials at present (Lei et al. 2014). Also there have been some theoretical approaches considered for decorated boron materials such as sheets (Wang et al. 2012; Qiu et al. 2014) and boron fullerenes (B_{80}) (Li et al. 2012).

Number of other materials have been considered for hydrogen storage applications such as heavy metals and their alloy, zeolites, and metal organic frameworks (MOFs). They have been summarized in a recent review by Bhattacharyya and Mohan (Bhattacharyya and Mohan 2015).

3.3 Hydrogen Conversion

Fuel cells are electrochemical devices that convert chemical energy intro electrical energy. They consist of anode, cathode and the electrolyte. Hydrogen can be used as a fuel for fuel cells. It reacts on anode, turning into an ion and an electron and as a result produces direct current. Typically, an ion is reunited with an electron at the cathode, and together with O_2 reacts to create water. There are various types of fuel cells, depending on the electrolyte they use for example: proton exchange membrane fuel cells (PEMFC), solid oxide fuel cells (SOFC), molten carbonate fuel cells (MCFC), and phosphoric acid fuel cells (PAFC) (Dodds et al. 2015). Schematic illustration of a PEMFC is presented in Fig. 6. Fuel cells are promising technology for sustainable energy, since they emit no CO_2 on H_2, and less than 50% CO_2 on hydrocarbons, and also they do not emit NO_X or SO_2 (Hussein 2015). MFCFs can even contribute to reduction of CO_2 emission (Chacartegui et al. 2013; Wee 2014). Main motivation for exploring utilization of nanomaterials and nanotechnology for fuel cells is possible cost reduction and efficiency improvement. Nanomaterials and nanotechnology can be utilized to improve membranes, electrodes and catalysts.

Carbon nanomaterials are believed to be promising approach to development of metal-free catalysts for fuel cells (Zhang and Dai 2012). It has been found that carbon nanofiber (CNF) and CNT based Pt catalysts have better durability than conventional carbon black catalysts for PEMFC application (Andersen et al. 2013). Progress of doped carbon nanomaterials intended for use as cathode catalysts in oxygen reduction reactions (ORR) has been recently reviewed by Yang et al.

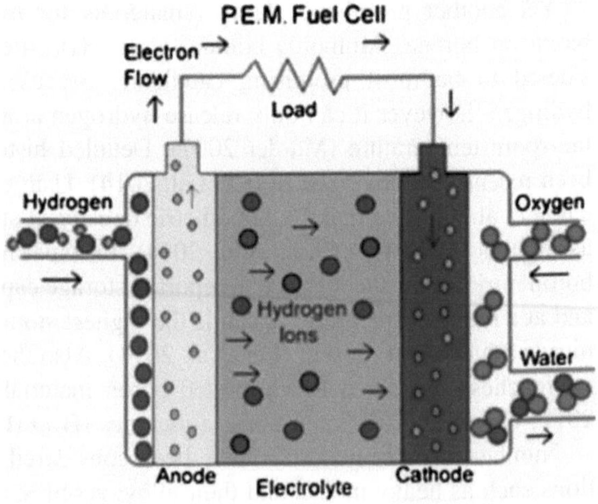

Fig. 6 Schematic illustration of the basic structure of a PEMFC. Reprinted with permission from (Mao and Chen 2007)

(2013). They concluded that present metal free doped carbon catalysts are yet far from the large scale practical applications.

Recently, novel electrodes based on CNF and Pt nanoparticles for PEMFC were developed by Salernitano et al. (2014). CNF with controlled morphology was grown of graphite paper, and Pt particles were deposited on CNF. It was found that these electrodes have better performance and stability than the commercial nanometric carbon powder, and that the electrocatalytic performance of CNF electrodes is strongly related to the fiber morphology and structure (Salernitano et al. 2014).

Nafion has been most frequently used as the material for proton exchange membrane (Chien et al. 2013). For example, Nafion membrane with nanosize TiO_2 and SnO_2 fillers was developed, and showed better performance than Nafion membranes without fillers (Abbaraju et al. 2008). Abbaraju et al. concluded that nanoparticle size i.e. surface area of nanopowders has a major role in water retention and performance of these membranes at temperatures of 80 and 120 °C (Abbaraju et al. 2008). Other approaches of utilizing Nafion other than as a main material for membrane have also been investigated. As an example, a composite membrane was developed using Nafion/histidine functionalized CNT for direct methanol fuel cell (DMFC) application (Asgari et al. 2013). These membranes exhibited higher proton conductivities at elevated temperatures when compared with conventional Nafion membranes.

Noble metal nanomaterials (NMN) as catalysts for fuel cells were reviewed by Guo et al. (Guo and Wang 2011). They concluded that improvement in fuel cell performance by the majority of such catalysts is limited to either activity or stability, but not to both.

In the last decade interest in biological fuel cells is resurging, where living cells or subcellular biological components are used as biocatalysts (Minteer et al. 2012). As an example, fabrication of Ni nanoparticles-dispersed micro-nanofiber for an electrode application in a microbial fuel cell (MFC) has been recently described by

Singh et al. in (Singh and Verma 2015). The authors reported that MFC showed high open circuit potential (710 mV), current density (1145 mA/m^2) and power density (1145 mW/m^2). They concluded that this novel electrodes could be potential alternative to electrodes that are based on noble metals.

Potential application of supercritical fluids have been investigated and summarized in review by Bozbag and Erkey (2012). They noted that electrocatalysts prepared by using supercritical fluids had higher electrochemical activities than commercial catalysts. They also concluded that supercritical fluids may have a role in development of non-carbon conductive nanostructured electrode materials.

Fairly recently nanofluidics was suggested for the first time for fuel cell application, in work by Lee et al. (2013). Nanofludic fuel cell shows rapid electrochemical kinetics and high active surface area, however with increasing current density high ohmic overpotentials gradually offset the gain from fast kinetics. Lee et al. suggest that this new fuel cell type could be optimized and have an important role in the future (Lee and Kjeang 2013).

MOFs have been also investigated as a possible hydrogen fuel cell material. It has been noted in the recent review that MOFs have been synthesized with proton conductivities that can rival that of the Nafion (Ren et al. 2013).

4 Nanotechnology for Batteries

Today's society is faced with a growing need for batteries, bearing in mind that the world becomes increasingly populated with mobile electronic products. Since its invention by Italian Count Alessandro Volta in 1799, their design and performance have evolved, but there is still a lot of room for further improvement. Actually, further development of existing products such as cell phones and laptop computers and in new applications such as electric car highly depend on improving current battery technology that needs upgrades in its life, power, recharging times, size and weight, beside other parameters. However, the development of batteries is relatively slow, mostly due to the lack of suitable electrode materials and electrolytes, together with difficulties in controlling the interfaces between them. For development of battery chemistry over the years see Fig. 7. It is well known, that batteries are inherently simple in concept, consisting of one or more electrochemical cells that convert stored chemical energy into electrical energy. The cells consist of a positive terminal, or *cathode*, and a negative terminal, or *anode*, while *electrolytes* enable ions to move between the electrodes and terminals, which in turn allows current to flow out of the battery to perform work. Two forms of batteries exists: (1) *primary* that are used once and discarded and (2) *secondary* batteries that can be recharged. Basically, we have two electrodes with different chemical potential and in case they are connected to an external device, electrons spontaneously flow from anode to cathode (Armand and Tarascon 2008). The charge balance is maintained by ions transported through electrolyte and the produced energy is tapped by the external circuit (Armand and Tarascon 2008). In rechargeable batteries, a larger

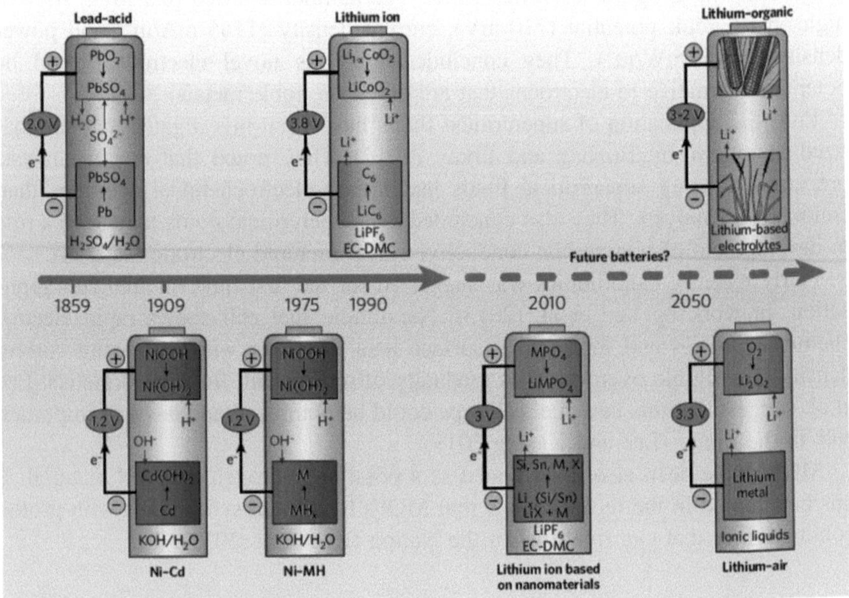

Fig. 7 Development of battery chemistry over the years. Reprinted with permission from (Armand and Tarascon 2008)

voltage applied in the opposite direction results in battery recharging (Armand and Tarascon 2008). The materials from which the batteries are composed have suffered many changes. Since their discovery, numerous solutions were proposed of which the most notable primary batteries are zinc–carbon and Zn-MnO_2, while lead-acid, Ni-Cd, NiMH, and Li-ion are the most common rechargeable batteries.

As it was mentioned above, growing electronic devices market and its wide application in everyday life pushed researchers to increase the investigations to find better solutions in rechargeable batteries, in order to provide higher specific density, higher cell voltage, longer shelf life, usable volumetric specific capacity, and to be relatively environmentally friendly.

Nowadays, the best performance of all rechargeable batteries with respect to energy density is exhibited by Li-ions batteries. They were suggested in the mid 1970's and the first time commercialized by Sonny in 1991. These batteries have a number of advantages in comparison to other rechargeable cells, such as nickel–cadmium (Ni-Cd), nickel metal hydride (Ni-MH), lead-acid, and lithium cells. For example, Ni-Cd and Ni-MH batteries provide a cell voltage of about 1.2 V, lead-acid batteries 2 V, while use of lithium in lithium ion cells provides a higher cell voltage of up to 4 V (Michael et al. 2009). In addition, lithium ion batteries have better self-discharging, a much longer shelf life, and also experience little or no memory effect. Also, because they do not contain lead, cadmium or mercury, they are considered relatively environmentally safer.

Basically, in Li-ion battery the small Li-ion can be inserted into an atomic structure or alloy with a metal of an electrode material. The commercial batteries consist of a carbon-based anode and $LiCoO_2$, or nickel and manganese substituted $LiCoO_2$, as the cathode. The open circuit cell voltage is 3.7 V that makes it more usable in comparison to other batteries. This comes because the Li-ion has the small radius, low density and its position at the very end of the electrochemical potential series (Philippe et al. 2013).

On other hand, current lithium ion battery has some disadvantages that need to be overcome in order to fulfill growing demand for improving their capacities. For example, the reversible capacity of the batteries rapidly fades over a number of charging and discharging cycles, despite lithium ion batteries having graphite anodes demonstrate a high initial capacity. It is believed that this is caused by exfoliation of graphite and/or irreversible electrochemical reduction of electrolyte that occurs to form a solid interphase layer at anode, which limits their applications where a large number of charging and discharging cycle is required (Michael et al. 2009). It is therefore obvious that the knowledge about different Li—containing materials and intercalation compounds opens up new areas within high and/or low, power applications.

Current research is focused on finding new high capacity anode and cathode materials. Bearing this in mind silicon comes as natural choice for studying replacement of the graphite anode (Philippe et al. 2012). It has been shown that silicon gives practical capacities of more than 2000 mAh/g that is actually more than five times larger than for graphite. However, silicon as a potential high-capacity anode material has high theoretical specific capacity of 4212 mAh g^{-1}, higher stability and safety in comparison to graphite (Yin et al. 2012), suffer from some drawbacks related to the drastic volume change (more than 300%) during the alloying/de-alloying reactions with Li (Huggins 1999), the intrinsic low electrical conductivity and unstable solid electrolyte interphase (Yin et al. 2012). In this respect, the research community is searching for the new solutions in the field of nanomaterials. This field was opened after revolutionary discovery of fullerenes and nanotubes and more recent graphene, because of their unique properties and a bit different physics lying behind in comparison to existed materials. Because of this, a completely new approach is taken in fabrication of materials by putting them on nano-scale. However there are some advantages but also disadvantages of nanomaterials for rechargeable Li-ion batteries. For example, they enable electrode reactions to occur, which cannot take place for micrometer-sized materials, but production of nanoparticles is not fully concurred and their dimensions can be difficult to control. Further, their small dimension increases the rate of lithium insertion/removal, because of the short distances for lithium-ion transport within the particles as well as electron transport within the particles (Bruce et al. 2008).

A high lithium-ion flux across the interface between a contact area and the electrolyte can be achieved by enlarging surface area that is the effect of their reduced dimensions.

However, this advantage, high electrolyte/electrode surface area may lead to more significant side reactions with the electrolyte, and more difficulty maintaining contact between particles (Bruce et al. 2008). It is worth to highlight that for the same mass of material the volume of the electrode increases reducing the volumetric energy density, as a consequence of the nanopowder less density than the density of same micrometer-sized material (Bruce et al. 2008). Nevertheless, a change of electrode potential, for very small particles, may modify the chemical potentials for lithium ions and electrons. Also, due to exertion associated with intercalation is often better accommodated for nanoparticles and the range of composition over which solid solutions exist is often more extensive (Bruce et al. 2008).

Many modifications of materials for electrodes as well as for electrolytes that use nanomaterials have been proposed. Different types of nanostructure have been investigated, such as nanoparticles, nanotubes/nanowires, graphene and nanoalloys. In this respect, among the first materials, the researchers started to optimize Si-based materials through diverse Si nanostructures, Si/C and Si/metal nanocomposites (Yin et al. 2012). It has been demonstrated that zero-dimensional (0D) silicon nanospheres exhibited a large specific capacity 3052 mAh cm^{-1} (current density 2000 mA cm^{-1}, electrolyte $LiPF_6$), and even after 48 cycles specific capacity was 1095 mAh cm^{-1} (Yin et al. 2012). In addition, excellent capacity retention showed interconnected Si hollow nanospheres electrode. Further, array of sealed Si one-dimensional (1D) nanowires or nanotubes, exhibited 3360 mAh cm^{-1} of charge capacity, which retention after 50 cycles were 81% (Yin et al. 2012). Additionally, Si thin films (2D) in principle can reduce volume variation and retain structural integrity, while cycling stability and Li accommodation capacity depend on film thickness (Yin et al. 2012). Despite pristine Si nanostructures exhibited improved capacity retention, they may lead to the risk of excessive side-reactions with electrolyte, besides the limitations of pure Si nano-materials (low electrical conductivity of Si) (Guo et al. 2008). In this respect, macroporous Si was coated with Ag, as metal with good electronic conductivity forming 3D structure, which delivered both a good reversible capacity and enhanced rate capability (Yu et al. 2010a).

Some of the research was focused on titanium oxides, $Li_4Ti_5O_{12}$, to find a replacement for graphite in order to link low cost, low toxicity, and the ability to be produced at a nanoscale level delivering fast lithium insertion/removal, with inherent protection against lithium deposition (Bruce et al. 2008). This material is environmentally safer and, produced at nanoparticle level, provides high rates of lithium insertion/removal owing to the short diffusion distances in the nanoparticles (Bruce et al. 2008). The produced prototype (Reale et al. 2004), did not demonstrated significant advantage since its capacity to store lithium is only half that of graphite, the cell voltage is reduced because of the increased potential of the negative electrode, which together leads to a reduced energy density (Bruce et al. 2008). Schematic representation of lithium insertion/de-insertion mechanism for current rechargeable lithium battery is presented in the Fig. 8.

However, these materials are very good candidates for application in high power batteries for application in hybrid electric vehicles and systems that require high power, because of their high reversible capacity, the high power density and the safety reasons (Goriparti et al. 2014).

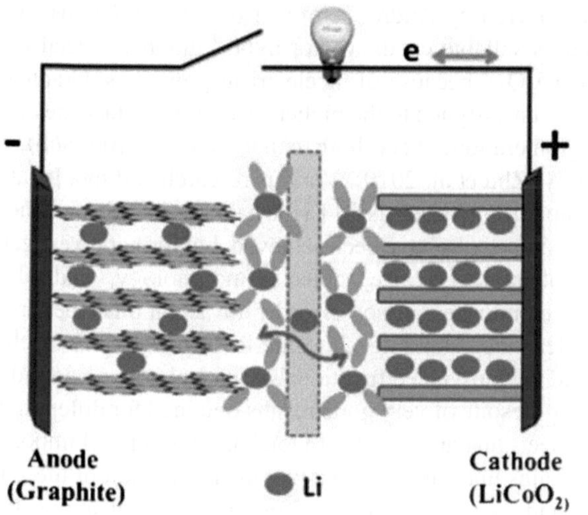

Fig. 8 Schematic representation of lithium insertion/de-insertion mechanism for current rechargeable lithium battery. Reprinted with permission from (Goriparti et al. 2014)

Active anode material	Theoretical capacity ($mAh\ g^{-1}$) [Reference]	Advantages	Common issues
Insertion/de-insertion materials			
A. Carbonaceous a. Hard carbons b. CNTS c. Graphene	200-600 1116 780/1116	➢ Good working potential ➢ Low cost ➢ Good safety	❖ Low coulombic efficiency ❖ High voltage hysteresis ❖ High irreversible capacity
B. Titanium oxides a. $LiTi_4O_5$ b. TiO_2	175 330	➢ Extreme safety ➢ Good cycle life ➢ Low cost ➢ High power capability	❖ Very low capacity ❖ Low energy density
Alloy/de-alloy materials			
a. Silicon b. Germanium c. Tin d. Antimony e. Tin oxide f. SiO	4212 1624 993 660 790 1600	➢ Higher specific capacities ➢ High energy density ➢ Good safety	❖ Large irreversible capacity ❖ Huge capacity fading ❖ Poor cycling
Conversion materials			
a. Metal oxides(Fe_2O_3, Fe_3O_4, CoO, Co_3O_4, Mn_xO_y, Cu_2O/CuO, NiO, Cr_2O_3, RuO_2, MoO_2/MoO_3 etc.) b. Metal phoshides/sulfides/nitrides (MXy: M = Fe, Mn, Ni, Cu, Co etc. and X = P, S, N)	500–1200 500–1800	➢ High capacity ➢ High energy ➢ Low cost ➢ Environmentally compatibility ➢ High specific capacity ➢ Low operation potential and Low polarization than counter oxides	❖ Low coulumbic efficiency ❖ Unstable SEI formation ❖ Large potential hysteresis ❖ Poor cycle life ❖ Poor capacity retention ❖ Short cycle life ❖ High cost of production

Fig. 9 Most common anode materials used for lithium ion batteries. Reprinted with permission from (Goriparti et al. 2014)

Another group of materials that have been investigated because of its promising properties, such as specific capacities that exceed that of the conventional graphite anode, are lithium metal alloys, Li_xM_y, *but their lifetime is limited to only a few* charge–discharge cycles. In Fig. 9 a review of the most common anode materials used for lithium batteries is given.

Recently discovered by Andre Geim and Konstantin Novoselov, graphene was also considered as suitable candidate for hybrid graphene/metal anodes (graphene with SnO_2 and Fe_2O_3), because of its electrical properties. Graphene's remarkable properties have been assigned to the higher accessible surface area (~2600 m^2/g due to lack of agglomeration), very high conductivity (~100 S/m), and very good chemical stability (Zhu et al. 2010). Various research and modification of graphene nanosheets as supporting materials for Li ion batteries have been done (Sun and Wang 2014) and promising results have been achieved, but further investigation is needed to reach control of porosity, size, defects, morphology, and alignment of the graphene-based composites. Also, in order to accomplish better electrochemical performance a clear understanding of the interfacial interactions mechanisms between the graphene and the introduced functional second-phase compounds is needed (Sun and Wang 2014). Despite of very good results obtained for different modifications of graphene nanosheets that are used for Li ion batteries, a lot of important features and mechanisms remains unclear. For example, the production of these materials is still very expensive and therefore limits their commercial applications. Further, the basic lithium storage mechanism in graphene-based composites is not fully clear, which is especially important for their surface functionalities/defects and hierarchical electrode structures (Sun and Wang 2014). In addition, a relatively high voltage range for lithium insertion and extraction in some of graphene-supported composites such as transitional metal oxides and metal sulphides may affect the stable output of the energy and reduce the power density of the battery (Sun and Wang 2014). Notwithstanding, materials for lithium-ion batteries based on grapheme electrode are very promising for future applications in many area such as electronics, electric vehicles, hybrid electric vehicles, and scalable energy storage (Sun and Wang 2014).

Generally, the use of nanomaterials in lithium-ion batteries can bring advantages such as higher electron transfer rates, enhanced power capability, improved triggering of some of electrochemical reactions in bulk materials and increased specific capacity because of intensification of the presence of active sites for lithium storage that is consequence of the active materials production, which have a high surface to volume ratio (Goriparti et al. 2014).

Beside the extensive research that has been done in application of nanostructured materials and its modifications as anode materials for lithium-ion batteries, some breakthrough has been made in their applications in conventional liquid electrolytes used in rechargeable lithium batteries as well as for cathodes. The nanopowders addition, such as ZrO_2, SiO_2 and Al_2O_3 to non-aqueous electrolytes may increase the conductivity by a factor of six (Bruce et al. 2008). It is also possible to produce nanostructured positive electrodes of a range of dimensions, and in general the performance is enhanced in comparison to bulk materials, which is especially refered to the rate capability (Bruce et al. 2008).

5 Other Applications of Nanomaterials for Sustainable Energy—Transportation and Electrical Energy Transmission

Generated energy must be distributed to the consumers. This problem raises many challenges, namely in transportation and transmission.

For example, energy can be conserved if transportation is reduced. Some authors suggest that there is possibility, when nanotechnology reaches the stage of personal nanofactories, that need for transport of the raw materials will be reduced (Ramsden 2014). They also noted that somewhere between 30 and 70% of total energy consumption is estimated to amounts only to transportation (Ramsden 2014). Also, lesser use of transportation implies smaller CO_2 emission. U.S. Environmental Protection Agency notes that about 32% of total U.S. CO_2 emission is because of transportation (US EPA 2015). Currently, about 25% of worldwide CO_2 emissions are attributed to transport according to International Energy Agency (OECD/IEA 2009). Lubricants are important in transportation. Addition of nanoparticles to conventional lubricants can lead to more environmentally friendly lubricants (Alves et al. 2013). Also, it can lead to less energy loses during operation of lubricated elements, thus potentially lowering fuel usage and CO_2 emission.

Transmission of energy through the electrical conductors is also accompanied by energy loses. Transmitted energy dissipates as heat because of the resistance of electrical conductors. CNTs are considered to be possible candidates for future electrical cabling (Ramsden 2014) and electronic applications in general (Aguilar et al. 2010).

There are number of challenges in the fabrication of CNT wires. CNTs must be assembled into macroscopic structures, in such a way that they do not lose their excellent mechanical and electrical properties (Lekawa-Raus et al. 2014). CNT yarns (also called threads or wires) are of particular relevance for this application (Lekawa-Raus et al. 2014). CNTs can be fabricated in the form of yarn by spinning, for example see references (Zhang et al. 2004; Atkinson et al. 2007; Miao 2013; Tran 2014), and Fig. 10. These CNT yarns show good electrical conductivity

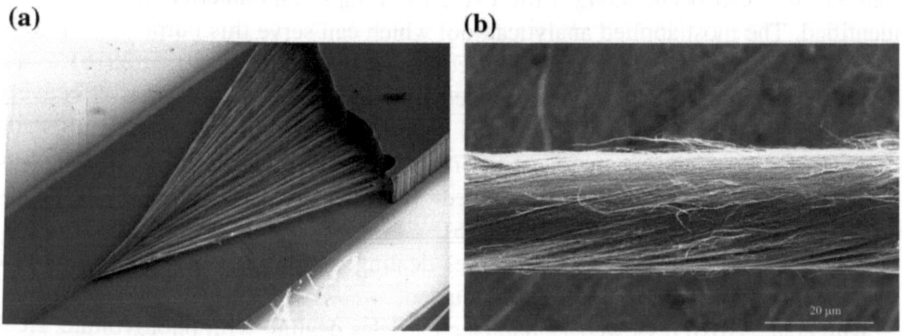

Fig. 10 Formation of CNT yarn by twist insertion (**a**) and SEM image of a CNT yarn (**b**). Reprinted with permission from (Miao 2013)

comparable with that of Cu and Al (Ramsden 2014). CNT yarns also demonstrated use as superconducting wires and battery and fuel cell electrodes (De Volder et al. 2013). It has been demonstrated that CNT wires can withstand salty and highly acidic conditions without surface protection and with no apparent degradation in their properties (Janas et al. 2013). However, most electrical wires need to have isolation. The inexpensive method for insulation that does not compromise electrical and mechanical performance of CNT wires was proposed in (Lekawa-Raus et al. 2014).

It has been reported that CNT wires and ribbons could be viable replacement for metallic conductors, which has been demonstrated on coaxial, USB and Ethernet cables (Jarosz et al. 2011).

Except good electrical conductivity, CNT materials have other potential benefits e.g. reduction in cable mass per unit length (Jarosz et al. 2011). Reduction in mass per unit length has significant implications. As summarized in (Lekawa-Raus et al. 2014) the use of lightweight CNT wires in the overhead power lines could lead to significant benefits, such as less strain on the supporting pylons and in the wires themselves. Lekawa-Raus et al. also noted that there could be potential application in transportation industry, since CNT wires could lead to reduction in weight of transportation machines (e.g. automobiles and airplanes) which in turn leads to reduction of fuel needed to propel these machines, thus making transportation more economic and environmentally friendly (Lekawa-Raus et al. 2014).

6 Life Cycle Assessment for Nanotechnologies Used in Systems for Sustainable Energy Production and Storage

Beside previously mentioned advantages of integration of nanomaterials and nanostructures in sustainable energy technologies, possible negative effects from their production, use and end of life, i.e. disposal and/or recycling should also be analyzed. In other words, all possible impacts on the human health and the environment that can occur along a life cycle of certain nanomaterial should also be identified. The most applied analytical tool which can serve this purpose is the Life Cycle Assessment (LCA) (Pieragostini et al. 2012; Perić et al. 2015, 2016).

Life cycle assessment can be defined as "an analytical tool that effectively creates a mass balance over an industrial system by analyzing all of the inputs and outputs of a product system over its entire life cycle" (Perić et al. 2016). The product of the system is usually »followed« from its »cradle«, extraction till its use and disposal—»grave« (Perić et al. 2016).

So far, certain number of LCA studies dealing with use of nano-materials in a sustainable energy sector has been conducted.

Kim et al. (2016) reviewed 72 published articles dealing with photovoltaic, LCA and nanotechnology, and indicated that only 8 of them are dealing with all three categories at the same time, i.e. actually consider the application of LCA on nanomaterials integrated in photovoltaic (PV) technology (Greijer et al. 2001; Fleischer and Grunwald 2008; Espinosa et al. 2011; Şengül and Theis 2011; Hsu et al. 2012; Kim et al. 2012b; Kim and Fthenakis 2013; Mater et al. 2013). The focus of these articles is on the production and use phase of nanomaterials, whereas no articles considering LCA of nanomaterials in the end-of-life phase have been identified.

Kim and colleagues summoned five studies as the best examples of the carbon footprint assessment of modern thin-film PV technologies, more precisely of the amorphous silicon (a-Si), cadmium telluride (CdTe) and copper indium gallium diselenide (CIGS) and harmonized results through alignment of efficiency, irradiation, performance ratio, balance of system, and lifetime (Kim et al. 2012b). Under given conditions (ground-mount application under southwestern United States irradiation of 2,400 kilowatt-hours per square meter per year (kWh/m^2/year), a performance ratio of 0.8, and a lifetime of 30 years), the resulting estimates for carbon footprints (g CO_2-eq/kWh) for CIGS were 26 g, for CdTe were 20 g and for a-Si 17 g of CO_2 and after the harmonization (for the roof-top PV systems with a performance ratio of 0.75 and the same irradiation) carbon footprint estimates were 27, 21 and 14 g CO_2-eq/kWh, respectively.

Similar analysis was performed for crystalline silicon (c-Si) by Hsu et al. (2012). Prior to harmonization, the median estimation of life cycle GHG emissions was 57 g carbon dioxide equivalent per kilowatt-hour (g CO_2-eq/kWh) whereas after the harmonization median estimate decreased to 45 g (Hsu et al. 2012).

LCA of a nanocrystalline dye sensitized solar cell (ncDSC) was performed by Greijer et al. (2001). This study analyzes only impacts from the generation of electricity from the solar cell system, where the transformation, the storing and systems for utilizing stored energy were omitted. The materials, the production of materials, the end of life of every component, i.e. recycling, landfill and incineration and an average process energy for production of the solar cell modules are included in this study (Greijer et al. 2001). The highest impact in ncDSC system comes from the process energy for the production of the solar cell module, then from substrate glass, frame and junction box. The most energy demanding steps in the manufacturing processes of the monolithic solar cells are the deposition and sintering of the porous layers. High impact from substrate glass is caused by the recycling stage in which large fraction of the lead is emitted. According to weighted methods applied, the most significant emissions are emissions of sulphur dioxide and carbon dioxide. The comparison of carbon footprint from electricity generation of a ncDSC system and a natural gas/combined cycle power plant designates ncDSC system as a less environmentally harmful due to emission of only 19–47 g CO_2 per kWh compared to 450 g CO_2/kWh occurring from a natural gas/combined cycle power plant. Higher solar cells efficiency, development of production technologies that uses low energy demanding processes, minimization of the material in the solar cell systems with maximization of the components recycling are proposed measures for further development and improvement of these systems.

Kim and Fthenakis reviewed 22 LCA studies and categorized them into few groups: studies of nanostructured (single) materials, nanostructured composites, nanophotovoltaics, nanocoatings, and other technologies (Kim and Fthenakis 2013). The emphasis was put on greenhouse gas (GHG) emissions and energy-related impacts and indicators (e.g., primary energy). They concluded that nanomaterials have higher cradle-to-grave energy demand per functional unit, and thus higher global warming impact, than their conventional counterparts. In order to produce 1 kg of carbon-based nanoparticles (carbon nanofibers, carbon nanotubes, and fullerenes), 1–900 gigajoules (GJ) of primary energy is required (depending on the synthesis method), whilst around 200 megajoules per kilogram (MJ/kg) of primary energy is required for aluminum production. Such a high energy requirements are due to the energy-intensive synthesis process of nanomaterials or due to the involvement of an additional mechanical process for particle size reduction. However, compared with conventional technologies, the "cradle-to-grave" energy demand and global warming impact from nanotechnologies at a device level, are actually lower because the amount of nanomaterials used in these devices is usually very small but sufficient to improve functionality and energy-efficiency (Kim and Fthenakis 2013).

Maybe one of the most comprehensive LCA studies for nanostructured PV conducted so far is the study done by Şengül and Theis (Şengül and Theis 2011). This study includes the analyze of CdSe quantum dot photovoltaic module (QDPV) life cycle (from raw materials acquisition, manufacturing to its use phase), comparison of QDPV with other PV modules (Ribbon multicrystalline-silicon (Ribbon-Si), Multicrystalline-silicon (Multi-Si), Monocrystalline-silicon (Mono-Si), Compound semiconductor CdTe/CdS, Copper indium gallium diselenide, Copper indium diselenide (CIGS/CIS), Dye sensitized PV (DSPV)) and comparison of QDPV with other energy sources, both renewable and nonrenewable (coal, oil, lignite, natural gas, diesel, nuclear, wind and hydropower). The energy output of QDPV modules during the use phase and four environmental impacts over the life cycle of PV modules that were investigated are CED, GWP, aquatic acidification potential, and heavy metal emissions. The GWP of CdSe QDPV is calculated to 25 kg CO_2-eq/m^2 or 5 g CO_2-eq/kWh at 14% efficiency, where the biggest contribution comes from the production stage of the quantum dot solar cell (70% of the total GWP), and the rest comes from the production of the PV module. Compared to other PV modules, QDPV modules exhibit shorter Energy Pay Back Time (EPBT) (except CdTe PV modules), lower Global Warming Potential (GWP), SOx and NOx emissions, but higher heavy metal emissions. Heavy metal emissions originate from the quantum dot solar cell production phase, due to hazardous waste disposal, aluminum foil, 1-butanol, ITO and methanol and electricity production. Compared to carbon-based energy sources, QDPV demonstrate better performance in all analyzed impact categories, except having the longer EPBT and higher GWP than wind and hydropower (Şengül and Theis 2011).

The results from the study conducted by Peng and colleagues coincides with previously mentioned study (Peng et al. 2013). Since QDPV modules were not included in the analysis, the results of the comparison of life cycle assessments of mono-Si, multi-Si, CdTe, CIS and amorphous silicon (a-Si), determines CdTe as the most environmental friendly solution for the application in PV systems in the terms of EPBT and GWP. For thin film PV systems (a-Si, CdTe and CIS), the life cycle total energy input ranges from 710 to 1990 MJ/m^2, the EPBT ranges from 0.75 to 3.5 years and GHG emission ranges from 10.5 to 50 g CO_2-eq./kWh. Advanced PV modules (high-concentration and dye-sensitized) were also reviewed. It is revealed that EBPT of high-concentration PV system is almost the same as for CIS and CdTe, where CO_2 emission rate of dye-sensitized PV system is a little higher. Authors suggest that the environmental performance of PV technologies will improve in the near future (by the advancement of the production processes, reduction of silicon and other raw materials consumption, increasing material recycling rates, etc.) (Peng et al. 2013).

The results from the previous study conducted by Sengül and Theis (Sengül and Theis 2009) which investigates environmental impacts during the synthesis of quantum dot and considers raw material use, energy use, and emissions to air, water, and soil, identifies the acquisition of raw materials as a potentially major contributor to environmental impacts. Considering conventional route of synthesis of CdSe quantum dots the biggest contribution to the cumulative emissions comes from air emissions, then from water and soil emissions (Sengül and Theis 2009).

The environmental assessment of carbon nanotubes (CNTs) used in the renewable energy installations was also conducted by various research groups (Köhler et al. 2008; Healy et al. 2008; Tan et al. 2012a; Amarakoon et al. 2013; Kim and Fthenakis 2013; Pereira and Coelho 2015).

Three SWNT production processes were investigated by Healy et al. (2008): arc ablation (arc), chemical vapor deposition (CVD), and high-pressure carbon monoxide (HiPco). The production of SWNTs is divided into the four separate steps: synthesis, purification, inspection, and packaging. Among all considered impacts the most dominant impact categories in these processes appear to be airborne inorganics, climate change and acidification. The results indicates that the highest impacts for all three SWNT production processes and in all three impact categories, come from the synthesis step, predominately due to electricity consumption (99% from the total impact) (Healy et al. 2008). These results coincides with the results published by Isaacs et al. (2006), Pereira and Coelho (2015). They confirmed that from all three SWNT production processes, the least environmentally burdensome process is HiPco, where the most environmentally burdensome process is CVD. The highest environmental impacts from CVD process come from the large electricity load due to higher number of batch processes needed to achieve the same production volume as the other production processes (Isaacs et al. 2006).

Köhler and colleagues investigated possible release of CNTs throughout all phases of a Li-ion battery life cycle (Köhler et al. 2008). They concluded that the emissions of CNTs may occur during all phases of the battery life cycle with higher risk for manufacturing and recycling and disposal phases where battery cells are either unsealed or damaged (Köhler et al. 2008). Also, Muller and Nowack stated that around 50% of CNTs can be released during battery end-of-life phase (Mueller and Nowack 2008). In the use phase risk of CNTs emission is much lower but not impossible and can be caused by the faulty operation by the battery user (Köhler et al. 2008).

Amarakoon and colleagues compared production of SWCNT anode and the battery-grade graphite anode (Amarakoon et al. 2013). Considering laboratory modeling data, energy requirements for the production of the SWCNT anode as well as abiotic depletion potential, global warming potential, acidification potential, eutrophication potential, ozone depletion potential, photochemical oxidation potential, ecological toxicity potential, human toxicity potential and occupational cancer hazard are much higher compared to the battery-grade graphite anodes production. In order to obtain potential benefits from the replacement of graphite anode with SWCNT anode, authors suggest that electricity consumption during SWCNT manufacture has to be reduced to 11 kWh per kWh capacity. In that case all impacts on the human health and the environment would be comparable to the ones occurring from the graphite anode production, except the occupational non-cancer hazard impacts, since SWCNT-based anode has a hazard value that is much higher than the geometric mean hazard of all chemical feedstocks in the impact category. On the contrary, ozone depletion potential is practically negligible (Amarakoon et al. 2013).

When analyzing environmental impacts from the production, use and the end-of-life of nanomaterials, of great importance is to evaluate toxicity. Unfortunately, questions about the potential toxicity of nanomaterials mainly stay unanswered.

Certain number of studies trying to answer this question has been published so far. Under certain conditions, using primary hepatocytes, Derfus and colleagues proved cytotoxicity of CdSe-core QDs which is caused by the liberation of Cadmium ions (Derfus et al. 2004). Surface oxidation through a variety of pathways can led to the formation of reduced Cd on the QD surface which can cause the release of free cadmium ions that correlates with cell death. Application of surface coatings such as ZnS and BSA are suggested to significantly reduce, but not eliminate QD cytotoxicity (Derfus et al. 2004).

Zhang and Monteiro-Riviere state that quantum dot can penetrate through skin abrasions (Zhang and Monteiro-Riviere 2008) and Holbrook and colleagues state that quantum dots can be transferred through a small food chain (Holbrook et al. 2008).

Genotoxicity of CNTs has been studied in many studies. Cytotoxic and genotoxic properties of SWCNT have been verified in cells of the human gastrointestinal tract (Pelka et al. 2013). The mitotic spindle aberrations of human airway epithelial cells in workers exposed to SWCNTs are observed by Sargent et al. (2012). The increased proliferation and chromosome aneuploidy observed in these cells can lead to both tumorigenesis and tumor progression (Sargent et al. 2012). Van Berlo and colleagues showed that short-term inhalation exposure to pure carbon nanoparticles can trigger oxidative stress responses in the lungs of mice but not in their brains (van Berlo et al. 2014).

TiO_2 nanoparticles can also induce genotoxic effects both in vivo and in vitro tests. Results of the different TiO_2 genotoxicity tests are often contradictory. Lindberg and colleagues observed no significant effect on the level of DNA damage in lung epithelial cells of mice, after 5-day inhalation exposure to nanosized TiO_2

(Lindberg et al. 2012) whereas Huerta-García and colleagues showed that TiO_2 nanoparticles can enter directly into the brain through the olfactory bulb and can be deposited in the hippocampus region, damaging rat and human glial cells (Huerta-García et al. 2014).

7 Summary and Conclusion

This survey has given an extensive overview of nanotechnology implementations for production, storage and transportation of electrical energy in a more sustainable manner. Nevertheless it is far from being really comprehensive, and it is not intended to be. There is a huge number of researchers dealing with problems in these domains, and exiting results are published on, almost, daily bases.

Nanostructured materials offer promising advantages for PV applications, especially in terms of improved efficiency and reduced production costs. Nevertheless, none of the mentioned PV devices based on nanomaterials has entered a large scale production, yet. Apart from good performance in controlled environment, there are some general issues concerning, more or less, all of the discussed nano-based PVs that need to be resolved before their successful commercialization. A leap from 1 cm^2 nanostructured solar cell to production-ready module is a huge one. Long-term environmental stability, robustness and adequate manufacturing technologies for mass production are still challenges that need to be properly addressed. At this point in time, hybrid solar cells based on organic thin films (conductive polymers) and inorganic nanoparticles and tandem solar cells with embedded QWs and QDs seem to be technologies that are closest to commercial exploitation. Other mentioned approaches are all, more or less, at earlier stages of research.

Nanotechnology also enabled new approaches for engineering the next generation of materials for lithium batteries that can be applied as anode, cathode or materials for electrolytes. However, further investigations are needed for the successful commercialization of such produced batteries. One very important objective, in these terms, is the investigation of the mechanisms, which control the interaction between lithium and the nanostructured forms of investigated materials together with the electron transport properties at the electrode/electrolyte interface. In order to obtain applicable and economically viable product it is necessary to master fabrication for large scale synthesis of nanostructured materials, which will achieve both higher energy and higher power density.

A review of hydrogen economy was also given, specifically in regard to hydrogen production, storage and conversion. On the whole, the process can be envisioned starting with conversion of solar energy into hydrogen by means of photocatalysis. Then, using the adequate nanomaterial or nanostructured material, hydrogen could be stored for later use. The final step is conversion of hydrogen into electric energy using a fuel cell. The substance that goes into and out of this process is water (water being the waste product of a fuel cell), thus this is clean and renewable energy economy.

There are many challenges in the production of hydrogen. The photocatalysis is considered not efficient enough. Many possible research paths are proposed for dealing with the production of hydrogen. For example, doping of the standard photocatalytic material (e.g. TiO_2), examining different morphologies of materials (e.g. whiskery, particulate etc.), and investigating the effects of various means of photocatalytic materials fabrication (e.g. by ball milling).

The most efficient way to store hydrogen is to bond it to a solid-state material. However this leads to hydrogen atoms that are either too strongly bounded, or a storage material that has a very low storage capacity at the room temperature. Different materials and effects of fabrication on storage capacity have been investigated. Nanoporous carbon materials could be a good candidate for hydrogen storage application.

As for materials related to hydrogen conversion, carbon is considered to be a promising material. Investigated alternatives for standard fuel cells are biological fuel cells. Also, as it was outlined in the article, there could be viable potential in using nanofluidics for hydrogen conversion.

So far, considerable number of LCA studies identifying possible environmental risks from production, use and end-of-life of nanomaterials installed in the sustainable energy systems has been conducted. Most of these studies consider only energy requirements during the production and use of nanomaterials and the global warming potential. Other potential environmental impacts such as acidification, eutrophication, ozone depletion, etc., are often omitted. Basically, all these LCA studies are not actually "cradle-to-grave" since they are only focused on impacts during production and use of nanomaterials while the impacts from the end-of-life are generally omitted due to lack of data. There are no data regarding impacts of nanowaste since the life cycle of nanomaterials are very long—some of the first manufactured nanomaterials are still in use. New nanoproducts are being produced while the potential environmental impacts occurring along the life cycle of the existing ones have not yet been determined.

The biggest concern regarding the use of nanoparticles is their potential toxicity. Even though there is no consensus among different toxicity studies conducted, possible citotoxicity and genotoxicity of certain number of nanoparticles have been confirmed. Nevertheless, scientist agree that the development of new technologies for the production of nanomaterials, together with the improvements of the existing ones, will reduce toxicity of nanomaterials and thus make application of nanomaterials in systems for sustainable energy production even more suitable.

Conflict of Interests Authors declare that there is no conflict of interest that they are aware of.

References

Abbaraju RR, Dasgupta N, Virkar AV (2008) Composite nafion membranes containing nanosize TiO[sub 2]/SnO[sub 2] for proton exchange membrane fuel cells. J Electrochem Soc 155: B1307

Abdelraouf OAM, Allam NK (2016) Towards nanostructured perovskite solar cells with enhanced efficiency: coupled optical and electrical modeling. Sol Energy 137:364–370. doi:10.1016/j.solener.2016.08.039

Adams JGJ, Elder W, Stavrinou P et al (2009) Efficiency enhancement in strain-balanced quantum well solar cells via anisotropic emission. In: Proceedings of 24th European photovoltaic solar energy conference, pp 19–21

Aguilar JO, Bautista-Quijano JR, Avilés F (2010) Influence of carbon nanotube clustering on the electrical conductivity of polymer composite films. Express Polym Lett 4:292–299. doi:10.3144/expresspolymlett.2010.37

Ahmad H, Kamarudin SK, Minggu LJ, Kassim M (2015) Hydrogen from photo-catalytic water splitting process: a review. Renew Sustain Energy Rev 43:599–610. doi:10.1016/j.rser.2014.10.101

Akimov YA, Koh WS (2011) Design of plasmonic nanoparticles for efficient subwavelength light trapping in thin-film solar cells. Plasmonics 6:155–161. doi:10.1007/s11468-010-9181-4

Alivisatos AP (1996) Semiconductor clusters, nanocrystals, and quantum dots. Science 271 (80):933–937

Alves SM, Barros BS, Trajano MF et al (2013) Tribological behavior of vegetable oil-based lubricants with nanoparticles of oxides in boundary lubrication conditions. Tribol Int 65:28–36. doi:10.1016/j.triboint.2013.03.027

Amarakoon S, Smith J, Segal B (2013) Application of life-cycle assessment to nanoscale technology : lithium-ion batteries for electric vehicles, pp 1–119

Aminorroaya S, Ranjbar A, Cho YH et al (2011) Hydrogen storage properties of Mg-10 wt% Ni alloy co-catalysed with niobium and multi-walled carbon nanotubes. Int J Hydrogen Energy 36:571–579. doi:10.1016/j.ijhydene.2010.08.103

Andersen SM, Borghei M, Lund P et al (2013) Durability of carbon nanofiber (CNF) and carbon nanotube (CNT) as catalyst support for proton exchange membrane fuel cells. Solid State Ionics 231:94–101. doi:10.1016/j.ssi.2012.11.020

Andrews J, Shabani B (2012) Where does hydrogen fit in a sustainable energy economy? Procedia Eng 49:15–25. doi:10.1016/j.proeng.2012.10.107

Arici E, Sariciftci NS, Meissner D (2003) Hybrid solar cells based on nanoparticles of $CuInS_2$ in organic matrices. Adv Funct Mater 13:165–171. doi:10.1002/adfm.200390024

Armand M, Tarascon J-M (2008) Building better batteries. Nature 451:652–657. doi:10.1038/451652a

Asgari MS, Nikazar M, Molla-Abbasi P, Hasani-Sadrabadi MM (2013) Nafion®/histidine functionalized carbon nanotube: high-performance fuel cell membranes. Int J Hydrogen Energy 38:5894–5902. doi:10.1016/j.ijhydene.2013.03.010

Atkinson KR, Hawkins SC, Huynh C et al (2007) Multifunctional carbon nanotube yarns and transparent sheets: fabrication, properties, and applications. Phys B Condens Matter 394: 339–343

Attia NF, Lee SM, Kim HJ, Geckeler KE (2013) Nanoporous carbon-templated silica nanoparticles: preparation, effect of different carbon precursors, and their hydrogen storage adsorption. Microporous Mesoporous Mater 173:139–146. doi:10.1016/j.micromeso.2013.02.016

Atwater HA, Polman A (2010) Plasmonics for improved photovoltaic devices. Nat Mater 9: 205–213

Aubry A, Lei DY, Fernández-Domínguez AI et al (2010) Plasmonic light-harvesting devices over the whole visible spectrum. Nano Lett 10:2574–2579. doi:10.1021/nl101235d

Barghi SH, Tsotsis TT, Sahimi M (2014) Chemisorption, physisorption and hysteresis during hydrogen storage in carbon nanotubes. Int J Hydrogen Energy 39:1390–1397. doi:10.1016/j.ijhydene.2013.10.163

Barnes J, Nelson J, Barnham KWJ et al (1996) Characterization of GaAs/InGaAs quantum wells using photocurrent spectroscopy

Barnham KWJ, Ballard I, Connolly JP et al (2002) Quantum well solar cells. Phys E Low-Dimension Syst Nanostruct 14:27–36. doi:10.1016/S1386-9477(02)00356-9

Barnham KWJ, Braun B, Nelson J et al (1991) Short-circuit current and energy efficiency enhancement in a low-dimensional structure photovoltaic device

Barnham KWJ, Duggan G (1990) A new approach to high-efficiency multi-band-gap solar cells

Bhattacharyya R, Mohan S (2015) Solid state storage of hydrogen and its isotopes: an engineering overview. Renew Sustain Energy Rev 41:872–883. doi:10.1016/j.rser.2014.09.004

Boden TA, Marland G, Andres RJ (2012) Global fossil-fuel CO_2 emissions

Bozbag SE, Erkey C (2012) Supercritical fluids in fuel cell research and development. J Supercrit Fluids 62:1–31. doi:10.1016/j.supflu.2011.09.006

Bruce PG, Scrosati B, Tarascon J-M (2008) Nanomaterials for rechargeable lithium batteries. Angew Chem Int Ed Engl 47:2930–2946. doi:10.1002/anie.200702505

Cabria I, López MJ, Alonso JA (2011) Simulation of the hydrogen storage in nanoporous carbons with different pore shapes. Int J Hydrogen Energy 36:10748–10759. doi:10.1016/j.ijhydene.2011.05.125

Cao G (2004) Nanostructures and nanomaterials: synthesis, properties and applications. Imperial College Press, London

Catchpole KR, Lin KL, Green MA et al (2002) Thin semiconducting layers and nanostructures as active and passive emitters for thermophotonics and thermophotovoltaics. Phys E Low-Dimension Syst Nanostruct 14:91–95. doi:10.1016/S1386-9477(02)00363-6

Catchpole KR, Mokkapati S, Beck F et al (2011) Plasmonics and nanophotonics for photovoltaics. MRS Bull 36:461–467. doi:10.1557/mrs.2011.132

Chacartegui R, Monje B, Sánchez D et al (2013) Molten carbonate fuel cell: towards negative emissions in wastewater treatment CHP plants. Int J Greenh Gas Control 19:453–461. doi:10.1016/j.ijggc.2013.10.007

Chainarong S, Sikong L, Pavasupree S, Niyomwas S (2011) Synthesis and characterization of nitrogen-doped TiO_2 nanomaterials for photocatalytic activities under visible light. Energy Procedia 9:418–427. doi:10.1016/j.egypro.2011.09.046

Chen D, Yu Y, Wang Y et al (2009) Cooperative energy transfer up-conversion and quantum cutting down-conversion in Yb^{3+}: TbF_3 nanocrystals embedded glass ceramics. J Phys Chem C 113:6406–6410. doi:10.1021/jp809995f

Chen X, Li C, Grätzel M et al (2012) Nanomaterials for renewable energy production and storage. Chem Soc Rev. doi:10.1039/c2cs35230c

Chiarello GL, Aguirre MH, Selli E (2010) Hydrogen production by photocatalytic steam reforming of methanol on noble metal-modified TiO_2. J Catal 273:182–190

Chien HC, Tsai LD, Lai CM et al (2013) Characteristics of high-water-uptake activated carbon/Nafion hybrid membranes for proton exchange membrane fuel cells. J Power Sources 226:87–93. doi:10.1016/j.jpowsour.2012.10.017

Choi JJ, Wenger WN, Hoffman RS et al (2011) Solution-processed nanocrystal quantum dot tandem solar cells. Adv Mater 23:3144–3148. doi:10.1002/adma.201100723

Chou C-C, Chen B-H, Lee D-J (2014) hydrogen storage in a chemical hydride fuel system containing ammonia borane and Ni-Co/r-GO catalyst. Energy Procedia 61:142–145. doi:10.1016/j.egypro.2014.11.925

Collavini S, Völker SF, Delgado JL (2015) Understanding the outstanding power conversion efficiency of perovskite-based solar cells. Angew Chemie Int Ed 54:9757–9759. doi:10.1002/anie.201505321

Conibeer G, Zhang B, Yao Y, Patterson R, Shrestha S, Green MA (2011) Electrical properties of conductive Ge nanocrystal thin films fabricated by low temperature in situ growth. Nanotechnology 22:125204

Conibeer G, Green MA, König D et al (2011b) Silicon quantum dot based solar cells: addressing the issues of doping, voltage and current transport. Prog Photovolt Res Appl 19:813–824. doi:10.1002/pip.1045

Conibeer G, Green M, Corkish R et al (2006) Silicon nanostructures for third generation photovoltaic solar cells. Thin Solid Films 511:654–662. doi:10.1016/j.tsf.2005.12.119

Conibeer GJ, Jiang C-W, König D et al (2008) Selective energy contacts for hot carrier solar cells. Thin Solid Films 516:6968–6973. doi:10.1016/j.tsf.2007.12.031

Corma A, Atienzar P, Garcia H, Chane-Ching J-Y (2004) Hierarchically mesostructured doped CeO_2 with potential for solar-cell use. Nat Mater 3:394–397

Cortecchia D, Dewi HA, Yin J et al (2016) Lead-free MA2CuClxBr 4–x hybrid perovskites. Inorg Chem 55:1044–1052. doi:10.1021/acs.inorgchem.5b01896

De Volder MFL, Tawfick SH, Baughman RH, Hart a J (2013) Carbon nanotubes: present and future commercial applications. Science 339(80):535–539. doi:10.1126/science.1222453

Derfus AM, Chan WCW, Bhatia SN (2004) Probing the cytotoxicity of semiconductor quantum dots. Nano Lett 4:11–18

Dodds PE, Staffell I, Hawkes AD et al (2015) Hydrogen and fuel cell technologies for heating: a review. Int J Hydrogen Energy 40:2065–2083

Drees M, Davis RM, Heflin JR (2004) Thickness dependence, in situ measurements, and morphology of thermally controlled interdiffusion in polymer-C_{60} photovoltaic devices. Phys Rev B 69:165320

Drees M, Premaratne K, Graupner W et al (2002) Creation of a gradient polymer-fullerene interface in photovoltaic devices by thermally controlled interdiffusion

EERE (2014) Hydrogen storage—current technology. Department of Energy. http://energy.gov/eere/fuelcells/hydrogen-storage-current-technology. Accessed 19 Feb 2015

Ekins-Daukes NJ, Adams J, Ballard IM et al (2009) Physics of quantum well solar cells. Proc SPIE 7211:72110L–72110L-11

Ekins-Daukes NJ, Barnes JM, Barnham KWJ et al (2001) Strained and strain-balanced quantum well devices for high-efficiency tandem solar cells. Sol Energy Mater Sol Cells 68:71–87. doi:10.1016/S0927-0248(00)00346-9

Eperon GE, Stranks SD, Menelaou C et al (2014) Formamidinium lead trihalide: a broadly tunable perovskite for efficient planar heterojunction solar cells. Energy Environ Sci 7:982–988. doi:10.1039/C3EE43822H

Espinosa N, Garcia-Valverde R, Urbina A, Krebs FC (2011) A life cycle analysis of polymer solar cell modules prepared using roll-to-roll methods under ambient conditions. Sol Energy Mater Sol Cells 95:1293–1302. doi:10.1016/j.solmat.2010.08.020

Fan J, Li Z, Zhou W et al (2014) Dye-sensitized solar cells based on TiO_2 nanoparticles/nanobelts double-layered film with improved photovoltaic performance. Appl Surf Sci 319:75–82. doi:10.1016/j.apsusc.2014.07.054

Fan Z, Razavi H, Do J et al (2009) Three-dimensional nanopillar-array photovoltaics on low-cost and flexible substrates. Nat Mater 8:648–653

Fleischer T, Grunwald A (2008) Making nanotechnology developments sustainable. A role for technology assessment? J Clean Prod 16:889–898. doi:10.1016/j.jclepro.2007.04.018

Franceschetti A, An JM, Zunger A (2006) Impact ionization can explain carrier multiplication in PbSe quantum dots. Nano Lett 6:2191–2195. doi:10.1021/nl0612401

Fujishima A, Honda K (1972) Electrochemical photolysis of water at a semiconductor electrode. Nature 238:37–38

Garnett EC, Brongersma ML, Cui Y, McGehee MD (2011) Nanowire solar cells. Annu Rev Mater Res 41:269–295. doi:10.1146/annurev-matsci-062910-100434

Gjessing J, Marstein ES, Sudbø A (2010) 2D back-side diffraction grating for improved light trapping in thin silicon solar cells. Opt Express 18:5481–5495. doi:10.1364/OE.18.005481

Goriparti S, Miele E, De Angelis F et al (2014) Review on recent progress of nanostructured anode materials for Li-ion batteries. J Power Sources 257:421–443

Grätzel M (2004) Conversion of sunlight to electric power by nanocrystalline dye-sensitized solar cells. J Photochem Photobiol A Chem 164:3–14. doi:10.1016/j.jphotochem.2004.02.023

Green MA, Emery K, Hishikawa Y et al (2016) Solar cell efficiency tables (version 47). Prog Photovolt Res Appl 24:3–11. doi:10.1002/pip.2728

Greenham NC, Peng X, Alivisatos AP (1996) Charge separation and transport in conjugated-polymer/semiconductor-nanocrystal composites studied by photoluminescence quenching and photoconductivity. Phys Rev B 54:17628–17637

Greijer H, Karlson L, Lindquist S-E, Hagfeldt A (2001) Environmental aspects of electricity generation from a nanocrystalline dye sensitized solar cell system. Renew Energy 23:27–39. doi:10.1016/S0960-1481(00)00111-7

Gribbin J, Gribbin M (1998) Richard Feynman: a life in science

Griffin PR, Barnes J, Barnham KWJ et al (1996) Effect of strain relaxation on forward bias dark currents in GaAs/InGaAs multiquantum well p–i–n diodes

Guo S, Wang E (2011) Noble metal nanomaterials: controllable synthesis and application in fuel cells and analytical sensors. Nano Today 6:240–264. doi:10.1016/j.nantod.2011.04.007

Guo YG, Hu JS, Wan LJ (2008) Nanostructured materials for electrochemical energy conversion and storage devices. Adv Mater 20:2878–2887. doi:10.1002/adma.200800627

Hagfeldt A, Graetzel M (1995) Light-induced redox reactions in nanocrystalline systems. Chem Rev 95:49–68. doi:10.1021/cr00033a003

Halls JJM, Walsh CA, Greenham NC et al (1995) Efficient photodiodes from interpenetrating polymer networks. Nature 376:498–500

Han W, Zang C, Huang Z et al (2014) Enhanced photocatalytic activities of three-dimensional graphene-based aerogel embedding TiO_2 nanoparticles and loading MoS_2 nanosheets as Co-catalyst. Int J Hydrogen Energy 39:19502–19512. doi:10.1016/j.ijhydene.2014.09.043

Hao F, Stoumpos CC, Cao DH et al (2014) Lead-free solid-state organic-inorganic halide perovskite solar cells. Nat Phot 8:489–494

Hao J-Y, Wang Y-Y, Tong X-L et al (2013) SiC nanomaterials with different morphologies for photocatalytic hydrogen production under visible light irradiation. Catal Today 212:220–224. doi:10.1016/j.cattod.2012.09.023

He M, Zheng D, Wang M et al (2014) High efficiency perovskite solar cells: from complex nanostructure to planar heterojunction. J Mater Chem A 2:5994–6003. doi:10.1039/C3TA14160H

He Y, Li Y (2011) Fullerene derivative acceptors for high performance polymer solar cells. Phys Chem Chem Phys 13:1970–1983. doi:10.1039/C0CP01178A

Healy ML, Dahlben LJ, Isaacs JA (2008) Environmental assessment of single-walled carbon nanotube processes. J Ind Ecol 12:376–393. doi:10.1111/j.1530-9290.2008.00058.x

Hetsch F, Xu XQ, Wang HK et al (2011) Semiconductor nanocrystal quantum dots as solar cell components and photosensitizers: material, charge transfer, and separation aspects of some device topologies. J Phys Chem Lett. doi:10.1021/jz200802j

Hochbaum AI, Yang P (2010) Semiconductor nanowires for energy conversion. Chem Rev 110:527–546. doi:10.1021/cr900075v

Hodes G, Cahen D (2014) Photovoltaics: perovskite cells roll forward. Nat Phot 8:87–88

Holbrook RD, Murphy KE, Morrow JB, Cole KD (2008) Trophic transfer of nanoparticles in a simplified invertebrate food web. Nat Nanotechnol 3:352–355

Hou H, Wang L, Gao F et al (2014) Fabrication of porous titanium dioxide fibers and their photocatalytic activity for hydrogen evolution. Int J Hydrogen Energy 39:6837–6844. doi:10.1016/j.ijhydene.2014.02.151

Hsu DD, O'Donoughue P, Fthenakis V et al (2012) Life cycle greenhouse gas emissions of crystalline silicon photovoltaic electricity generation: systematic review and harmonization. J Ind Ecol. doi:10.1111/j.1530-9290.2011.00439.x

Huerta-García E, Pérez-Arizti JA, Márquez-Ramírez SG et al (2014) Titanium dioxide nanoparticles induce strong oxidative stress and mitochondrial damage in glial cells. Free Radic Biol Med 73:84–94

Huggins RA (1999) Lithium alloy negative electrodes. J Power Sources 81–82:13–19. doi:10.1016/S0378-7753(99)00124-X

Hussein AK (2015) Applications of nanotechnology in renewable energies—a comprehensive overview and understanding. Renew Sustain Energy Rev 42:460–476. doi:10.1016/j.rser.2014.10.027

Huynh WU, Dittmer JJ, Alivisatos AP (2002) Hybrid nanorod-polymer solar cells. Science 295 (80):2425–2427

Im J-H, Chung J, Kim S-J, Park N-G (2012) Synthesis, structure, and photovoltaic property of a nanocrystalline 2H perovskite-type novel sensitizer ($(CH_3CH_2NH_3)PbI_3$. Nanoscale Res Lett 7:1–7. doi:10.1186/1556-276X-7-353

Im J-H, Lee C-R, Lee J-W et al (2011) 6.5% efficient perovskite quantum-dot-sensitized solar cell. Nanoscale 3:4088–4093. doi:10.1039/C1NR10867K

Inoue T, Watanabe K, Toprasertpong K et al (2015) Enhanced light trapping in multiple quantum wells by thin-film structure and backside grooves with dielectric interface. IEEE J Photovolt 5:697–703

Isaacs JA, Tanwani A, Healy ML (2006) Environmental assessment of SWNT production. In: Proceedings of the 2006 IEEE international symposium on electronics and the environment, 2006. IEEE, pp 38–41

Janas D, Vilatela AC, Koziol KKK (2013) Performance of carbon nanotube wires in extreme conditions. Carbon N Y 62:438–446. doi:10.1016/j.carbon.2013.06.029

Jarosz P, Schauerman C, Alvarenga J et al (2011) Carbon nanotube wires and cables: near term applications and future perspectives. Nanoscale 3:4542–4553

Jiang C, Yan Y, Moutinho HR et al (2011) Phosphorus and boron doping effect on nanocrystalline formation in hydrogenated amorphous and nanocrystalline mixed-phase silicon thin films. MRS Proc. doi:10.1557/PROC-1153-A17-07

Jones MR (2009) File: Dye_Sensitized_Solar_Cell_Scheme @ commons.wikimedia.org. https://commons.wikimedia.org/wiki/File:Dye_Sensitized_Solar_Cell_Scheme.png

Jyothirmayee Aravind SS, Ramanujachary K, Mugweru A, Vaden TD (2015) Molybdenum phosphide-graphite nanomaterials for efficient electrocatalytic hydrogen production. Appl Catal A Gen 490:101–107. doi:10.1016/j.apcata.2014.11.003

Kahane SV, Sasikala R, Vishwanadh B et al (2013) CdO-CdS nanocomposites with enhanced photocatalytic activity for hydrogen generation from water. Int J Hydrogen Energy 38:15012–15018. doi:10.1016/j.ijhydene.2013.09.077

Kallmann H, Pope M (1959) Photovoltaic effect in organic crystals

Kanamori Y, Hane K, Sai H, Yugami H (2001) 100 nm period silicon antireflection structures fabricated using a porous alumina membrane mask

Kandjani SA, Mirershadi S, Nikniaz A (2015) Inorganic–organic perovskite solar cells. Sol Cells—New Approaches Rev 223–246. doi:10.5772/58970

Kearns D, Calvin M (1958) Photovoltaic effect and photoconductivity in laminated organic systems

Khataee A, Arefi-Oskoui S, Fathinia M et al (2015) Synthesis, characterization and photocatalytic properties of Er-doped PbSe nanoparticles as a visible light-activated photocatalyst. J Mol Catal A Chem 398:255–267. doi:10.1016/j.molcata.2014.11.009

Kim H-S, Lee C-R, Im J-H et al (2012a) Lead iodide perovskite sensitized all-solid-state submicron thin film mesoscopic solar cell with efficiency exceeding 9%. Sci Rep 2:591

Kim HC, Fthenakis V (2013) Life cycle energy and climate change implications of nanotechnologies. J Ind Ecol 17:528–541. doi:10.1111/j.1530-9290.2012.00538.x

Kim HC, Fthenakis V, Choi JK, Turney DE (2012b) Life cycle greenhouse gas emissions of thin-film photovoltaic electricity generation: systematic review and harmonization. J Ind Ecol. doi:10.1111/j.1530-9290.2011.00423.x

Kim J, Rivera JL, Meng TY et al (2016) Review of life cycle assessment of nanomaterials in photovoltaics. Sol Energy 133:249–258. doi:10.1016/j.solener.2016.03.060

Köhler AR, Som C, Helland A, Gottschalk F (2008) Studying the potential release of carbon nanotubes throughout the application life cycle. J Clean Prod 16:927–937. doi:10.1016/j.jclepro.2007.04.007

Kojima A, Teshima K, Shirai Y, Miyasaka T (2009) Organometal halide perovskites as visible-light sensitizers for photovoltaic cells. J Am Chem Soc 131:6050–6051. doi:10.1021/ja809598r

König D, Casalenuovo K, Takeda Y et al (2010) Hot carrier solar cells: principles, materials and design. Phys E Low-Dimension Syst Nanostruct 42:2862–2866. doi:10.1016/j.physe.2009.12.032

Law M, Greene LE, Johnson JC et al (2005) Nanowire dye-sensitized solar cells. Nat Mater 4:455–459. doi:10.1038/nmat1387

Lee JW, Kjeang E (2013) Nanofluidic fuel cell. J Power Sources 242:472–477. doi:10.1016/j.jpowsour.2013.05.129

Lee MM, Teuscher J, Miyasaka T et al (2012) Efficient hybrid solar cells based on meso-superstructured organometal halide perovskites. Science 338(80):643–647

Lei W, Zhang H, Wu Y et al (2014) Oxygen-doped boron nitride nanosheets with excellent performance in hydrogen storage. Nano Energy 6:219–224. doi:10.1016/j.nanoen.2014.04.004

Lekawa-Raus A, Kurzepa L, Peng X, Koziol K (2014) Towards the development of carbon nanotube based wires. Carbon N Y 68:597–609. doi:10.1016/j.carbon.2013.11.039

Li H, Yang Q, Chen X, Shore SG (2014) Ammonia borane, past as prolog. J Organomet Chem 751:60–66. doi:10.1016/j.jorganchem.2013.08.044

Li JL, Hu ZS, Yang GW (2012) High-capacity hydrogen storage of magnesium-decorated boron fullerene. Chem Phys 392:16–20. doi:10.1016/j.chemphys.2011.08.017

Lindberg HK, Falck GC-M, Catalán J et al (2012) Genotoxicity of inhaled nanosized TiO(2) in mice. Mutat Res 745:58–64

Luther JM, Law M, Beard MC et al (2008) Schottky solar cells based on colloidal nanocrystal films. Nano Lett 8:3488–3492. doi:10.1021/nl802476m

Mao SS, Chen X (2007) Selected nanotechnologies for renewable energy applications. Int J Energy Res 31:619–636

Mao SS, Shen S, Guo L (2013) Nanomaterials for renewable hydrogen production, storage and utilization. Prog Nat Sci Mater Int 22:522–534. doi:10.1016/j.pnsc.2012.12.003

Marder TB (2007) Will we soon be fueling our automobiles with ammonia-borane? Angew Chemie—Int Ed 46:8116–8118. doi:10.1002/anie.200703150

Martí A, Cuadra L, Luque A (2000) Quantum dot intermediate band solar cell. In: Twenty-eighth IEEE photovoltaic specialists conference, pp 940–943

Martí A, López N, Antolín E et al (2006) Novel semiconductor solar cell structures: the quantum dot intermediate band solar cell. Thin Solid Films 511–512:638–644. doi:10.1016/j.tsf.2005.12.122

Mater A, Università S, Xxv C et al (2013) Scienze ambientali: tutela e gestione delle risorse naturali a case study on nano-TiO_2

Mazzer M, Grünbaum E, Barnham KWJ et al (1996) Study of misfit dislocations by EBIC, CL and HRTEM in GaAs/InGaAs lattice-strained multi-quantum well p-i-n solar cells. Mater Sci Eng, B 42:43–51. doi:10.1016/S0921-5107(96)01681-9

McDonald SA, Konstantatos G, Zhang S et al (2005) Solution-processed PbS quantum dot infrared photodetectors and photovoltaics. Nat Mater 4:138–142

McLeskey Jr JM, Qiao Q (2010) Nanostructured organic solar cells. In: Nanotechnology for photovoltaics. CRC Press, pp 147–185

Miao M (2013) Yarn spun from carbon nanotube forests: production, structure, properties and applications. Particuology 11:378–393. doi:10.1016/j.partic.2012.06.017

Michael I, Wagner J, Rendek LJ, et al (2009) United States Patent. 2

Minteer SD, Atanassov P, Luckarift HR, Johnson GR (2012) New materials for biological fuel cells. Mater Today 15:166–173. doi:10.1016/S1369-7021(12)70070-6

Mueller NC, Nowack B (2008) Exposure Modeling of Engineered Nanoparticles in the Environment. Environ Sci Technol 42:4447–4453

Nelson J, Paxman M, Barnham KWJ et al (1993) Steady-state carrier escape from single quantum wells. IEEE J Quantum Electron 29:1460–1468

Nikitin A, Li X, Zhang Z et al (2008) Hydrogen storage in carbon nanotubes through the formation of stable C–H bonds. Nano Lett 8:162–167. doi:10.1021/nl072325k

Noel NK, Stranks SD, Abate A et al (2014) Lead-free organic-inorganic tin halide perovskites for photovoltaic applications. Energy Environ Sci 7:3061–3068. doi:10.1039/C4EE01076K

Noh JH, Im SH, Heo JH et al (2013) Chemical management for colorful, efficient, and stable inorganic-organic hybrid nanostructured solar cells. Nano Lett 13:1764–1769. doi:10.1021/nl400349b

Nozik AJ, Parsons CA, Dunlavy DJ et al (1990) Dependence of hot carrier luminescence on barrier thickness in GaAs/AlGaAs superlattices and multiple quantum wells. Solid State Commun 75:297–301. doi:10.1016/0038-1098(90)90900-V

O'Regan B, Gratzel M (1991) A low-cost, high-efficiency solar cell based on dye-sensitized colloidal TiO_2 films. Nature 353:737–740

OECD/IEA (2009) Transport, energy and CO2

Ohsawa Y, Saji T (1992) Electrochemical detection of C606- at low temperature. J Chem Soc Chem Commun 781–782. doi:10.1039/C39920000781

Park Y, Drouard E, El Daif O et al (2009) Absorption enhancement using photonic crystals for silicon thin film solar cells. Opt Express 17:14312–14321. doi:10.1364/OE.17.014312

Pei F, Liu Y, Xu S et al (2013) Nanocomposite of graphene oxide with nitrogen-doped TiO_2 exhibiting enhanced photocatalytic efficiency for hydrogen evolution. Int J Hydrogen Energy 38:2670–2677. doi:10.1016/j.ijhydene.2012.12.045

Peiro AM, Ravirajan P, Govender K et al (2006) Hybrid polymer/metal oxide solar cells based on ZnO columnar structures. J Mater Chem 16:2088–2096. doi:10.1039/B602084D

Pelka J, Gehrke H, Rechel A et al (2013) DNA damaging properties of single walled carbon nanotubes in human colon carcinoma cells. Nanotoxicology 7:2–20

Peng J, Lu L, Yang H (2013) Review on life cycle assessment of energy payback and greenhouse gas emission of solar photovoltaic systems. Renew Sustain Energy Rev 19:255–274

Pereira SR, Coelho MC (2015) Can nanomaterials be a solution for application on alternative vehicles?—a review paper on life cycle assessment and risk analysis. Int J Hydrogen Energy 1–11. doi:10.1016/j.ijhydene.2014.12.132

Perić M, Hut I, Pelemiš S, Matija L (2015) Possible approaches to LCA methodology for nanomaterials in sustainable energy production. Contemp Mater (Renew Energy Sources) VI-2:160–169. doi:10.7251/COMEN1502160P

Perić M, Komatina M, Bugarski B, Antonijević D (2016) Best practices of biomass energy life cycle assessment and possible applications in serbia-review paper. Croat J For Eng 37:375–390

Petala A, Tsikritzis D, Kollia M et al (2014) Synthesis and characterization of N-doped TiO_2 photocatalysts with tunable response to solar radiation. Appl Surf Sci 305:281–291. doi:10.1016/j.apsusc.2014.03.062

Philippe B, Dedryveìre R, Gorgoi M et al (2013) Improved performances of nanosilicon electrodes using the salt LiFSI: a photoelectron spectroscopy study. J Am Chem Soc 135:9829–9842. doi:10.1021/ja403082s

Philippe B, Dedryvère R, Allouche J et al (2012) Nanosilicon electrodes for lithium-ion batteries: Interfacial mechanisms studied by hard and soft X-ray photoelectron spectroscopy. Chem Mater 24:1107–1115. doi:10.1021/cm2034195

Pieragostini C, Mussati MC, Aguirre P (2012) On process optimization considering LCA methodology. J Environ Manage 96:43–54. doi:10.1016/j.jenvman.2011.10.014

Pistidda C, Bergemann N, Wurr J et al (2014) Hydrogen storage systems from waste Mg alloys. J Power Sources 270:554–563. doi:10.1016/j.jpowsour.2014.07.129

Preethi V, Kanmani S (2013) Photocatalytic hydrogen production. Mater Sci Semicond Process 16:561–575. doi:10.1016/j.mssp.2013.02.001

Qiu J, Qiu Y, Yan K et al (2013) All-solid-state hybrid solar cells based on a new organometal halide perovskite sensitizer and one-dimensional TiO_2 nanowire arrays. Nanoscale 5:3245–3248. doi:10.1039/C3NR00218G

Qiu NX, Tian ZY, Guo Y et al (2014) A first-principle study of calcium-decorated BC_2N sheet doped by boron or carbon for high hydrogen storage. Int J Hydrogen Energy 39:9307–9320. doi:10.1016/j.ijhydene.2014.04.063

Ramsden JJ (2014) Applied nanotechnology. Elsevier

Reale P, Panero S, Scrosati B et al (2004) A safe, low-cost, and sustainable lithium-ion polymer battery. J Electrochem Soc 151:A2138

Ren Y, Chia GH, Gao Z (2013) Metal-organic frameworks in fuel cell technologies. Nano Today 8:577–597. doi:10.1016/j.nantod.2013.11.004

Robledo CB, Rojas MI, Cámara O, Leiva EPM (2014) First-principles studies concerning optimization of hydrogen storage in nanoporous reduced graphite oxide. Int J Hydrogen Energy 39:4396–4403. doi:10.1016/j.ijhydene.2013.12.206

Rodriguez I, Ramiro-Manzano F, Atienzar P et al (2007) Solar energy harvesting in photoelectrochemical solar cells. J Mater Chem 17:3205–3209. doi:10.1039/B618065E

Rosenwaks Y, Hanna MC, Levi DH et al (1993) Hot-carrier cooling in GaAs: quantum wells versus bulk. Phys Rev B 48:14675–14678

Street RA (2005) Hydrogenated amorphous silicon. Cambridge University Press, Cambridge

Salernitano E, Giorgi L, Dikonimos Makris T (2014) Direct growth of carbon nanofibers on carbon-based substrates as integrated gas diffusion and catalyst layer for polymer electrolyte fuel cells. Int J Hydrogen Energy 39:15005–15016. doi:10.1016/j.ijhydene.2014.07.060

Saliba M, Matsui T, Seo J-Y et al (2016) Cesium-containing triple cation perovskite solar cells: improved stability, reproducibility and high efficiency. Energy Environ Sci 9: doi:10.1039/C5EE03874J

Sargent LM, Hubbs AF, Young S-H et al (2012) Single-walled carbon nanotube-induced mitotic disruption. Mutat Res 745:28–37

Sariciftci NS, Braun D, Zhang C et al (1993) Semiconducting polymer-buckminsterfullerene heterojunctions: diodes, photodiodes, and photovoltaic cells

Savenije TJ, Warman JM, Goossens A (1998) Visible light sensitisation of titanium dioxide using a phenylene vinylene polymer. Chem Phys Lett 287:148–153. doi:10.1016/S0009-2614(98)00163-8

Schur V, Tarasov P, Yu S et al (2002) The prospects for using of carbon nanomaterials as hydrogen storage systems. Int J Hydrogen Energy 27:1063–1069. doi:10.1016/S0360-3199(02)00009-5

Seabold JA, Shankar K, Wilke RHT et al (2008) Photoelectrochemical properties of heterojunction $CdTe/TiO_2$ electrodes constructed using highly ordered TiO_2 Nanotube Arrays. Chem Mater 20:5266–5273. doi:10.1021/cm8010666

Sengül H, Theis TL (2009) Life cycle inventory of semiconductor cadmium selenide quantum dots for environmental applications. Nanotechnol Appl Clean Water 561–582. doi:10.1016/B978-0-8155-1578-4.50047-0

Şengül H, Theis TL (2011) An environmental impact assessment of quantum dot photovoltaics (QDPV) from raw material acquisition through use. J Clean Prod 19:21–31. doi:10.1016/j.jclepro.2010.08.010

Seo M-S, Jeong I, Park J-S et al (2016) Vertically aligned nanostructured TiO_2 photoelectrodes for high efficiency perovskite solar cells via a block copolymer template approach. Nanoscale 8:11472–11479. doi:10.1039/C6NR01010E

Serdiukova I, Monier C, Vilela MF, Freundlich A (1999) Critical built-in electric field for an optimum carrier collection in multiquantum well p-i-n diodes

Serrano E, Rus G, García-Martínez J (2009) Nanotechnology for sustainable energy. Renew Sustain Energy Rev 13:2373–2384. doi:10.1016/j.rser.2009.06.003

Shao H, Asano K, Enoki H, Akiba E (2009) Fabrication, hydrogen storage properties and mechanistic study of nanostructured $Mg_{50}Co_{50}$ body-centered cubic alloy. Scr Mater 60: 818–821

Shao H, Ma W, Kohno M et al (2014) Hydrogen storage and thermal conductivity properties of Mg-based materials with different structures. Int J Hydrogen Energy 39:9893–9898. doi:10.1016/j.ijhydene.2014.02.063

Shao H, Xin G, Zheng J et al (2012) Nanotechnology in Mg-based materials for hydrogen storage. Nano Energy 1:590–601. doi:10.1016/j.nanoen.2012.05.005

Shi H, Zhou M, Song D et al (2014) Highly porous SnO_2/TiO_2 electrospun nanofibers with high photocatalytic activities. Ceram Int 40:10383–10393. doi:10.1016/j.ceramint.2014.02.124

Singh S, Verma N (2015) Fabrication of Ni nanoparticles-dispersed carbon micro-nanofibers as the electrodes of a microbial fuel cell for bio-energy production. Int J Hydrogen Energy 40:1145–1153. doi:10.1016/j.ijhydene.2014.11.073

Sun W, Wang Y (2014) Graphene-based nanocomposite anodes for lithium-ion batteries. Nanoscale 6:11528–11552. doi:10.1039/C4NR02999B

Sun Y-Y, Shi J, Lian J et al (2016) Discovering lead-free perovskite solar materials with a split-anion approach. Nanoscale 8:6284–6289. doi:10.1039/C5NR04310G

Švrček V, Slaoui A, Muller J-C (2004) Silicon nanocrystals as light converter for solar cells. Thin Solid Films 451–452:384–388. doi:10.1016/j.tsf.2003.10.133

Tan CW, Tan KH, Ong YT et al (2012a) Energy and environmental applications of carbon nanotubes. Environ Chem Lett 10:265–273. doi:10.1007/s10311-012-0356-4

Tan H, Santbergen R, Smets AHM, Zeman M (2012b) Plasmonic light trapping in thin-film silicon solar cells with improved self-assembled silver nanoparticles. Nano Lett 12:4070–4076. doi:10.1021/nl301521z

Taniguchi N (1974) On the basic concept of nanotechnology

Terrés E, Panella B, Hayashi T et al (2005) Hydrogen storage in spherical nanoporous carbons. Chem Phys Lett 403:363–366. doi:10.1016/j.cplett.2005.01.032

Tibbetts GG, Meisner GP, Olk CH (2001) Hydrogen storage capacity of carbon nanotubes, filaments, and vapor-grown fibers. Carbon N Y 39:2291–2301. doi:10.1016/S0008-6223(01)00051-3

Tran C-D (2014) Nanotube superfiber materials. Elsevier

Tsai H, Nie W, Blancon J-C et al (2016) High-efficiency two-dimensional Ruddlesden-Popper perovskite solar cells. Nature 536:312–316

Tsakalakos L (2010) Nanotechnology for photovoltaics. CRC Press, Taylor & Francis, Boca Raton, FL

Tsao J, Lewis N, Crabtree G (2006) Solar FAQs. US Dep Energy 1–24

Turner S, Mokkapati S, Jolley G et al (2013) Periodic dielectric structures for light-trapping in InGaAs/GaAs quantum well solar cells. Opt Express 21:A324–A335. doi:10.1364/OE.21.00A324

Tylianakis E, Dimitrakakis GK, Martin-Martinez FJ et al (2014) Designing novel nanoporous architectures of carbon nanotubes for hydrogen storage. Int J Hydrogen Energy 39:9825–9829. doi:10.1016/j.ijhydene.2014.03.011

US EPA (2015) Carbon dioxide emissions. http://www.epa.gov/climatechange/ghgemissions/gases/co2.html. Accessed 3 Mar 2015

van Berlo D, Hullmann M, Wessels A et al (2014) Investigation of the effects of short-term inhalation of carbon nanoparticles on brains and lungs of c57bl/6j and p47(phox-/-) mice. Neurotoxicology 43:65–72

Vlachopoulos N, Liska P, Augustynski J, Graetzel M (1988) Very efficient visible light energy harvesting and conversion by spectral sensitization of high surface area polycrystalline titanium dioxide films. J Am Chem Soc 110:1216–1220. doi:10.1021/ja00212a033

Wang H, Gao Q, Hu J, Chen Z (2009) High performance of nanoporous carbon in cryogenic hydrogen storage and electrochemical capacitance. Carbon N Y 47:2259–2268. doi:10.1016/j.carbon.2009.04.021

Wang YS, Wang F, Li M et al (2012) Theoretical prediction of hydrogen storage on Li decorated planar boron sheets. Appl Surf Sci 258:8874–8879. doi:10.1016/j.apsusc.2012.05.107

Wee JH (2014) Carbon dioxide emission reduction using molten carbonate fuel cell systems. Renew Sustain Energy Rev 32:178–191. doi:10.1016/j.rser.2014.01.034

Westland DJ, Ryan JF, Scott MD et al (1988) Hot carrier energy loss rates in GaInAs/InP quantum wells. Solid State Electron 31:431–434. doi:10.1016/0038-1101(88)90311-5

Xu GR, Wang JN, Li CJ (2013) Template directed preparation of TiO_2 nanomaterials with tunable morphologies and their photocatalytic activity research. Appl Surf Sci 279:103–108. doi:10.1016/j.apsusc.2013.04.043

Yang SJ, Jung H, Kim T, Park CR (2012) Recent advances in hydrogen storage technologies based on nanoporous carbon materials. Prog Nat Sci Mater Int 22:631–638

Yang Z, Nie H, Chen X et al (2013) Recent progress in doped carbon nanomaterials as effective cathode catalysts for fuel cell oxygen reduction reaction. J Power Sources 236:238–249. doi:10.1016/j.jpowsour.2013.02.057

Yin YX, Wan LJ, Guo YG (2012) Silicon-based nanomaterials for lithium-ion batteries. Chin Sci Bull 57:4104–4110. doi:10.1007/s11434-012-5017-2

Yu G, Gao J, Hummelen JC et al (1995) Polymer photovoltiac cells: Enhanced efficiencies via a network of internal donor-acceptor heterojunctions. Science 270(80):1789

Yu Y, Gu L, Zhu C et al (2010a) Reversible storage of lithium in silver-coated three-dimensional macroporous silicon. Adv Mater 22:2247–2250. doi:10.1002/adma.200903755

Yu Z, Raman A, Fan S (2010b) Fundamental limit of light trapping in grating structures. Opt Express 18:A366–A380. doi:10.1364/OE.18.00A366

Zeng L, Bermel P, Yi Y et al (2008) Demonstration of enhanced absorption in thin film Si solar cells with textured photonic crystal back reflector. Appl Phys Lett. doi 10(1063/1):3039787

Zhang B, Xie W, Xiang Y (2015) Development and prospect of nanoarchitectured solar cells. Int J Photoenergy. doi:10.1155/2015/382389

Zhang C, Li X, Shang A et al (2014) Design of dual-diameter nanoholes for efficient solar-light harvesting. Nanoscale Res Lett 9:481. doi:10.1186/1556-276X-9-481

Zhang LW, Monteiro-Riviere NA (2008) Assessment of quantum dot penetration into intact, tape-stripped, abraded and flexed rat skin. Skin Pharmacol Physiol 21:166–180

Zhang M, Atkinson KR, Baughman RH (2004) Multifunctional carbon nanotube yarns by downsizing an ancient technology. Science 306:1358–1361

Zhang M, Dai L (2012) Carbon nanomaterials as metal-free catalysts in next generation fuel cells. Nano Energy 1:514–517. doi:10.1016/j.nanoen.2012.02.008

Zhang Q, Dandeneau CS, Zhou X, Cao G (2009) ZnO nanostructures for dye-sensitized solar cells. Adv Mater 21:4087–4108. doi:10.1002/adma.200803827

Zhang Y, Zhang W, Wang A et al (2007) $LiBH_4$ nanoparticles supported by disordered mesoporous carbon: hydrogen storage performances and destabilization mechanisms. Int J Hydrogen Energy 32:3976–3980

Zhu J, Cheng J, Dailly A et al (2014) One-pot synthesis of Pd nanoparticles on ultrahigh surface area 3D porous carbon as hydrogen storage materials. Int J Hydrogen Energy 39:14843–14850. doi:10.1016/j.ijhydene.2014.07.002

Zhu Y, Murali S, Stoller MD et al (2010) Microwave assisted exfoliation and reduction of graphite oxide for ultracapacitors. Carbon N Y 48:2118–2122. doi:10.1016/j.carbon.2010.02.001

Integration of Nanostructured Thermoelectric Materials in Micro Power Generators

D. Dávila, A. Tarancón and L. Fonseca

Abstract The demand for portable power generation required by miniaturized systems with long-lasting autonomous operation is large and expanding. In order to fulfill this demand, high-energy density storage and generation devices are required. Currently, batteries and micro fuel cells represent the technologies available for this purpose. However, true energy autonomy would eventually be better enabled by energy harvesting devices, either by themselves or in combination with storage devices (batteries or capacitors). In this respect, thermoelectric generators may lead to a quite enabling energy autonomy solution in those particular application scenarios in which waste heat or hot surface are available. This chapter serves as a brief introduction to the general principles of thermoelectricity and the state-of-the-art of thermoelectric materials. An overview of the different efforts performed by the research community to develop thermoelectric devices and the implementation approaches followed to fabricate thermoelectric microgenerators based on nanostructured materials is presented.

Keywords Thermoelectricity · Nanostructured materials · Thermoelectric micro generators · Energy harvesting · Micro/Nanodevices

D. Dávila (✉)
Science & Technology Department, IBM Research - Zurich Lab, Säumerstrasse 4, 8803 Rüschlikon, Switzerland
e-mail: dda@zurich.ibm.com

A. Tarancón
Catalonia Institute for Energy Research (IREC),
Jardins de Les Dones de Negre 1, 2da. Pl. 08930 Barcelona, Spain
e-mail: atarancon@irec.cat

L. Fonseca
Institute of Microelectronics of Barcelona (IMB-CNM),
Campus Universitat Autònoma de Barcelona (UAB), 08193
Cerdanyola Del Vallès, Barcelona, Spain
e-mail: luis.fonseca@imb-cnm.csic.es

D. Brabazon et al. (eds.), *Commercialization of Nanotechnologies–A Case Study Approach*, DOI 10.1007/978-3-319-56979-6_4

1 Introduction

The increasing world's demand for ubiquitous power generation required by miniaturized systems is driving the development of new materials and technologies able to extract energy from the environment and to enable long-lasting autonomous operation without cords or even batteries. For ambient energy sources such as vibrational, solar or thermal, the power density that can be harvested is high enough (in the range of tens of microwatts to miliwatts per cubic centimeter) for powering small electronic devices, nodes in wireless sensor networks or remote actuators (Tan and Panda 2010). These applications usually involve scaling down the power source to the microscale by using MEMS technology, therefore limiting the materials selection to those compatible with this technology. Furthermore, recent advances in the integration of one-dimensional nanostructures (Huang et al. 2001; Bachtold et al. 2001) with ultra-low power consumption call for power sources compatible with nano- or microelectronic technologies to take advantage of their size (Wang et al. 2007; Wang 2012).

Among other energy harvesters, thermoelectric generators (TEGs) able to convert naturally occurring thermal gradients into electricity are promising candidates to lead this small-power revolution. Due to their high reliability, scalable size, and relatively low cost, these solid-state devices have the potential to become a viable power source for low-power miniaturized systems in appropriate scenarios.

This chapter presents a brief overview of the fundamentals of thermoelectricity and the state-of-the-art of thermoelectric materials. The various advances in enhancing the performance of these materials through nanostructuring approaches and the achievements in the fabrication of thermoelectric microgenerators are reviewed with particular attention to silicon.

2 The Thermoelectric Effects

The thermoelectric effects (Seebeck, Peltier and Thomson) define the interaction between heat and electricity in metals and semiconductors (Rowe 2006b). If a circuit formed from two dissimilar conductors (A and B), also known as thermocouple, is connected electrically in series and thermally in parallel and subjected to a temperature gradient an electromotive force will develop between its hot and cold ends (Fig. 1). Since the kinetic energy of charge carriers is temperature dependent, the charge carriers (electrons for n-type and holes for p-type semiconductors) at the hot end will diffuse to the cold end resulting in a charge accumulation at the cold end and producing an electrostatic potential (ΔV). This property, known as the Seebeck effect, is the basis for thermoelectric power generation:

$$V = \alpha \cdot \Delta T, \tag{1}$$

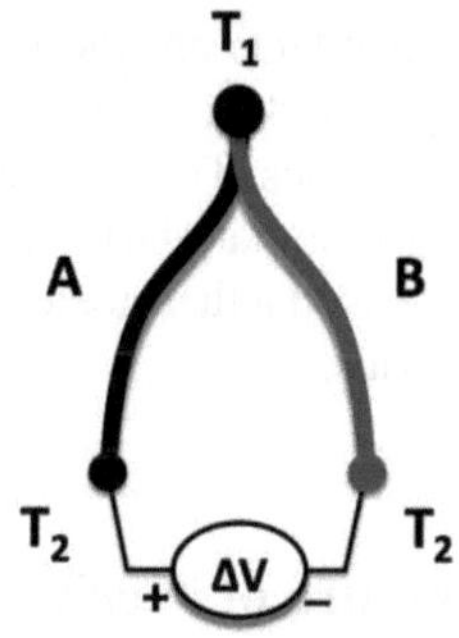

Fig. 1 Schematic of a basic thermocouple. Two dissimilar conductors, material A and material B, produce a voltage difference, ΔV, across their terminals if their junctions are held at different temperatures when connected in an open circuit. The produced voltage is directly proportional to the difference between the hot (T_1) and cold (T_2) junction temperatures

where α is the differential Seebeck coefficient between the materials A and B and ΔT is the temperature difference. The Seebeck coefficient depends on the conductor's absolute temperature, material and molecular structure. Any electrically conductive material with a finite Seebeck coefficient and subjected to a temperature gradient develops a voltage difference between points of different temperatures.

Conversely, when an electric current I flows between two dissimilar conducting materials held at constant temperature, the carriers attempt to return to the equilibrium that existed before the current was applied by absorbing heat at one connector of the thermocouple and releasing it at the other, this phenomenon is known as the Peltier effect. The heat is directly proportional to the magnitude of the current flowing across the junction and can be decomposed into a contribution from each material. The ratio of heat to current for each material defines the Peltier coefficient:

$$Q = \Pi \cdot I = \alpha \cdot T \cdot I, \tag{2}$$

where I is the electric current magnitude and Q is the heat flux at the interface of the two materials. The Peltier coefficient, Π, is the product of T at the interface and the difference in α between the two materials that compose the interface.

The relation between the Peltier and the Seebeck effect is given by the Thomson effect. If an electric current flows through a single conductor subjected to a temperature difference ΔT, heat is generated or absorbed.

$$Q = \beta \cdot I \cdot \Delta T, \tag{3}$$

where β is the Thomson coefficient and is given as

$$\beta = \frac{d\alpha}{dT} \tag{4}$$

While the Seebeck and Peltier effects are the main thermoelectric effects, the Thomson effect is not of primary importance in thermoelectric devices but it should not be neglected in detailed calculations. The Joule effect (the heat generated when a voltage is applied across a resistive material), although not a reversible thermoelectric effect and not generally termed a thermoelectric effect, is an important non-ideality of any thermoelectric device.

3 The Thermoelectric Figure of Merit

Thermoelectric devices are normally made from alternating *n*-type (electron-charge carrier) and *p*-type (hole-charge carrier) semiconductor elements joined by metallic connectors (Fig. 2). Because two dissimilar materials have different Seebeck coefficients, charge carriers flow through the *n*-type element, cross a metallic interconnect, and pass into the *p*-type element. When a heat source is provided, the thermoelectric device functions as a power generator (Fig. 3). A typical thermoelectric power generator contains several *n*- and *p*-type thermocouples electrically connected in series to add all Seebeck voltages of the individual thermoelectric legs and thermally in parallel to convert heat (temperature gradients) into electrical energy. The heat flow

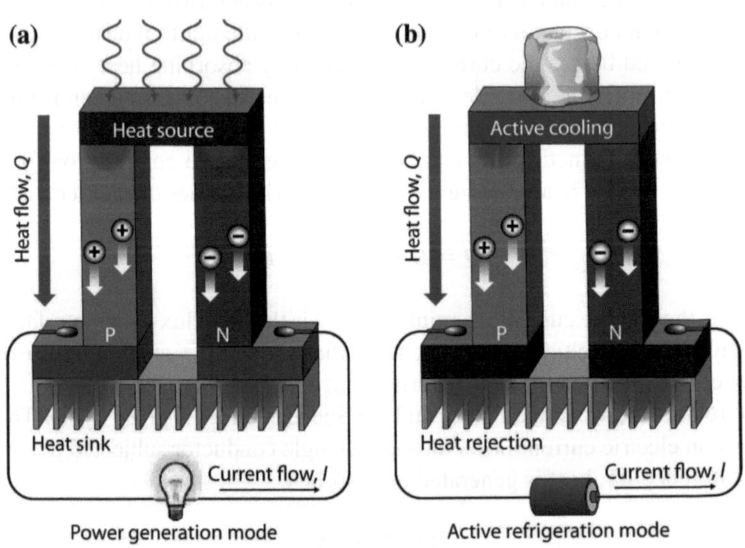

Fig. 2 Schematic illustrations of a thermoelectric module for **a** power generation (Seebeck effect) and **b** active refrigeration (Peltier effect). **a** The charge carriers in the material (electrons or holes) diffuse from the hot side to the cold side when a temperature difference is applied, which results in a current flow through the circuit. **b** The generated heat is absorbed at the lower junction and dissipated at the upper junction when a current flows through the circuit. Reproduced with permission from (Li et al. 2010)

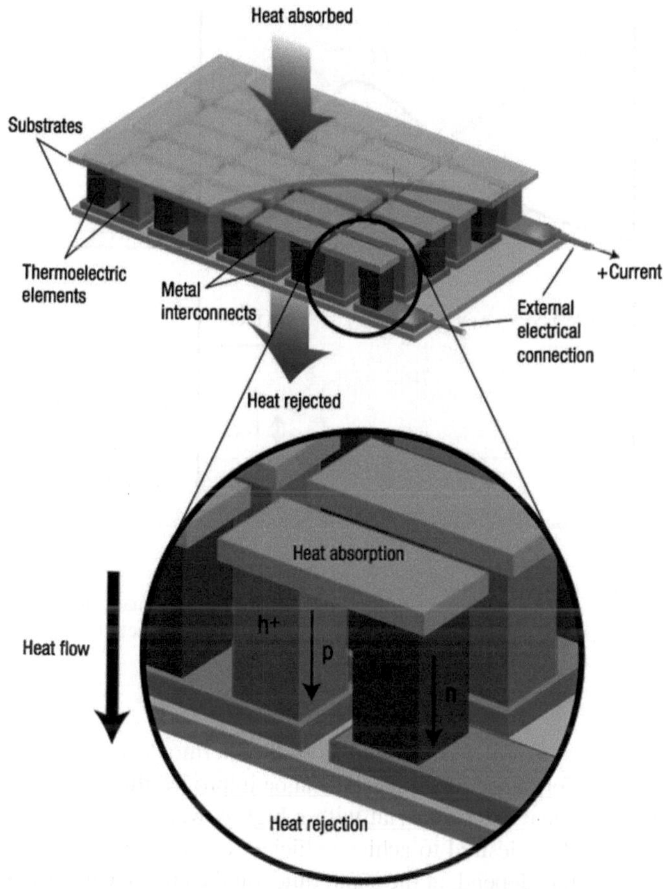

Fig. 3 Schematic of a thermoelectric module. The image shows the direction of charge flow on both cooling and power generation. Reproduced with permission from (Snyder and Toberer 2008)

across the legs drives the electrical current used to power an external circuit. Conversely, cooling can be achieved in a similar structure when forcing a heat flow by applying an electrical current. The maximum efficiency of a thermoelectric material for both power generation and cooling is determined by its dimensionless thermoelectric figure of merit *ZT* (Rowe 2006b) given by:

$$ZT = \frac{\alpha^2 \cdot \sigma}{\kappa} T, \tag{5}$$

where α is the Seebeck coefficient, σ is the electrical conductivity, κ is the thermal conductivity, and T is the absolute temperature. The κ of a material has two components: a component due to lattice vibrations (phonons propagation through the lattice, κ_L) and a component due to the random kinetic energy transported by

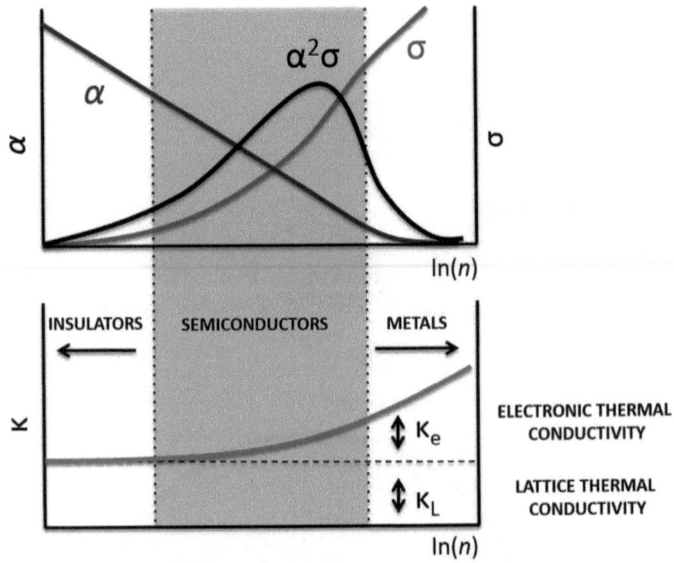

Fig. 4 Dependence of *ZT* components on carrier concentration. Schematic dependence of electrical conductivity, Seebeck coefficient, power factor, and thermal conductivity on concentration of free carriers (n). Adapted from (Rowe 2006b)

electrons (κ_e). The term $\alpha^2\sigma$ is referred as to the thermoelectric power factor. The thermoelectric performance of a material can be improved through the independent adjustment of α, σ and κ. A material with a high power factor and a low thermal conductivity would be desired to achieve efficient energy conversion, a. Nevertheless, these parameters depend, at the same time, on the charge carrier concentration (n) leading to a significant trade-off between them (Fig. 4). First, the Seebeck coefficient of the material decreases with increasing σ. Therefore, in order to increase the power factor, the carrier concentration cannot simply be increased to saturation: a peak of $\alpha^2\sigma$ occurs prior to the saturation of n. Second, as the carrier concentration in the material increases, the electronic contribution to the thermal conductivity κ_e also increases. An optimal carrier concentration to attain high *ZT* values is found in highly doped semiconductors, making them the best candidates for thermoelectric materials. Nevertheless, only those materials with a $ZT > 0.5$ are usually regarded as good thermoelectric materials. It is important to notice that, in real thermoelectric devices, apart from the fundamental material losses, there are additional parasitic losses that affect the efficiency of the device. The thermal conduction of the assembly materials and insulation as well as the contact and wiring resistances introduce significant non-idealities (imperfect impedance matching and non-zero thermal contact resistance) that have to be considered for final designs.

4 State-of-the-art of Thermoelectric Materials

The boost of thermoelectric materials began in the 1950s when the basic science behind this field was established. It was during these years that the definition of the efficiency of a thermoelectric material was first described through the thermoelectric figure of merit, *ZT*, and the promising thermoelectric material bismuth telluride (Bi_2Te_3) was discovered and with it, the thermoelectric industry was triggered. Few advances were made in increasing *ZT* and up to 1990 three material families dominated thermoelectrics: Bi_2Te_3-based materials for applications around room temperature (up to 450 K), PbTe-based materials for use in an intermediate temperature range (up to 850 K) and SiGe for use at the highest temperatures (up to 1350 K), primarily in Radioisotope Thermoelectric Generators (RTGs) used to power spacecrafts (Rowe 2006a; Vining 2007). Even though these classical materials are the ones mainly used in current commercial thermoelectric applications, there have been important advances since the 1990s regarding the synthesis of new materials and the fabrication of material structures with improved thermoelectric performance. It was at the beginning of the 1990s, that funding agencies became interested in the potential of thermoelectrics. This led to the development of two new different research lines in this field: one focused on synthesizing new bulk materials with complex crystalline structures and other focused on smart structuring of traditional ones to achieve low-dimensional materials systems.

Figure 5 gives an overview of the state-of-the-art of bulk thermoelectric materials and their *ZT* values. The search for new bulk materials has led to the development of a huge number of families and is mainly motivated by the suggestion made by Slack and Tsoukala (1994) based on the search for the so-called phonon glass-electron crystal (PGEC), which implies that an optimum thermoelectric material should have a low lattice thermal conductivity as in a glass, and a high electrical conductivity as in a crystal. In this type of new materials, such as skutterudites (Yang et al. 2004; Sales et al. 1996) and clathrates (Nolas et al. 1998), the thermal conductivity can be reduced while maintaining the electrical conductivity at a high level. This is due to the strong phonon scattering effect induced by the vibrating motion of loosely bonded atoms within a large cage, which has less impact on the electron transport. Effects on phonons such as a decrease of their mean-free-path, an increased disorder or an increased diffusion can be produced by engineering different thermoelectric materials with complex nanostructures, vacancies, impurities or grain mixtures (Slack 1995; Ebling 2009; Li et al. 2010).

4.1 Low-Dimensional Thermoelectric Materials

The research focused on low-dimensional material systems began in 1993, when Hicks and Dresselhaus discovered their potential working on two theoretical papers that predicted the enhanced thermoelectric properties of quantum wells (Hicks and Dresselhaus 1993b) and quantum wires (Hicks and Dresselhaus 1993a). These

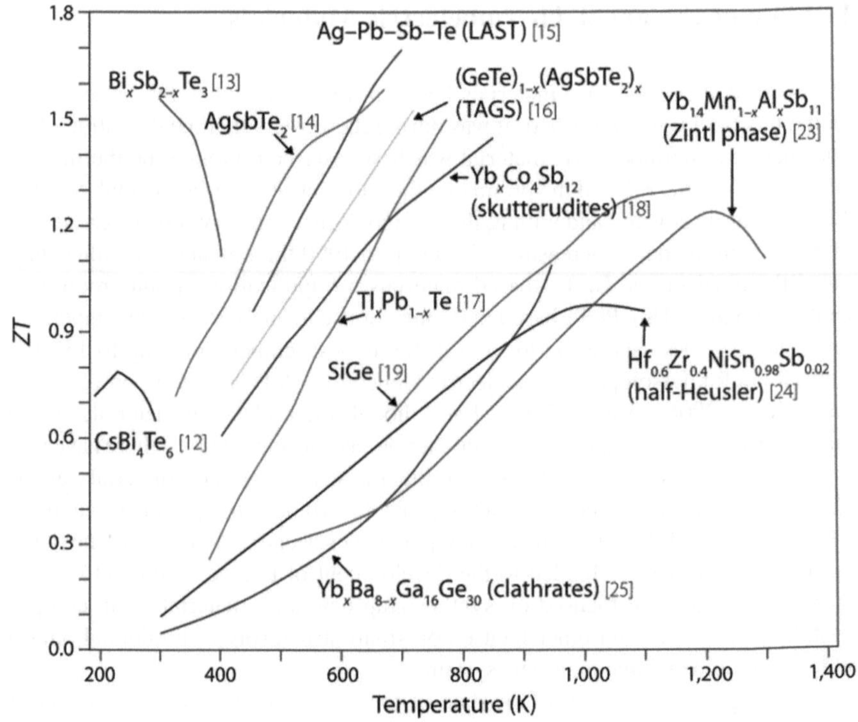

Fig. 5 State-of-the-art of thermoelectric materials. Figure of merit (*ZT*) of high-performance bulk thermoelectric materials as a function of applied temperature. The high performance of most of these materials, which are complex alloys with dopants, is related to nanostructure engineering. Reproduced with permission from (Li et al. 2010)

low-dimensional structures provide a route for achieving significantly improved thermoelectric figures of merit with an enlarged Seebeck coefficient, over that of the bulk, due to a modified density of states when quantum confinement of the electron charge carriers takes place (Rowe 2006a; Dresselhaus et al. 2007). When the dimensionality of the system is decreased from 3D crystalline solids to 2D (quantum wells) to 1D (quantum wires) and finally to 0D (quantum dots) and approaches a scale similar to the feature length of electron behavior (e.g. mean-free-path, wavelength) in any direction, the density of electronic states (DOS) is increased significantly due to quantum confinement leading to sharp changes in the DOS (Fig. 6). The electron energy bands in quantum-confined structures are progressively narrower as the confinement increases and the dimensionality decreases. These narrow bands should produce high effective masses (electrons behave like free particles with an effective mass that is different from the free electron mass in vacuum) and therefore enhanced Seebeck coefficients. Electron motion is thus limited to high-energy electrons causing a filtering effect, which can dramatically increase the Seebeck coefficient with a relatively modest decrease in electrical conductivity. This electron filtering effect

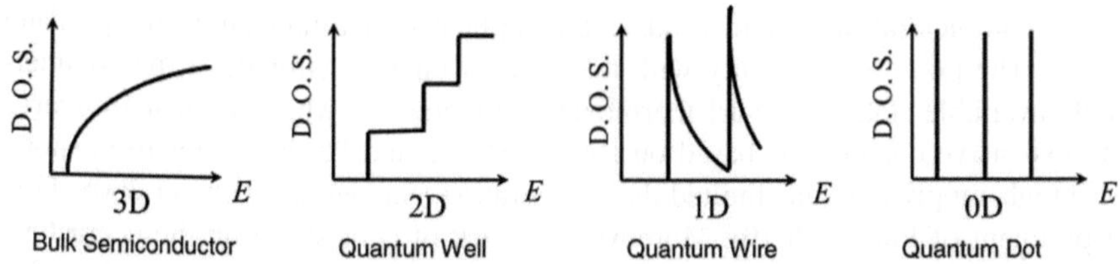

Fig. 6 Schematic representation of the electronic density of states. Electronic density of states for a bulk 3D crystalline semiconductor, a 2D quantum well, a 1D nanowire or nanotube, and a 0D quantum dot. Since α is related to the derivative of the DOS, at the Fermi level, through the Mott relation, reducing the dimensionality of the material allows increasing α and hence ZT. Reproduced with permission from (Dresselhaus et al. 2007)

allows to decouple the Seebeck coefficient and the electrical conductivity enabling an enhancement of the *ZT* of the material.

Although all these theoretical predictions anticipate the potential to improve $\alpha^2\sigma$ by quantum confinement in low-dimensional structures, there is no convincing experimental evidence to validate such predictions (Singh 2008). Most of the improvements in *ZT* have been mainly achieved due to an enhancement of phonon scattering by increasing the presence of interfaces, a phenomenon that reduces κ (in the denominator of *ZT*).

Ideally, the thermal conductivity κ and the electrical conductivity σ of a material can be tuned independently by nanostructuring approaches since phonons and electrons have different mean-free-paths (hundreds and tens of nanometers respectively) (Weber and Gmelin 1991; Esfarjani et al. 2011; Feser et al. 2012). This means that it is possible to restrict the movement of phonons in a material by scattering effects at surfaces and interfaces decreasing κ as the surface area to volume ratio of the material increases. On the other hand, σ is less sensitive to these nanostructuring effects, which means that electron mobility is kept unaffected, although it will degrade through the scattering of charge carriers at interfaces at sufficiently small sizes (Yu and Roca i Cabarrocas 2010). Scattering the propagation of phonons by nanostructuring of materials leads, therefore, to an enhanced thermoelectric figure of merit. In this direction, nanowires represent a promising low-dimensional material that enable easy pathways to electrons while restricting the flow of phonons by scattering effects, allowing in this way a directional thermal transport (Iyengar 2010).

4.2 *Nanostructured Silicon: A Case Study*

In the last decade, the scientific community has been mainly focused on the development of thermoelectric materials with an enhanced figure of merit by different approaches and many materials with *ZT* >1.5 have been reported. Nevertheless, no high-*ZT* devices demonstrating this efficiency have been shown. The integration of low-dimensional thermoelectric materials is a challenging process that usually entails complex fabrication methods and complicated measurement techniques and setups. The poor compatibility with mainstream microelectronics of the commercially available state-of-the-art thermoelectric materials, which are usually scarce and expensive compounds based on Bi, Te, Sb, Se and Pb that suffer from short-term high supply risk, has limited their integration (Snyder and Toberer 2008; U.S. Department of Energy 2010). Moreover, the lack of consistency in the characterization methods employed and systematic inaccuracies in the measurements yields to unreliable results difficult to reproduce. Certainly, the use of materials traditionally employed in microelectronics could simplify part of these integration issues. Nonetheless, the poor thermoelectric properties of these materials limits their use for thermoelectric applications.

Silicon, as the second most common element on the Earth's crust and with a well-established and mature technology represents the ideal material for the development of any CMOS device. Nanostructuring silicon into nanowires, nanomeshes or porous membranes could overcome the poor bulk thermoelectric performance of this material with a $ZT \approx 0.01$. Independent works of Boukai et al. and Hochbaum et al. showed an astonishing enhancement of the thermoelectric properties for the particular case of single silicon nanowires (Fig. 7) related to the low dimensionality of the material (Boukai et al. 2008; Hochbaum et al. 2008). The work performed by Boukai et al. showed that *ZT* values representing an approximately 100 fold improvement over bulk Si could be achieved over a broad temperature range by varying the nanowire size and impurity doping levels, including $ZT \approx 1$ at 200 K. Moreover, Hochbaum et al. demonstrated that it is possible to achieve $ZT = 0.6$ at room temperature in rough Si nanowires of ≈50 nm diameter, attributing this *ZT* enhancement to efficient scattering throughout the phonon spectrum by the introduction of nanostructures at different length scales (diameter, roughness and point defects). However, while the thermal conductivity of silicon nanowires has been shown to decrease from its bulk value for a wide range of nanowire diameters (Hochbaum et al. 2008; Boukai et al. 2008; Chen et al. 2008; Li et al. 2003; Ponomareva et al. 2007), there is only evidence of the enhancement of the Seebeck coefficient for the smallest diameters (≈20 nm) (Boukai et al. 2008). In addition, this dimensional reduction has an insignificant effect on the electrical conductivity of the material which is close to the bulk value. Feser et al. have reported thermal conductivities of highly roughened nanowires significantly below the boundary scattering limit predicted and below the previously measured for smooth single nanowires of similar diameter (100–160 nm) (Feser et al. 2012). Although when synthesized bottom-up silicon nanowires are usually obtained in large numbers, their enhanced thermoelectric properties have been mainly measured in ad hoc complex test structures. Stranz et al. have used a less complex approach to measure the thermoelectric properties of top-down vertical arrays of silicon nanowires by stacking them with bulk silicon (Stranz et al. 2011a, b, 2012, 2013). Their results show a reduction of the thermal conductivity with measured values of 9 ± 2 W/m-K (less than 10% of the value for bulk silicon, i.e. 150 W/m−K) and a Seebeck coefficient of -370 ± 50 μV/K, which is comparable to that of bulk silicon. Yu et al. have fabricated a silicon nanomesh thin film patterned with periodic spacings that are comparable to, or shorter than, the phonon mean-free-path

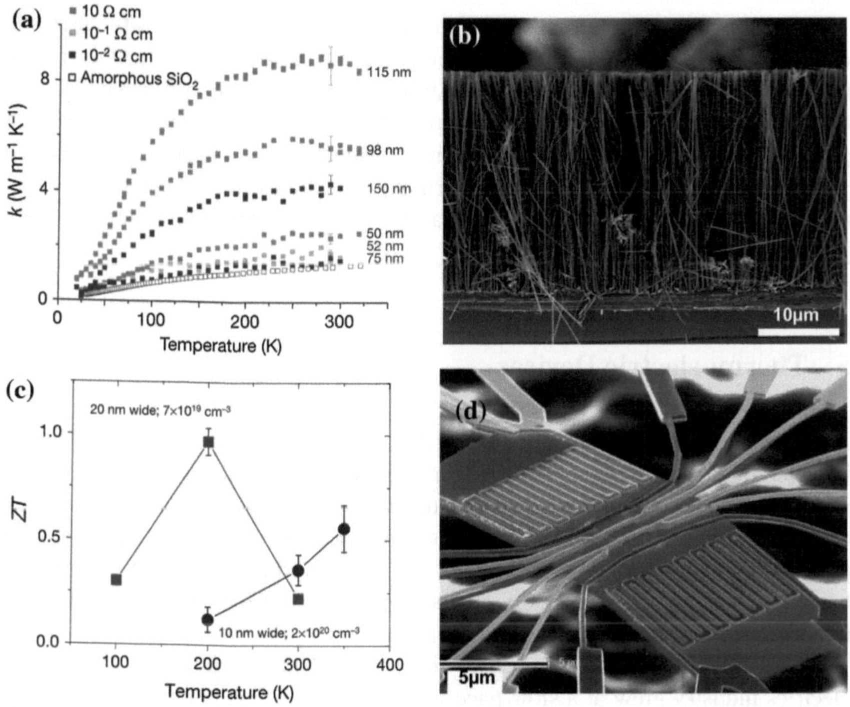

Fig. 7 Silicon nanowires as efficient thermoelectric materials. **a** Thermal conductivity values of Si nanowires measured by Hochbaum et al., nanowires were fabricated from wafers of different resistivities: 10 $\Omega \cdot$ cm (*red squares*), 10^{-1} $\Omega \cdot$ cm (*green squares*), and 10^{-2} $\Omega \cdot$ cm (*blue squares*) using an aqueous electroless etching (EE) method. **b** Cross-sectional SEM image of an EE Si nanowire array fabricated by Hochbaum et al., reproduced with permission from (Hochbaum et al. 2008). **c** Temperature dependence of *ZT* for two different groups of nanowires and **d** SEM image of the structure of the device used by Boukai et al. to measure Si nanowire arrays. By using this type of suspended structures, the thermopower and electrical and thermal conductivity of Si nanowire arrays were measured. Reproduced with permission from (Boukai et al. 2008)

exhibiting substantially lower thermal conductivities than an equivalent prepared array of silicon nanowires while preserving bulk-like electrical conductivity (Yu et al. 2010). Nanostructuration does not restrict to low dimensional structures themselves but it can also have a role in more bulky materials. For instance, porous silicon nanomeshes can be fabricated in films of tens of microns (Boor et al. 2012). Furthermore, nanocrystalline bulk silicon has also been shown to be a promising thermoelectric material. Bux et al. first applied the ball milling and hot pressing approach to fabricate *n*-type nanocrystalline bulk silicon with a very competitive figure of merit of $ZT \approx 0.7$ at 1200 K and a thermal conductivity of 6.3 W/m−K at room temperature (Bux et al. 2009). Schierning et al. have obtained similar results by using nanopowder from a gas phase process and compact the raw material by a current-activated pressure assisted sintering technique (Schierning et al. 2011; Schierning 2014).

All these significant findings open up the opportunity to use nanostructured silicon for thermoelectric applications although when compared to state-of-the-art thermoelectric materials, high density nanostructures are needed to overcome the trade-off of silicon being an enabling technological material with modest thermoelectric properties. Nevertheless, those high density nanostructures must be obtained in a way that assures their electrical and thermal connection with low electric and thermal contact resistances in real devices avoiding cumbersome and time-consuming procedures.

5 Thermoelectric Devices

As mentioned previously, thermoelectric devices allow to convert waste heat directly into electrical current (Seebeck effect) or to convert electrical current into a temperature gradient (Peltier effect). Clear examples of applications of this type of devices can be observed in thermocouples and in the cooling systems of electronic components (Winder et al. 1996). Although not considered here, in addition to energy harvesting and heat management, another application of thermoelectric devices is temperature sensing, both in contact (hot surfaces) or remotely (infrared radiation).

After the basis of thermoelectrics was established in the 1950s, the thermoelectrics industry grew at a slow pace whereas the research community gave little attention to the field. Few products were developed during the first decades of thermoelectrics until the beginning of the 1990s, when a picnic basket cooler was brought to market introducing the first significant consumer product. Nevertheless, the most relevant commercial application of themoelectric devices emerged in the automotive sector, when Amerigon placed thermoelectric coolers/heaters in the passenger seats of automobiles (Vining 2007). The automotive sector is still an important animator of thermoelectric. While environmental concerns are driving more stringent regulations, two thirds of energy in combustion engines is lost as waste heat, 40% of which through exhaust gases. Recovering part of this lost heat and converting it into electricity would translate into a reduction of fuel consumption and emissions. This application, though, is best suited to bulk thermoelectrics (nanostructured or not) and the high temperatures existing in the engine/exhaust need to be considered from a material and system reliability perspective.

On the other hand, heat management of microprocessor-based electronic systems is an interesting niche for small size thermoelectric devices. Current CMOS technologies used for the fabrication of microelectronic devices are steadily moving into the few nanometers regime dramatically increasing the number of minute transistors integrated per unit of area and their switching speed. As a result, this increase of computing capability density is reaching a serious limit in the power that can be dissipated during computation. There is a real and continuous need of efficiently cooling down the heat produced at present and future computing and graphics microprocessors.

Continuous monitoring scenarios and distributed intelligence paradigms (Internet of Things, Trillion Sensors) are another interesting application for small thermoelectric devices. Energy autonomy is still one of the most desired enabling functionalities for autonomous applications, such as wireless sensor networks. In many of these off-grid applications, batteries are normally used due to the fact that wired power is not feasible. Nevertheless, replacing batteries is not a practical solution (economically, environmentally, and logistically) not only for sensor networks in remote places or harsh environments, but also for applications where the number of nodes increases exponentially. Harvesting energy, tapping into environmentally available sources, may be a good solution in man-made scenarios applications. Thermoelectrics is an open door to energy autonomy in those particular scenarios in which waste heat or hot surfaces are available.

Thermoelectric devices possess several advantages, they are silent, reliable and scalable, they do not comprise moving parts, which reduces their maintenance to a minimum and makes them more reliable and long-lasting; and it is easy to switch from cooling to heating mode, which makes them very adaptable depending on the application required. Hence, thermoelectrics are ideal for a broad range of applications requiring thermal management. For example, thermoelectrics could substitute conventional compressor-based cooling systems for tight geometric space or low weight applications, or they could be used as power generators converting waste heat into energy in remote locations.

Commercially available thermoelectric modules are conventional modules manually assembled by connecting a large number of thermocouples or pellets electrically in series and thermally in parallel, which are then "sandwiched" between two high thermal conductivity but low electrical conductivity ceramic plates (Fig. 8). So far, these modules have been designed to operate in the Peltier mode since cooling applications have been more successful in the past years, whereas power conversion applications have not reached a widely spread usage due to the low conversion efficiencies (in the range of 5–10% for semiconductors compared to the maximum theoretically predicted value of 20%) (Rowe 2002; Rowe and Min 1996). Nevertheless, the recent search for alternative energy sources is driving the fabrication of new modules with tailored geometries and materials to achieve reliable thermoelectric generators that

Fig. 8 Image of a typical thermoelectric module. This type of device is composed of thermoelectric couples (*n*- and *p*-type semiconductor legs) that are connected electrically in series, thermally in parallel and, fixed by soldering, sandwiched between two ceramic plates. The latter form the hot and cold thermoelectric sides

outweigh the low conversion efficiency by finding niche applications in the market. Thermoelectric technology offers a green/environmentally friendly method for converting waste heat into electrical power. Nevertheless, the cost and functionality of energy harvesters are important factors for their acceptance by industry and for moving them into mass production. Reduction of the cost can be achieved by using micro and nanotechnologies for high throughput processing (Su et al. 2010b).

5.1 Thermoelectric Microgenerators Based on Nanostructured Materials

In the last decade a lot of effort has been invested in the development of thermoelectric microgenerators. Most of these attempts, which will be shortly reviewed, have tried to follow the approach of scaling down the typical architecture of a macro thermoelectric generator, i.e. the "out-of-the-plane" leg configuration, which better adapts the naturally occurring thermal gradients. From the materials point of view, microgenerators move from bulky macrostructures like pellets to nanostructures like thin films, superlattices or nanowires that enable higher material efficiencies and lower costs. Scaling down allows to achieve higher power densities, automation and mass production if standard microtechnologies are involved (Whalen et al. 2008).

Regarding the "out-of-the-plane" (vertical) architecture, such geometry is difficult to replicate using microelectronics technology. The reason for this is strictly process-related since microelectronic processes have historically been developed to fabricate lateral structures on silicon wafers, and, in addition, vertical distances between layers have been usually kept to a minimum. In this planar configuration, the thermocouple legs are patterned on a substrate and the interconnections between them are fabricated on the same surface level, making the fabrication process simpler than the vertical approximation, but consuming more area. This horizontal or "in-the-plane" architecture is typically used in the fabrication of silicon-based thermopiles for most commercial temperature or infrared sensors. If a traditional vertical or "out-of-the-plane" configuration should be replicated, the legs of the thermocouple have to be implemented through the substrate material or structured on it, and some type of suspended interconnections are then needed. This kind of 3D arrangement is hard to achieve monolithically at the microscale in a single substrate. Some alternatives to overcome this issue involve attaching two complementary dies face to face employing 3D hyperintegration strategies.

The first functional micro thermoelectric device (μTEG) was a miniature thermoelectric generator based on CMOS technology developed by Rowe et al. and was composed by miniaturized thermocouples (*p*- and *n*-type Si) as the ones used in infrared detectors (Rowe et al. 1989). Following works on this direction were mainly focused on the integration of these thermocouples to achieve high densities and consequently high output voltages. Strasser et al. developed thermoelectric devices based on doped poly-Si and poly-SiGe for wearable electronics applications

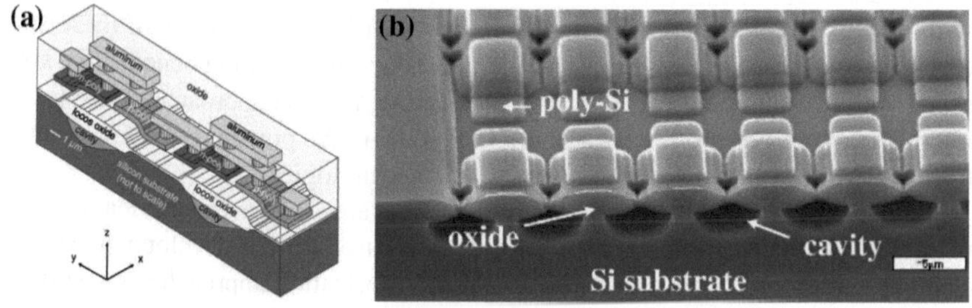

Fig. 9 Miniaturized thermoelectric generators developed by Strasser et al., the devices are based on poly-Si and Poly-SiGe surface micromachining. **a** Schematic view of two thermoelectric couples of a BiCMOS realization using standard materials. **b** SEM micrograph showing the cavities etched into a Si-substrate in order to optimize the vertical heat flux direction within the microgenerator. Reproduced with permission from (Strasser et al. 2002)

(Fig. 9), with a thermoelectric microgenerator formed by 16000 vertical thermocouples in an area of 7 mm^2 to generate 0.112 μW (ΔT = 10 K) (Strasser et al. 2002). Xie et al. fabricated a device based on heavily doped *p*- and *n*-type polysilicon thin films with an area of 1 cm^2 able to develop an open-circuit voltage of 16.7 V and an output power of 1.3 μW (ΔT = 5 K) under matched load resistance. Nevertheless, all these devices based on heavily doped *p*- and *n*-type silicon thermocouples possess high contact resistances that yield to output powers lower than 10 μW/cm^2 (Glosch et al. 1999; Strasser et al. 2003; Huesgen et al. 2008; Xie et al. 2010).

Higher power outputs have been achieved by combining MEMS technology with thin films of the state-of-the-art V-VI semiconductors, e.g. Bi_2Te_3. Böttner et al. developed a sputtering method based on $Bi_{2+x}Te_{3-x}$ integrated at a wafer-level to produce thermoelectric devices compatible with modern semiconductor industry and mass-production methods (Böttner et al. 2007). A high voltage thin-film thermogenerator developed by this group is commercially available and capable of delivering 0.14 V/K (Fig. 10). Venkatasubramanian et al. developed a technology based on a Bi_2Te_3-Sb_2Te_3 superlattice, producing a high voltage thermoelectric generator with output power levels of 1.5 mW and an open circuit voltage of 0.25 V at a ΔT of 10 K (0.025 V/K) (Venkatasubramanian et al. 2006). These groups have achieved devices

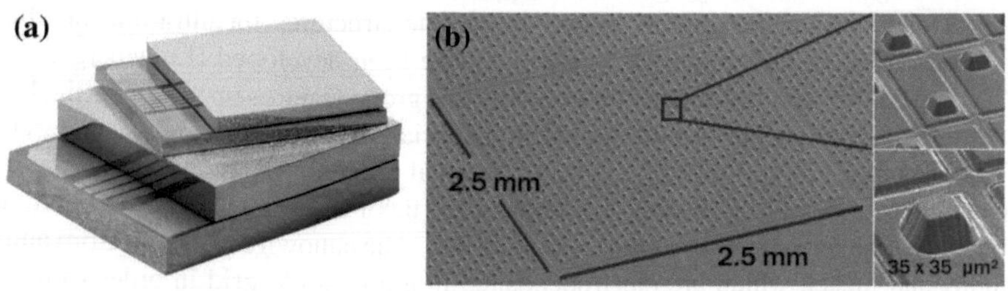

Fig. 10 **a** High voltage thin-film thermogenerator developed by Bttner et al. (Böttner et al. 2007). **b** With this technology more than 50 vertical leg pairs are possible on a square millimeter. Reproduced with permission from ©Micropelt

with strongly reduced electrical resistances, which is the key factor of these successful prototypes that have reached general production providing benefits like size, speed and heat pumping capacity (W/cm^2). Some of these approaches have led to microgenerators being offered for some years now in the portfolio of a few companies, e.g. Micropelt and Nextreme (now Laird Technologies).

To achieve energy harvesting devices with larger areas and overcome feature sizes of square millimeters some groups have been focused on the development of low cost fabrication processes alternative to MEMS fabrication approaches by working on large-area polymer-based flexible TEGs. Nevertheless, the performance of some of these devices has been below their small-sized counterparts based on MEMS technology (Itoigawa et al. 2005; Hasebe et al. 2004; Pasold et al. 2011). One of the first examples of polymer-based TEG devices can be found in the work developed by Stordeur et al. (Stordeur and Stark 1997) and Stark et al. (Stark and Stordeur 1999) based on V-VI semiconductors ($(Bi, Sb)_2Te_3/Bi_2(Se,Te)_3$) where a thin film of the material was sputtered in a polyimide-foil substrate integrating 2250 thermocouples in 64 mm^2 and generating 1.6 μW ($\Delta T = 5$ K) (Stordeur and Stark 2003). Glatz et al. have developed a flexible TEG based on the electrochemical deposition of Copper and Nickel into a polymer mold of SU-8 with an easy device integration achieving 12.0 ± 1.1 nW/cm^2 for a ΔT of 0.12 K (Glatz et al. 2006). Further research performed by this group achieved a 20 fold improvement factor when replacing the thermoelectric material with *n*- and *p*-type $Bi_{2+x}Te_{3-x}$ pillars (Glatz et al. 2009). Several fabrication approaches to find an alternative cost-effective process to deposit V-VI semiconductors such as spark plasma sintering or screen printing have been developed in the last years with little improvements.

Application-tailored devices such as the ones developed by Su et al. (Su et al. 2010a) and van Andel et al. (van Andel et al. 2010) for energy scavenging from the human body have been able to harvest enough energy to power wearable wireless medical devices using thermal energy, like an electrocardiogram (ECG) in a shirt or a solar/thermal-hybrid powered electro-encephalogram (EEG) with hybrid energy scavenger. They have also developed wrist thermoelectric generators that allow to power different types of wireless sensors (Leonov et al. 2009). Their activity mainly comprises the fabrication of both planar and vertical micromachined modules using poly-SiGe thermocouples (Fig. 11).

Although not an energy generation application, it is worth to review the work developed by Lindeberg et al., where thermopile structures for ultra-thin flexible IR sensors using thermoelectric nanowires have been developed (Lindeberg et al. 2008). The thermopile structure proposed by this group features up to 224 vertically-arranged thermocouple legs deposited in a specially prepared polyimide material. The thermopile is optimized for infrared thermal radiation detection and its fabricated using a flexible printed circuit board-like (flex PCB-like) process embedding bundles of metal nanowires as thermocouple legs. The nanowires are built from antimony and nickel, which are electrodeposited in a polyimide grid in order to form the nanowires. This device shows a clever thermoelectric application for nanowires, however, this technology is limited by the metals that can be electrodeposited, and the low Seebeck coefficient of metals in general.

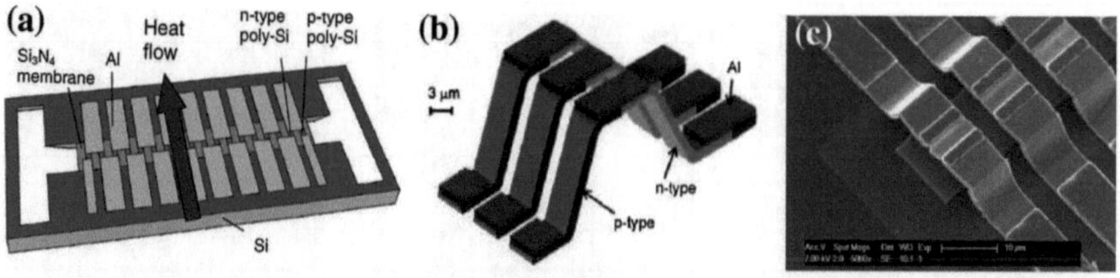

Fig. 11 Micromachined thermoelectric energy harvesters developed at the IMEC research center. **a** Membrane-less in-plane bulk-micromachined thermopiles for energy harvesting, the schematic layout shows the supporting membrane. **b** Schematic of the design of the thermocouple employed in a thermoelectric energy harvester fabricated by stepper. **c** SEM image of the released thermocouples of the design in (**b**). Reproduced with permission from (Su et al. 2010b; van Andel et al. 2010)

5.2 Integration of Silicon Nanostructures in Thermoelectric Microgenerators

Low-dimensional structures, although reported to have promising thermoelectric characteristics, have been barely successfully integrated in devices. Most of the developments performed to integrate such structures in thermoelectric microgenerators have followed the vertical "out-of-plane" configuration by electrodepositing V-VI semiconductor materials into nanopore templates but no relevant result, apart from an uncontrolled process, has been shown (Wang et al. 2005). In this direction, silicon offers both bottom-up and top-down approaches to achieve controlled integration of nanostructures. Moreover, being silicon technology the one used for the fabrication of microsensors, it makes technological and economical sense prospecting the use of this same technology and (compatible) materials for building microenergy solutions of similar robustness able to power such sensors during their whole lifetime.

Li et al. have developed a high-density silicon nanowire-based thermoelectric generator fabricated following a top-down CMOS-compatible technique (Li et al. 2011). The 25 mm^2 TEG is comprised by densely packed alternating *n*- and *p*-type silicon nanowire (Si NW) bundles (Fig. 12) with each wire having a diameter of 80 nm and a heights of 1 µm (each bundle serving as an individual thermoelectric element with 540×540 wires). The fabricated TEG is able to deliver an open circuit voltage of 1.5 mV and a short circuit current of 3.79 µA with an estimated temperature gradient across the device of 0.12 K. The work developed by Li et al. represents the first attempt to develop a Si NW-based TEG using a top-down approximation for the nanowire growth. Nevertheless, the complexity of this "out-of-plane" configuration added to a thermal gradient restricted by the nanowire height limits the maximum power output attainable to 1.5 nW (6 nW/cm^2). Further work developed by this group with the same device configuration yielded maximum generated power outputs of 4.6 nW by improving the thermal gradient across the device although the thermal parasitic losses of the device requires design optimization to truly enhance its performance (Li et al. 2012a, b).

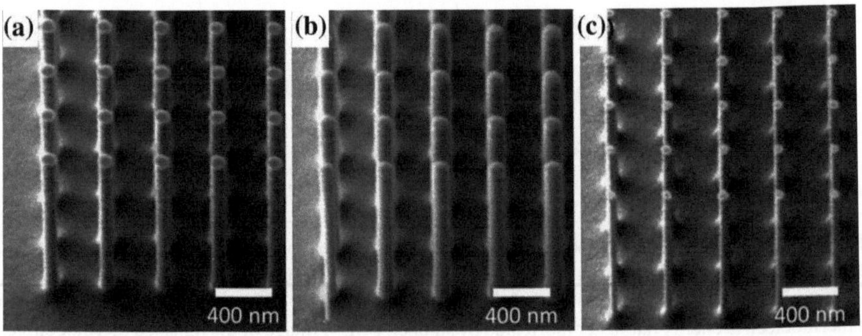

Fig. 12 SEM images of a SiNW array: **a** as etched, **b** after thermal oxidation to reduce the SiNW diameter, and **c** after removal of SiO_2 to confirm the reduction in diameter of the SiNWs. Reproduced with permission from (Li et al. 2012a)

Another top-down approach to integrate vertical arrays of silicon nanowires was developed by Curtin et al. (Curtin and Bowers 2012; Curtin et al. 2012). This group has fabricated top-down silicon nanowires by etching heavily doped n-type substrates. The nanowires, with diameters of 80–90 nm and 1 μm-height, are embedded in spin-on-glass (SOG) to form a dense and robust composite material showing a x16 decrease of the thermal conductivity over the one of bulk silicon. Although not a complete thermoelectric generator was reported, power generation from a combined *n*-type Si NW/SOG and substrate device was presented with a maximum generated power of 29.3 μW for a ΔT of 56 K.

Dávila et al. have demonstrated a planar or "in-the-plane" approach for integrating monolithically horizontal aligned arrays of bottom-up silicon nanowires into microfabricated structures (Dávila et al. 2011a, b, 2012). This group has implemented multiple electrically connected dense arrays of silicon nanowires (Si NWs) into microfabricated structures to develop thermoelectric microgenerators. Suspended silicon structures have been designed and fabricated by silicon micromachining to naturally generate thermal gradients in planar microthermoelements. The hot and cold parts of the device are linked with horizontal arrays of Si NWs grown by the Vapour-Liquid-Solid (VLS) bottom-up process (Fig. 13) after having seeded with gold nanoparticles the exposed (111) vertical trenches. It is important to notice that the attachment of VLS SiNWs to both silicon surfaces is quasi-epitaxial so that very low thermal and electric contact resistances are expected. In order to improve the thermal behavior of this device as energy harvester, multiple arrays of silicon nanowires have been successively linked, increasing the thermoelement "height", to generate larger temperature differences while keeping small internal thermoelement resistances (Dávila et al. 2013). The fabricated thermoelements have generated power output densities up to 1.44 mW/cm^2 for a ΔT of 300 °C and, working as energy harvesters, a power density of 9 μW/cm^2 for a ΔT of 27 °C (across the nanowires) in a single thermoelement. The fabricated microgenerator, taking advantage of the simple planar geometry and the 3D lateral growth of VLS Si NWs enables obtaining an effective arbitrary long Si NW-based metamaterial while preserving its electrical connectivity. Although the results presented by this group correspond to a single-leg

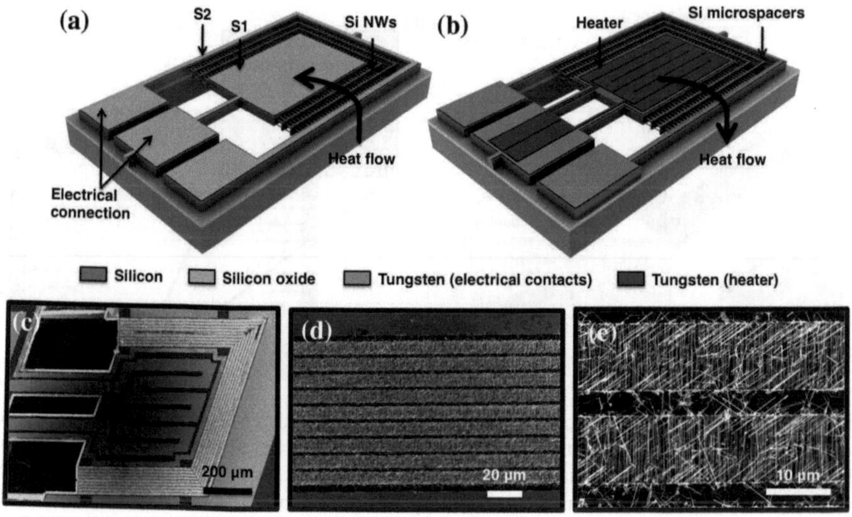

Fig. 13 Sketch showing the design of the basic structure developed by Davila et al. **a** A thermally isolated suspended silicon mass (*S1*) is connected to the surrounding silicon bulk (*S2*) through silicon nanowires (*Si NWs*) allowing a planar temperature difference to be achieved. In this way, a thermoelement in which the *Si NWs* act as a nanostructured thermoelectric material is formed; **b** sketch of a similar device that shows the heater that was integrated and used both as heat source and sensor for controlling and characterizing the temperature gradients that were reached in the structure. Heat flow is reversed in this configuration. **c** SEM image of a device with an integrated heater and silicon microspacers linking nine 10 μm-long nanowire arrays; **d** top view of nine silicon nanowire arrays connected through silicon microspacers, a detail is shown in (**e**). Reproduced with permission from (Dávila et al. 2012)

thermocouple structure, the loss of conversion efficiency in this one-leg device is compensated by a simpler architecture and a full technological compatibility with standard microfabrication processes that allows a large number of these thermocouples to be integrated and appropriately connected in a single chip for the generation of higher power devices.

Recently, Perez-Marín et al. (2014) have fabricated ultrathin single-crystalline Si membranes (100 nm in thickness) with embedded *n*- and *p*-type doped regions electrically connected in series and thermally in parallel as the active elements of a thermoelectric microgenerator (Fig. 14). The proof-of-concept planar device is able to deliver an output power density of 4.5 μW/cm^2 for a ΔT of 5.5 K. Moreover, if the membrane is pattern into Si nanowires or its thickness reduced to few nm, the thermal conductivity of the structure could be further reduced, which could yield substantial improvements.

Although a fair comparison of all these devices is difficult to achieve due to the different device architectures, geometries and characterization methods employed, Table 1 attempts to summarize the different results obtained by the aforementioned research groups. Power densities were calculated by dividing the output power values by the area of the device in each case.

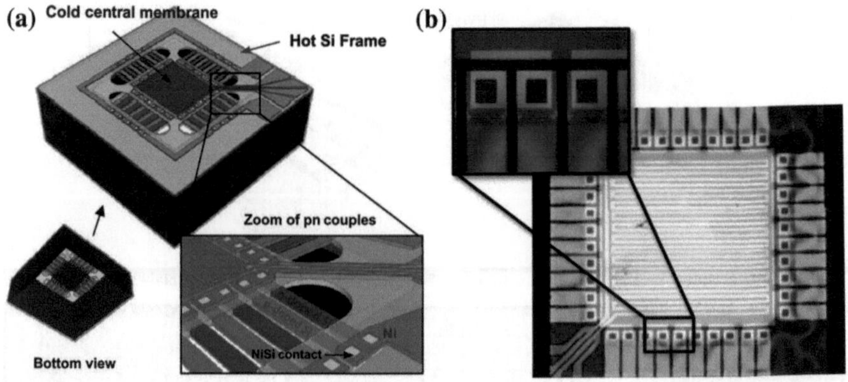

Fig. 14 Schematics of the device fabricated by Perez-Marín et al. **a** Main draw of the device. The *red* and *blue* regions represent the *n*- and *p*-couples defined in the Si membranes and *white* the NiSi intermetallic at the open via contacts. **b** Optical microscope images of the fabricated device. A detailed view of the *n*- and *p*-regions and the open vias used for contacts are shown in the inset. Reproduced with permission from (Perez-Marín et al. 2014)

Table 1 Thermoelectric microgenerators—comparison of thermoelectric microgenerators developed by different research groups

Reference	Thermoelectric material	Area (mm^2)	Number pn-couples	Power (mW/cm^2)	ΔT (K)
(Rowe et al. 1989)	*p*-Si/*n*-Si	1.8	5	$7.2e^{-4}$	25
(Stordeur and Stark 1997; Stark and Stordeur 1999)	$(Bi_{0.25}Sb_{0.75})Te_3$/ $Bi_2(Se_{0.1}Te_{0.9})_3$	67.5	2250	$8.6e^{-3}$	10
(Lim et al. 2002; Snyder et al. 2003)	Sb_2Te_3/Bi_2Te_3	2.89	63	$3.5e^{-2}$	2
(Böttner et al. 2004)	$(Bi, Sb)_2Te_3/Bi_2Te_3$	1.12	12	$5.9e^{-2}$	5
(Böttner et al. 2007)	$(Bi, Sb)_2Te_3/Bi_2Te_3$	33.4	1800	8.4	10
(Venkatasubramanian et al. 2006)	Sb_2Te_3/Bi_2Te_3	10.23	N/A	14.6	10
(Glatz et al. 2006)	Ni/Cu	14	90	$12e^{-6}$	0.12
(Glatz et al. 2009)	*p*-/*n*- Bi_2Te_3	14	99	0.12	48.4
(Strasser et al. 2002, 2003)	*p*-/*n*- poly-Si/-SiGe	7	16000	$1.6e^{-3}$	10
(Glosch et al. 1999)	Si/Al	16.5	1000	$7.3e^{-3}$	10
(Li et al. 2011, 2012a)	Top-down Si NWs	162	25	$9.3\ e^{-7}$	0.12
(Curtin and Bowers 2012; Curtin et al. 2012)	Top-down *n*- Si NWs	$2.5e^{-3}$	1	11.7	56
(Dávila et al. 2012, 2013)	Bottom-up *p*- Si NWs	1	1	$2.3\ e^{-2}$	50
(Perez-Marín et al. 2014)	*p*-Si/*n*-Si	0.64	20	$4.5\ e^{-3}$	5

6 Concluding Remarks

In summary, thermoelectric microgenerators are particularly interesting for portable devices although their implementation in microelectronics is not straightforward. Good thermoelectric materials cannot be easily integrated monolithically in microelectronics processes, and mainstream microelectronics materials result unattractive due to their poor thermoelectric properties. It is clear that the challenge to widespread commercial applications using these thermoelectric microdevices is highly related to the improvement of the thermoelectric figure of merit of the material, ZT, where current fundamental research in nanoscience is actually trying to push the limits, but it also evident that an extensive development on the integration of these materials must be carried out. With research groups having pursued thin-film thermoelectric devices and proving the feasibility of integrating thermoelectrics with microtechnology it seems reasonable to expect further size and cost reductions through mass production and utilization of modern semiconductor manufacturing technologies for introducing new products in the market.

Furthermore, the preliminary results achieved with nanostructured silicon could bring a new silicon-based thermoelectric generation technology closer to practical implementation of μTEGs in local energy supply applications such as wireless sensor networks and remote actuators or advanced micro/nanodevices. Although the technological limitations of the maximum thermoelement height achieved by nanostructuring silicon through top-down approaches can be overcome by switching to a bottom-up approach, further improvements need to be developed to achieve higher thermoelectric efficiencies and smart integration processes of both p- and n-type elements into a single device using this technology.

Overcoming the difficult implementation of thermoelectric generators with microelectronics materials and microelectronics processes still remains a big challenge for the thermoelectrics research community. A good combination of the thermoelectric properties shown by nanostructured materials and a suitable integration into microelectronics could yield a major breakthrough in the fabrication and application of thermoelectric microgenerators.

References

Bachtold A, Hadley P, Nakanishi T, Dekker C (2001) Logic circuits with carbon nanotube transistors. Science 294:1317–1320

Boor J, Kim D, Ao X, Becker M, Hinsche N, Mertig I, Zahn P, Schmidt V (2012) Thermoelectric properties of porous silicon. Appl Phys A 107:789–794

Böttner H, Nurnus J, Gavrikov A, Kühner G, Jägle M, Künzel C, Eberhard D, Plescher G, Schubert A, Schlereth K-H (2004) New thermoelectric components using microsystem technologies. J Microelectromech Syst 13(3):414–420

Böttner H, Nurnus J, Schubert A, Volkert F (2007) New high density micro structured thermogenerators for stand alone sensor systems. In: 26th international conference on thermoelectrics. IEEE, pp 306–309

Boukai A, Bunimovich Y, Tahir-Kheli J, Yu J-K, Goddard W, Heath J (2008) Silicon nanowires as efficient thermoelectric materials. Nature 451(7175):168–71

Bux S, Blair R, Gogna P, Lee H, Chen G, Dresselhaus M, Kaner R, Fleurial J (2009) Nanostructured bulk silicon as an effective thermoelectric material. Adv Func Mater 19:2445–2452

Chen R, Hochbaum A, Murphy P, Moore J, Yang P, Majumdar A (2008) Thermal conductance of thin silicon nanowires. Phys Rev Lett 101(10):1–4

Curtin B, Bowers J (2012) Thermoelectric properties of silicon nanowire array and spin-on glass composites fabricated with CMOS-compatible techniques. In MRS proceedings 1408:1–6

Curtin B, Fang E, Bowers J (2012) Highly ordered vertical silicon nanowire array composite thin films for thermoelectric devices. J Electron Mater

Dávila D, Tarancón A, Calaza C, Salleras M, Fernández-Regúlez M, San Paulo A, Fonseca L (2012) Monolithically integrated thermoelectric energy harvester based on silicon nanowire arrays for powering micro/nanodevices. Nano Energy 1:812–819

Dávila D, Tarancón A, Calaza C, Salleras M, Fernández-Regúlez M, San Paulo A, Fonseca L (2013) Improved thermal behavior of multiple linked arrays of silicon nanowires integrated into planar thermoelectric microgenerators. J Electron Mater 42(7):1918–1925

Dávila D, Tarancón A, Fernández-Regúlez M, Calaza C, Salleras M, San Paulo A, Fonseca L (2011a) Silicon nanowire arrays as thermoelectric material for a power microgenerator. J Micromech Microeng 21(10):104007

Dávila D, Tarancón A, Kendig D, Fernández-Regúlez M, Sabaté N, Salleras M, Calaza C, Cané C, Gràcia I, Figueras E, Santander J, San Paulo A, Shakouri A, Fonseca L (2011b) Planar thermoelectric microgenerators based on silicon nanowires. J Electron Mater 40(5):851–855

Dresselhaus M, Chen G, Tang M, Yang R, Lee H, Wang D, Ren Z, Fleurial J, Gogna P (2007) New directions for low-dimensional thermoelectric materials. Adv Mater 19(8):1043–1053

Ebling D (2009) Thermoelectrics for high temperatures—a survey of state of the art. In: Energy harvesting workshop

Esfarjani K, Chen G, Stokes H (2011) Heat transport in silicon from first-principles calculations. Phys Rev B 84(8):085204

Feser J, Sadhu J, Azeredo B, Hsu K, Ma J, Kim J, Seong M, Fang N, Li X, Ferreira P, Sinha S, Cahill D (2012) Thermal conductivity of silicon nanowire arrays with controlled roughness. J Appl Phys 112(11):114306

Glatz W, Muntwyler S, Hierold C (2006) Optimization and fabrication of thick flexible polymer based micro thermoelectric generator. Sens Actuators A Phys 132(1):337–345

Glatz W, Schwyter E, Durrer L, Hierold C (2009) Bi2Te3-based flexible micro thermoelectric generator with optimized design. J Microelectromech Syst 18(3):763–772

Glosch H, Ashauer M, Pfeiffer U, Lang W (1999) A thermoelectric converter for energy supply. Sens Actuators A Phys 74(1–3):246–250

Hasebe S, Ogawa J, Shiozaki M, Toriyama T, Sugiyama S, Ueno H, Itoigawa K (2004) Polymer based smart flexible thermopile for power generation. In: 17th IEEE international conference on micro electro mechanical systems (MEMS), pp 689–692

Hicks L, Dresselhaus M (1993a) Effect of quantum-well structures on the thermoelectric figure of merit. Phys Rev B 47(19):12727–12731

Hicks L, Dresselhaus M (1993b) Thermoelectric figure of merit of a one-dimensional conductor. Phys Rev B 47(24):16631–16634

Hochbaum A, Chen R, Delgado R, Liang W, Garnett E, Najarian M, Majumdar A, Yang P (2008) Enhanced thermoelectric performance of rough silicon nanowires. Nature 451(7175):163–167

Huang Y, Duan X, Cui Y, Lauhon L, Kim K, Lieber C (2001) Logic gates and computation from assembled nanowire building blocks. Science 294(November):1313–1317

Huesgen T, Woias P, Kockmann N (2008) Design and fabrication of MEMS thermoelectric generators with high temperature efficiency. Sens Actuators A Phys 145–146:423–429

Itoigawa K, Ueno H, Shiozaki M, Toriyama T, Sugiyama S (2005) Fabrication of flexible thermopile generator. J Micromech Microeng 15(9):S233–S238

Iyengar A (2010) Synthesis and characterization of micro/nano material for thermoelectric applications. PhD thesis, Case Western Reserve University
Leonov V, Van Hoof C, Vullers R (2009) Thermoelectric and hybrid generators in wearable devices and clothes. In: IEEE (ed) Sixth international workshop on wearable and implantable body sensor networks, 2009. BSN 2009. IEEE, pp 195–200
Li D, Wu Y, Kim P, Shi L, Yang P, Majumdar A (2003) Thermal conductivity of individual silicon nanowires. Appl Phys Lett 83(14):2934
Li J-F, Liu W-S, Zhao L-G, Zhou M (2010) High-performance nanostructured thermoelectric materials. NPG Asia Mater 2:152–158
Li Y, Buddharaju K, Singh N, Lo GQ, Lee SJ (2011) Chip-level thermoelectric power generators based on high-density silicon nanowire array prepared with top-down CMOS technology. IEEE Electron Device Lett 32(5):674–676
Li Y, Buddharaju K, Singh N, Lee S (2012a) Top-down silicon nanowire-based thermoelectric generator: design and characterization. J Electron Mater 41:989–992
Li Y, Buddharaju K, Tinh B, Singh N, Lee S (2012b) Improved vertical silicon nanowire based thermoelectric power generator with polyimide filling. IEEE Electron Device Lett 33(5):715–717
Lim, J., Snyder G, Huang C-K, Herman J, Ryan M, Fleurial J-P (2002) Thermoelectric microdevice fabrication process and evaluation at the Jet Propulsion Laboratory (JPL), pp 535–539
Lindeberg M, Yousef H, Rödjegård H, Martin H, Hjort K (2008) A PCB-like process for vertically configured thermopiles. J Micromech Microeng 18(6):065021
Nolas G, Cohn J, Slack G, Schujman S (1998) Semiconducting Ge clathrates: promising candidates for thermoelectric applications. Appl Phys Lett 73(2):178
Pasold G, Etlin P, Hahn M, Muster U, Nersessian V, Bonfrate D, Buser R, Cucinelli M, Gutsche M, Kehl M, Zäch N, Hazelden R (2011) Powering wireless sensors: microtechnology-based large-area thermoelectric generator for mass applications. Proceedings of IEEE sensors 1:1293–1296
Perez-Marín A, Lopeandía A, Abad L, Ferrando-Villaba P, Garcia G, Lopez A, Muñoz-Pascual F, Rodríguez-Viejo J (2014) Micropower thermoelectric generator from thin Si membranes. Nano Energy 4:73–80
Ponomareva I, Srivastava D, Menon M (2007) Thermal conductivity in thin silicon nanowires: phonon confinement effect. Nano Lett 7(5):1155–1159
Rowe D (2002) Recent concepts in thermoelectric power generation. In: Twenty-first international conference on thermoelectrics, 2002. Proceedings ICT '02, pp 365–374
Rowe D (2006a) Thermoelectric waste heat recovery as a renewable energy source. Int J Innov Energy Syst Power 1(1):13–23
Rowe D (2006b) Thermoelectrics handbook: macro to nano. CRC Press, Taylor & Francis Group
Rowe D, Min G (1996) Design theory of thermoelectric modules for electrical power generation. IEE Proc Sci Meas Technol 143(6):351–356
Rowe D, Morgan D, Kiely J (1989) Miniature low-power/high-voltage thermoelectric generator. Electron Lett 25(2):166–168
Sales B, Mandrus D, Williams R (1996) Filled skutterudite antimonides: a new class of thermoelectric materials. Science 272(5266):1325–1328
Schierning G (2014) Silicon nanostructures for thermoelectric devices: a review of the current state of the art. Phys Status Solidi (A) 211(6):1235–1249
Schierning G, Theissmann R, Stein N, Petermann N, Becker A, Engenhorst M, Kessler V, Geller M, Beckel A, Wiggers H, Schmechel R (2011) Role of oxygen on microstructure and thermoelectric properties of silicon nanocomposites. J Appl Phys 110(11):113515
Singh R (2008) Experimental characterization of thin film thermoelectric materials and film deposition via molecular beam epitaxy. PhD thesis, University of California Santa Cruz
Slack G (1995) New materials and performance limits for thermoelectric cooling, Chapter 34. CRC Press, Boca Raton, pp 407–440
Slack G, Tsoukala V (1994) Some properties of semiconducting IrSb3. J Appl Phys 76(3):1665

Snyder G, Lim J, Huang C-K, Fleurial J-P (2003) Thermoelectric microdevice fabricated by a MEMS-like electrochemical process. Nat Mater 2:528–531

Snyder G, Toberer E (2008) Complex thermoelectric materials. Nat Mater 7(2):105–114

Stark I, Stordeur M (1999) New micro thermoelectric devices based on bismuth telluride-type thin solid films. In: Eighteenth international conference on thermoelectrics. Proceedings, ICT'99 (Cat. No. 99TH8407), pp 465–472

Stordeur M, Stark I (1997) Low power thermoelectric generator—self-sufficient energy supply for micro systems. In: Proceedings of the 16th international conference on thermoelectrics, pp 575–577

Stordeur M, Stark I (2003) DTS: thin film thermoelectric generator systems GmbH. In: Proceedings of the 22nd international conference on thermoelectrics, vol 97, p 661

Stranz A, Kähler J, Merzsch S, Waag A, Peiner E (2012) Nanowire silicon as a material for thermoelectric energy conversion. Microsyst Technol 18:857–862

Stranz A, Sökmen Ü, Kähler J, Waag A, Peiner E (2011a) Measurements of thermoelectric properties of silicon pillars. Sens Actuators A Phys 171(1):48–53

Stranz A, Waag A, Peiner E (2011b) Thermal characterization of vertical silicon nanowires. J Mater Res 26(15):1958–1962

Stranz A, Waag A, Peiner E (2013) High-temperature performance of stacked silicon nanowires for thermoelectric power generation. J Electron Mater 42(7):2233–2238

Strasser M, Aigner R, Franosch M, Wachutka G (2002) Miniaturized thermoelectric generators based on poly-Si and poly-SiGe surface micromachining. Sens Actuators A Phys 97–98(1–2):535–542

Strasser M, Aigner R, Lauterbch C, Sturm T, Franosch M, Wachutka G (2003) Micromachined CMOS thermoelectric generators as on-chip power supply. In: TRANSDUCERS '03. 12th international conference solid-state sensors, actuators microsystems. IEEE

Su J, Leonov V, Goedbloed M, van Andel Y, de Nooijer M, Elfrink R, Wang Z, Vullers R (2010a) A batch process micromachined thermoelectric energy harvester: fabrication and characterization. J Micromech Microeng 20(10):104005

Su J, Vullers R, Goedbloed M, van Andel Y, Leonov V, Wang Z (2010b) Thermoelectric energy harvester fabricated by Stepper. Microelectron Eng 87(5–8):1242–1244

Tan Y, Panda S (2010) Review of energy harvesting technologies for sustainable wireless sensor network. InTech

U.S. Department of Energy (2010) Critical materials strategy, Technical report

van Andel Y, Jambunathan M, Vullers R, Leonov V (2010) Membrane-less in-plane bulk-micromachined thermopiles for energy harvesting. Microelectron Eng 87(5–8):1294–1296

Venkatasubramanian R, Watkins C, Caylor C, Bulman G (2006) Microscale thermoelectric devices for energy harvesting and thermal management. In: PowerMEMS 2006. The 6th international workshop on micro and nanotechnology for power generation and energy conversion applications, pp 1–4

Vining C (2007) ZT = 3.5: fifteen years of progress and things to come. In: The 5th European conference on thermoelectrics ECT 2007, pp 6–11

Wang W, Jia F, Huang Q, Zhang J (2005) A new type of low power thermoelectric micro-generator fabricated by nanowire array thermoelectric material. Microelectron Eng 77(3–4):223–229

Wang X (2012) Piezoelectric nanogenerators-harvesting ambient mechanical energy at the nanometer scale. Nano Energ 1(1):13–24

Wang X, Song J, Liu J, Wang Z (2007) Direct-current nanogenerator driven by ultrasonic waves. Science 316:102–106

Weber L, Gmelin E (1991) Transport properties of silicon. Appl Phys A 140:136–140

Whalen S, Apblett C, Aselage T (2008) Improving power density and efficiency of miniature radioisotopic thermoelectric generators. J Power Sources 180:657–663

Winder E, Ellis A, Lisensky G (1996) Thermoelectric devices: solid-state refrigerators and electrical generators in the classroom. J Chem Educ 73(10):940

Xie J, Lee C, Feng H (2010) Design, fabrication, and characterization of CMOS MEMS-based thermoelectric power generators. J Microelectromech Syst 19(2):317–324
Yang J, Chen Y, Peng J, Song X, Zhu W, Su J, Chen R (2004) Synthesis of CoSb3 skutterudite by mechanical alloying. J Alloys Compd 375(1–2):229–232
Yu J-K, Mitrovic S, Tham D, Varghese J, Heath J (2010) Reduction of thermal conductivity in phononic nanomesh structures. Nat Nanotechnol 5:718–721
Yu L, Roca i Cabarrocas P, (2010) Growth mechanism and dynamics of in-plane solid-liquid-solid silicon nanowires. Phys Rev B 81(8):1–11

Towards Voltage-Driven Nano-Spintronics: A Review

Jin Zhang, Eva Pellicer and Jordi Sort

Abstract Nano-spintronics or nano-spin-electronics is a highly emergent technology that has revolutionized information and communication technologies offering non-volatility in high-density recorded information and an increase of the data processing speed by exploiting both the fundamental charge of electrons as well as their spin degree of freedom in nanoscale devices. The utilization of electric currents, though, poses a challenge in terms of minimization of electric power consumption. Magnetic storage systems and magneto-electronic devices are conventionally controlled by means of magnetic fields (via electromagnetic induction) or using spin-polarized electric currents (spin-transfer torque). Both principles involve significant energy loss by heat dissipation (Joule effect). The replacement of electric current with voltage (or electric field) to control the processing of information would drastically reduce the overall power consumption. Strain-mediated magneto-electric coupling in piezoelectric-magnetostrictive bilayers might appear a proper strategy to achieve this goal. However, this approach is not so suitable in spintronics because of the clamping effects with the substrate, need of epitaxial interfaces and risk of fatigue-induced mechanical failure. The exciting possibility to control ferromagnetism of metals and semiconductors directly with electric field (without strain) has been reported in recent years, but most significant effects occur below 300 K and only in ultra-thin films and nanoparticles. Herein, we provide an overview of the progress in voltage-driven magneto-electric effects in different types of magnetic materials and systems at the nanoscale. The possibility to use these effects in real applications (e.g., electrically-assisted high-density recording media, magnetic random access memories and spin field effect transistors) is described. The ongoing progress in the understanding of these effects is likely to

J. Zhang · E. Pellicer
Facultat de Ciències, Departament de Física,
Universitat Autònoma de Barcelona, 08193 Bellaterra, Spain

J. Sort (✉)
Facultat de Ciències, Institució Catalana de Recerca i Estudis Avançats (ICREA) and Departament de Física, Universitat Autònoma de Barcelona, 08193 Bellaterra, Spain
e-mail: jordi.sort@uab.cat

D. Brabazon et al. (eds.), *Commercialization of Nanotechnologies–A Case Study Approach*, DOI 10.1007/978-3-319-56979-6_5

open new paradigms in the field of spintronics and will certainly have a high economic transcendence.

Keywords Nano-spintronics • Magneto-electric devices • Energy efficiency • Ferromagnetic alloys and semiconductors • MRAM • Spin field effect transistor

1 Introduction

Spintronics or "spin-electronics" is a rapidly emerging technology in which traditional electronics is combined with spin-dependent effects to deliver non-volatile magnetic storage devices that can be operated at ultra-fast processing rates (Zútic et al. 2004; Dieny et al. 2010). More specifically, spintronics exploits both the fundamental charge of electrons as well as their spin momentum to offer non-volatile magnetic storage and an enhanced data processing speed. Spintronics can involve the active manipulation of the spin degree of freedom in solid-state systems, not only by using magnetic fields (as in most conventional magneto-electronic devices) but also by means of electric current or, more recently, electric fields (or voltage).

The origins of spintronics can be traced back to the first utilization of the magneto-resistance effect. The ordinary magneto-resistance originates from the curved trajectory of electrons in a non-ferromagnetic metal, due to the Lorentz force, when an external magnetic field is applied (Pippard 2009). Magneto-resistance in metals is, however, not very pronounced in general. Nevertheless, a "giant magnetoresistance effect" was observed in magnetic multilayers consisting of two ferromagnetic layers separated by a non-ferromagnetic metal, typically Cu or Cr. The resistance of this tri-layered stack was found to be dependent on the relative orientation of the two ferromagnets, attaining maximum values when the two ferromagnetic layers are antiparallel aligned to each other. This effect, discovered separately by Fert's group (Baibich et al. 1988) and Grünberg's group (Binasch et al. 1989), led to the implementation of spin valves (Dieny et al. 1991), which are utilized nowadays in the magnetic recording industry. Due to the high transcendence of this discovery, Fert and Grünberg were awarded the Nobel Prize in Physics in 2007. Remarkably, a further enhancement of the magnetoresistance (up to a factor 600–800%) can be achieved if the non-magnetic metal is replaced by an oxide dielectric spacer (Al_2O_3, MgO or HfO_2) (Moodera et al. 1995) since electric conductivity takes place in this case only by quantum tunneling effect (leading to the so-called tunneling magnetoresistance and tunneling junction devices, which are conceptually similar to the aforementioned spin valves).

During the last decade, enormous progress has been made on the understanding of phenomena like spin polarization (generation of electric currents in which all spins possess a unique, up or down, spin orientation) (Wiesendanger and Güntherodt 1990), injection of spin-polarized currents from ferromagnetic to non-ferromagnetic metals and semiconductors (Wolf et al. 2001), giant magneto-resistance in magnetic multilayers (Baibich et al. 1988; Binasch et al. 1989), spin-transfer torque (change of the magnetization direction of a given material through the interaction with a spin-polarized current) (Albert et al. 2000; Slonczewski 1996) or spin-Hall effects (Valenzuela and Tinkham 2006). Pure spintronic effects are observed in experiments in which only the spin is propagating (spin current) rather than the electric charges themselves (Valenzuela and Tinkham 2006; Hoffmann 2013; Scheid et al. 2007). Spin injection can be achieved by (i) optical excitation (using circularly polarized light) (Ohno et al. 1999) or (ii) letting the electrons traverse a magnetized ferromagnet (Wolf et al. 2001). In turn, detection of spin current can be accomplished by (i) inverse spin Hall effect (Saitoh et al. 2006), (ii) Handle effect (which is conceptually similar to the Faraday effect in magneto-optics) (Lou et al. 2007), (iii) emission of light using a quantum well (Fiederling et al. 1999) or (iv) magneto-optic Kerr microscopy (Cormier et al. 2008). Based on these phenomena, spintronics has revolutionized information recording strategies and has triggered the development of a number of innovative devices: magnetic tunnel junctions (Moodera et al. 1995), magnetic tunneling transistors (Huang et al. 2004), magnetic random-access memories (MRAM) (Prenat et al. 2015), all-spin logic devices or spin quantum computation (Hu et al. 2015), the latter still in its infancy.

2 Why Voltage-Driven Magnetic/Spintronic Nanodevices?

Technological progress in industrialized countries has created huge demands for electric energy. Electricity represents almost half of the overall energy spent in Europe. Transmitting electricity from power plants to our homes involves considerable energy loss in the form of heat dissipation by Joule effect (about 3–5% of the generated energy is lost, which is equivalent to thousands of M€ every year). The use of high voltages (and low currents) in power transmission lines is specifically designed to reduce Joule heating losses, since the dissipated power is proportional to i^2, where i denotes the current intensity. Heating effects are also quite significant in electronic devices. Most of the electric power utilized by central processing units is converted to heat, although the rate of computation attained per Watt of power consumed (i.e., number of operations/s · W) has dramatically increased, by over a trillion times, since the first commercial computers (Scogland et al. 2013). This has been possible thanks to the progressive miniaturization of magnetic storage systems and the reading/writing heads, attaining lateral sizes of only few tens of nm. Nonetheless, the search for new methods to implement energy-efficient electronic nanodevices remains still one of the main goals of the Digital Agenda for Europe in the Horizon 2020 Programme (http://ec.europa.eu/digital-agenda/en/electronics). Besides the heat dissipation itself, there is also a large amount of energy spent in cooling systems or air conditioning (as, for example, in powerful servers).

Nano-spintronics has largely contributed to the enhancement of the hard disk capacity. However, power consumption in magneto-electronic devices continues to be an issue (Hu et al. 2011). The currents needed to create the magnetic fields able to switch magnetization in MRAM units are of the order of 10 mA; spin-transfer torque requires currents of around 0.5 mA, still a factor five times larger than the output currents delivered by highly miniaturized metal-oxide-semiconductor field-effect transistors (MOSFET) (Ralph et al. 2011). By analogy to electric power transmission, the use of applied voltages has been envisaged as a suitable strategy to reduce the current intensity in spintronic devices and hence their power consumption. This is particularly appealing in systems based on semiconductors which would be compatible with complementary metal oxide seminconductor (CMOS) architectures (Ohno et al. 2000; Yamada et al. 2011; Stolichnov et al. 2008). Furthermore, while magnetization reversal induced solely by an applied voltage (without external magnetic field) remains very challenging (Chiba et al. 2003; Maruyama et al. 2009), effects of the electric field on the magnetic anisotropy and magnetization orientation have been reported not only in diluted semiconductors but also in metallic nanoparticles and ultra-thin films, in some cases at room temperature (RT) (Weisheit et al. 2007; Cherifi et al. 2014; Kim et al. 2013; Maruyama et al. 2009; Shiota et al. 2011). This has opened new avenues for the development of low-power spintronic devices. If technically realizable, switching of the magnetization using voltage instead of spin-polarized electric current has been estimated to bring about an energy cost reduction factor of about 1/500 (Shiota et al. 2011). Remarkably, in spin-transfer torque MRAM devices, application of electric field can reduce the current needed to switch the state of one of the memory units by roughly two orders of magnitude (from 10^6 to 10^4 A/cm^2), thus representing a factor 10^4 in energy saving (Joule heating) in each magnetization reversal (Wang et al. 2012).

Finally, there is the interesting issue that voltage can be applied at a very local length scale, hence it is a very suitable actuation principle for the progress of nano-spintronics. This is opposite to magnetic field, which is not so local and can easily cause undesired reversal of neighboring bits and consequent loss of magnetic information.

3 The Magneto-Electric Effect: Piezoelectric-Magnetostrictive Composites and Single-Phase Multiferroics

Magneto-electricity (ME) refers to the influence of an external magnetic field on the electric polarization of certain materials or, conversely, the effect of external electric field on the intrinsic net magnetization. The effect was first conjectured by P. Curie in 1854 for asymmetric molecular bodies (Curie 1894) but it was not experimentally observed until 1960 (Astrov 1960). Hitherto, ME effects have been observed in

(i) *single-phase* "multiferroics" (a term coined by Schmid in 1994 to designate materials exhibiting simultaneously two or three types of ferroic order (Schmid 1994)) or (ii) piezoelectric-magnetostrictive or piezomagnetic-electrostrictive composite *heterostructures* (Wang et al. 2010). The primary requirement for ME in single-phase materials is the presence of adequate structural and symmetry conditions that permit the coexistence of spontaneous long-range magnetic and electric orders. Progress in this direction is currently hampered by the dearth of available materials and the reduced strength of magneto-electric coupling, even at low temperatures. In spite of intensive efforts, most single phase multiferroic materials are complex oxides that exhibit their properties only at low temperatures. Moreover, the strength of the ME coupling is generally too weak for practical applications. ME effects at RT have been reported for Cr_2O_3 and $BiFeO_3$ but these materials are both antiferromagnetic, with zero net magnetization (Ramesh and Spaldin 2007).

An alternative strategy for engineering enhanced voltage-driven ME effects at RT is to introduce indirect coupling, mediated via strain, between piezoelectric and magnetostrictive materials. In such systems, the applied voltage induces strain in the piezoelectric component, which is transmitted to the magnetostrictive material, causing changes in the orientation of its net magnetization. This effect has potential applications in magnetic actuators and transducers (Ramesh and Spaldin 2007). However, in spintronic devices, this approach is not easily applicable for several reasons. First, there is limited piezoelectric response due to mechanical clamping of the films with the substrate. Secondly, the strength of the coupling is typically rather low and affected by mechanical mismatch at the interface between the two materials. Finally, the endurance of devices is compromised by eventual mechanical breakdown.

Strain coupling requires an optimum mechanical matching between the two constituent phases. For this reason, the effects are only observed in epitaxial composite films or patterned structures where the crystallographic orientation and interfacial roughness are accurately controlled. Unfortunately, the changes in magnetization induced by voltage in thin films are typically rather low (<5%), due to clamping effects with the substrate, which strongly limit the available strains. This hinders the implementation of such composite films in spintronic devices. Furthermore, in case of magnetostrictive films directly grown onto piezoelectric substrates, the endurance of devices would be highly compromised by eventual fatigue-induced mechanical failure. Hence, further progress in the field of voltage-driven magnetic actuation relies on engineering new procedures, beyond the state-of-the-art, to induce large voltage effects at RT without involving strain.

4 Controlling Magnetism with Electric Field Without Strain-Mediated Effects

A number of exciting experiments performed in recent years have shown the possibility to modify the magnetic properties of diluted magnetic semiconductors and some specific metallic alloys (in the form of nanoparticles or ultra-thin films of just a few monolayers) directly with an applied electric field, without strain-mediated effects, via accumulation of electric charges at their surface (Ohno et al. 2000; Yamada et al. 2011; Stolichnov et al. 2008; Chiba et al. 2003; Weisheit et al. 2007; Cherifi et al. 2014; Kim et al. 2013; Maruyama et al. 2009; Shiota et al. 2011). Although the physical origin of the observed effects in semiconductors and metals is completely different, in both types of materials an electric field in the range 0.5–2 V/nm can induce significant variations in the values of coercivity (H_C) or Curie temperature (T_C) (Chiba et al. 2003, 2011; Weisheit et al. 2007; Wang et al. 2012; Bonell et al. 2011). Such electric fields can be created applying voltages of tens or hundreds of V, depending on the thickness of the dielectric layer. In a liquid electrolyte, lower voltages generate reasonably high electric fields due to the very small thickness of the electric double-layer that is formed.

From a physical point of view, the origin of voltage-driven magnetic effects is different depending on whether the actuated material is a semiconductor or a metal. In diluted semiconductors, magnetic effects are typically observed at low temperatures and they are induced by the voltage-driven modification of the concentration of holes/donors which mediate the exchange interactions responsible for the ferromagnetic response (Ohno et al. 2000; Yamada et al. 2011; Stolichnov et al. 2008; Chiba et al. 2003; Weisheit et al. 2007; Cherifi et al. 2014; Kim et al. 2013). Such changes occur in a rather thin channel, determined by the so-called Debye radius, which is of the order of 10–12 nm. In metallic alloys, electric field is more effectively screened and electrostatic charges accumulate within the first 0.5 nm from the interface between the metal and the dielectric (the Thomas-Fermi screening length is $\lambda_{TF} \sim 0.5$ nm). Yet, magnetic effects can propagate a few nm towards the interior of the alloy, within the spin-spin correlation length, which exceeds 20 nm in many metallic systems (Ovchinnikov and Wang 2009). The fundamental physical origin of the voltage-driven modification of the magnetic anisotropy in metallic alloys still remains controversial. It has been suggested that electric field modifies the number of unpaired *d* electrons with energy close to the Fermi level (Weisheit et al. 2007). Moreover, direct changes in the electronic band structure can also be introduced by the electric field (Nakamura et al. 2009). Very recently, Rashba spin-orbit coupling, caused by the surface electric field, has been suggested as an additional mechanism contributing to the change of perpendicular magnetic anisotropy observed in some ultra-thin metallic films by application of voltage (Barnes et al. 2014).

Among the magnetic effects caused by electric field at low temperatures one can mention the following: change of T_C in magnetic diluted semiconductors such as (Ga, Mn)As or (In, Mn)As (Chiba et al. 2008); increase of the superparamagnetic

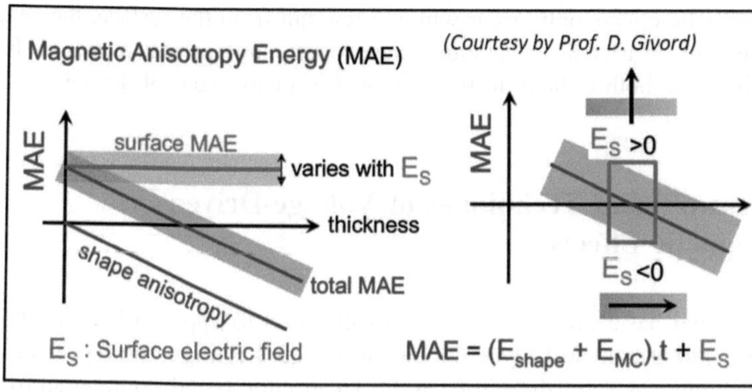

Fig. 1 Effect of an applied electric field (E) on the net magnetic anisotropy energy

blocking temperature in metallic nanoparticles (Kim et al. 2013); and change of coercivity (H_C) (Ohno et al. 2000) and, in some cases, orientation of the magnetic easy axis (Chiba et al. 2008; Zhernenkov et al. 2010). Figure 1 illustrates how an electric field can induce a reorientation of the magnetic easy axis of a ferromagnetic thin film. In the figure, a positive magnetic anisotropy energy (MAE) indicates that the film will have a perpendicular-to-plane magnetic easy axis, whereas a negative MAE means in-plane preferential orientation of the magnetization. In the absence of any external electric field, the MAE is given by the sum of the shape anisotropy plus the magnetocrystalline anisotropy energies ($E_{shape} \cdot t$ and $E_{MC} \cdot t$, respectively, where t is the film thickness). This energy is negative and its absolute value increases with the film thickness. The electric field provides an additional energy, E_S, which is positive. If the films are thick, the in-plane contribution (negative MAE) predominates. If the films are very thin, the magnetic easy axis will be perpendicular-to-plane due to the influence of the electric field. However, around a critical film thickness, the resulting $MAE = (E_{shape} + E_{MC}) \cdot t + E_S$ can become positive or negative depending on the sign of the electric field, thus eventually inducing a magnetic reorientation transition from in-plane towards perpendicular-to-plane.

Unfortunately, the effects of electric field on the magnetic properties at RT are still too weak to be used in relevant applications. Moreover, to observe such effects it is usually necessary to apply relatively high voltage pulses, a procedure which is not straightforward to be scaled up for industrial purposes. There are a number of reasons that make the observation of such effects difficult at RT. First, most magnetic diluted semiconductors exhibit $T_C \ll 300$ K and the increase of T_C obtained with the applied electric field is of just a few K (Ohno et al. 2000). Some studies have reported the existence of RT-ferromagnetism in "oxide diluted magnetic semiconductors" (ODMS), such as ZnO, TiO_2, In_2O_3 and SnO_2 doped with 3d metals (Matsumoto et al. 2001; Coey et al. 2005), but the results are still controversial and it is not clear whether the magnetic properties are indeed intrinsic or caused by transition-metal clusters. Furthermore, in the case of metallic alloys, since magnetic effects only occur within a few nm from the surface, the candidate materials are restricted to nanoparticles (often superparamagnetic at RT) or ultra-thin films, both difficult to be integrated in many types of devices.

5 Measurement Techniques of Voltage-Driven Magnetic Effects

There are two basic principles conventionally used to apply voltage in magnetic systems: (i) to use an electrolytic double layer (liquid configuration) or (ii) to apply voltage through a solid dielectric, as in a capacitor (solid configuration), more convenient for real spintronic applications.

Application of voltage in a metallic film immersed in suitable electrolytes is a rather straightforward way to generate an electrical field at its surface. In this configuration, the material to be electrically actuated does not need to be coated with a solid dielectric. Instead, the layer is itself used as an electrode (Fig. 2). A voltage is applied between the material of interest and a counter-electrode. As voltage is increased, ions in the electrolyte solution diffuse towards the surface of the electrode. The charges accumulated at the interface between the electrode and the electrolyte will form the electric double-layer which will act as an ionic dielectric, insulating the electrode from the rest of the electrolyte (Fig. 2). To minimize hydrogen evolution, specific non-aqueous electrolytes can be used, such as propylene carbonate (Weisheit et al. 2007). Magnetic measurements can be

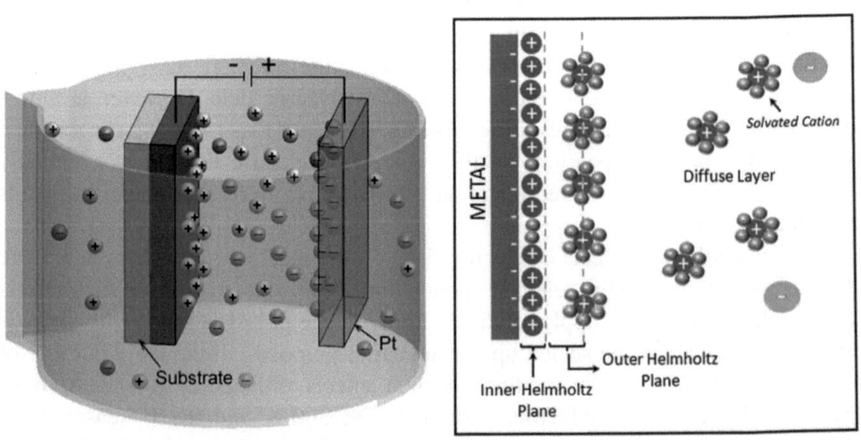

Fig. 2 (*Left*) Schematic representation of the "liquid electrolyte configuration", to induce charge accumulation at the surface of a metallic film via an electric double-layer; (*Right*) Schematic picture of the structure of an electrochemical double-layer. The negative charge of the surface is neutralized by the cations in the inner Helmholtz plane, which also contains adsorbed solvent molecules. The solvated cations form the outer Helmholtz plane and finally the diffuse layer extending into the bulk solution is found

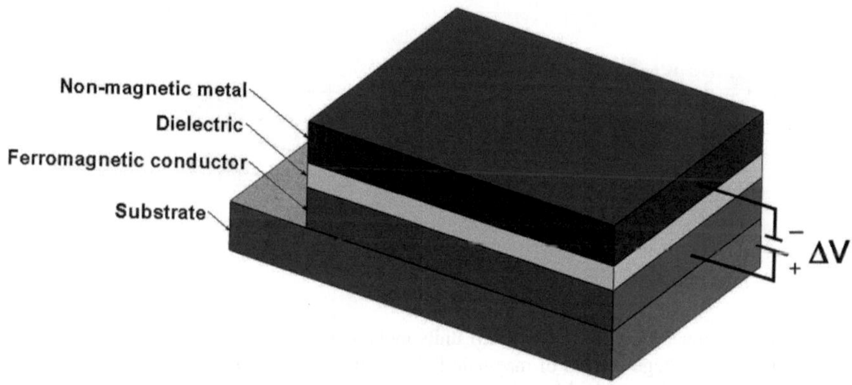

Fig. 3 Schematic drawing to illustrate how to induce accumulation of surface electric charges using a solid dielectric

performed by means of magneto-optic Kerr effect. Note that although this configuration is not so convenient for solid-state nanospintronic devices, it is very suitable for fundamental studies since it typically allows the observation of the effects at lower applied voltages due to the very small thickness of the double-layer.

Alternatively, one can use a solid configuration to investigate voltage effects in magnetic materials (Fig. 3). A layer of solid dielectric material is then grown on top of the magnetic material to be actuated. The other electrode can be a sputtered metallic film, such as Cu. Magnetic characterization in this case can be performed using extraordinary Hall Effect.

Suitable oxide materials that can act as dielectrics are Al_2O_3 or HfO_2. Reasonable dielectric strength is required to avoid current leakage effects while a high dielectric constant (like in HfO_2) is desirable to enhance the accumulated electric charge for a given applied voltage. An alternative to metal oxides can be insulating polymers. Some examples of these materials are polyethylene, polyimide, or polycarbonate, which exhibit high electric breakdown field, and polymerize at relatively low temperatures (Teyssedre and Laurent 2013). An advantage of these materials (as compared to piezoelectric-magnetostrictive composites) is that epitaxial interfaces between the conducting and dielectric components are not needed.

6 Examples of Voltage-Driven Nano-Spintronic Applications

Besides its fundamental interest, voltage-driven magnetic effects are of paramount importance for nano-spintronics applications. Among the diverse technological applications that one can envisage from these effects, the following three are

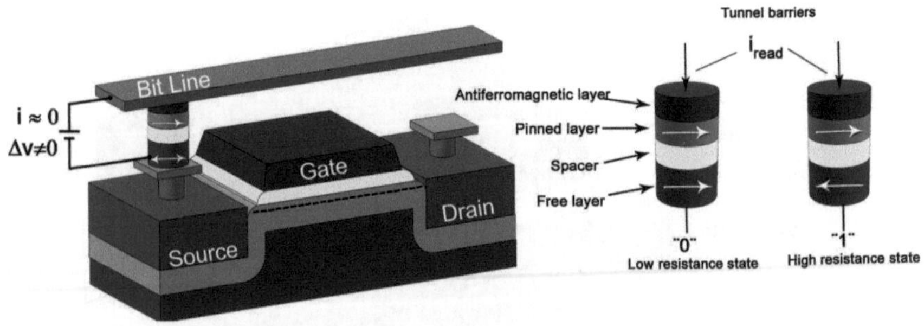

Fig. 4 (*Left*) Tunnel junctions for MRAM units including a free layer whose resistance is to be tuned applying a voltage instead of magnetic field or spin-transfer torque current, assisted by the magnetic stray field from the pinned layer; (*Right*) detail on how the resistance of the tunneling junctions changes depending on whether the pinned and free layers are parallel (low resistance) or antiparallel (high resistance) to each other

particularly appealing: (i) electrically-assisted magnetic recording media, (ii) voltage-driven magnetic switching in MRAM units, and (iii) voltage-assisted spin-field-effect transitors (FET).

One of the current problems in high-density recording media is the need to apply large magnetic fields to write the bits. High fields are not easy to generate in localized regions and these fields can eventually cause undesirable switching of neighboring bits. To overcome this problem, magnetic writing is often thermally-assisted. Local heating reduces the effective magnetic anisotropy of the media, so that information can be stored using lower magnetic field. However, heating is not energetically effective. Thus, tailoring the anisotropy by applying voltage instead of heat is much more cost-effective. This idea can be exploited using semi-hard ferromagnetic (or ferrimagnetic) materials (e.g., FePt or $CoFe_2O_4$).

Magnetic random access memories (MRAMs) combine an ensemble of magnetic storage units (the so-called magnetic tunnel junctions, MTJ) with standard CMOS transistors (see Figs. 4 and 5). The key features of MRAMs are: ability to store non-volatile information, ultra-fast processing rates, and high read and write endurance. Each MTJ comprises a fixed ferromagnetic layer (typically pinned via the direct exchange coupling with an antiferromagnet, Fig. 4), a thin dielectric tunnel barrier and a soft ferromagnetic layer (free layer), whose orientation can be changed by applying a magnetic field or, as demonstrated in recent years, a spin-polarized electric current (spin-transfer torque effect) (Huai 2008). When a current (i_{read}) traverses the MTJ, the resistance of each unit depends on the relative orientation of the free and pinned layers (Fig. 8). During the reading stage, the magnetoresistance is low (or high) when the two layers are parallel (or antiparallel) to each other.

The MTJ can operate along in-plane or perpendicular-to-plane directions. For the latter, a $[Pt/Co]_n$ stack coupled to an antiferromagnet (such as IrMn), grown by sputtering, can be used as pinned layer (Sort et al. 2005). While reading the

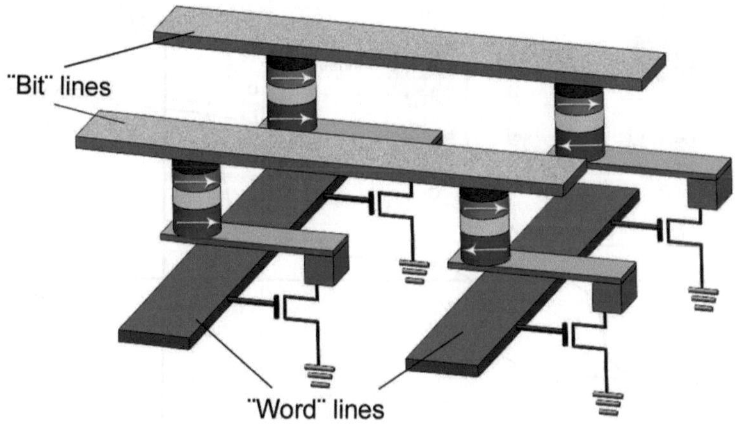

Fig. 5 Schematic drawing of an MRAM device in which each magnetic unit consists of a magnetic tunnel junction coupled to CMOS transistor

information stored in each MTJ of the MRAM is rather simple, the writing process is more complicated and remains not optimized.

Originally, the bit writing relied on two orthogonal magnetic fields generated by electric current flowing in the so-called "write or word lines". However, relatively high writing currents, of the order of 10 mA, were needed. This was not convenient in terms of energy efficiency and also because the high currents induced undesirable switching in neighbouring MTJs, particularly as MRAMs were progressively shrunk in size. Several alternatives have been proposed to overcome this issue, such as thermally-assisted switching (which works in an analogous way as thermally-assisted magnetic recording) (Prejbeanu et al. 2007) or spin-transfer MRAMs (Huai 2008), in which the switching of the free layer is controlled by a current of spin-polarized electrons that traverses the MTJs. However, in both cases, the power consumption remains exceedingly high and the requirement for moderate electric currents also precludes the reduction in size of the associated transistors, thus hindering the miniaturization of MRAMs. For these reasons, MRAMs have not succeeded in replacing other types of memories (like the "electrically erasable programmable read-only memory of flash devices), in spite of their obvious advantages in terms of long-term endurance.

Figure 6 depicts the operation principle of a voltage-driven MRAM. The use of voltage instead of magnetic field (or electric current) allows for a drastic reduction of the device power consumption, while permitting further miniaturization with respect to the state-of-the-art. The idea is to grow a soft magnetic film with a critical thickness that will result in a poorly defined magnetic easy axis, where the voltage-controlled perpendicular anisotropy will compete with the easy-plane magnetic anisotropy. To illustrate the switching process, let's consider a MRAM operating along the perpendicular-to-plane direction. Let us assume that the MTJ is resting in the high resistance configuration "a" in Fig. 6, corresponding to an

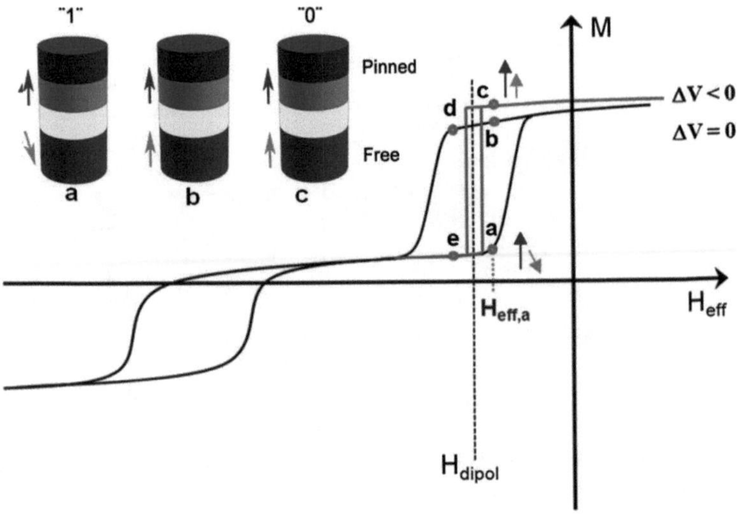

Fig. 6 Operation principle for a voltage-driven MRAM comprising magnetic tunnel junctions that operate along the perpendicular-to-film direction. The free layer is constituted by a film whose orientation is controlled with an external electric field, assisted by the magnetic dipolar field emanated from the pinned layer (H_{dipol}). The sub-loop at the *bottom-left* corresponds to the reversal of the pinned layer, whereas the one on the *upper right* corresponds to the reversal of the free layer. In *black* are the loops in absence of external electric field and in *red* is the sub-loop of the free layer upon application of a negative voltage

effective magnetic field $H_{eff,a}$. The magneto-resistance (MR) is high (state "1") because the two layers are roughly antiparallel to each other (although the free layer is not completely aligned along the pillar axis because of the competing anisotropies). If a negative electric field is applied, the coercivity of the free layer will decrease, similar to the observations made in ultra-thin FePt (Weisheit et al. 2007). In the new energy landscape, the free layer is forced to relax to state "b" for the given effective magnetic field $H_{eff,a}$. Once the voltage is removed, the two layers will remain parallel to each other since no external magnetic field is applied to vary H_{eff} (state "c"). Hence, keeping constant the $H_{eff,a}$ value, the MTJ will change from a high resistance state "1" to a low resistance state "0". Similarly, the resistance state could be reversed from "0" to "1" applying an analogous procedure from configuration "d" to "e".

The field-effect transistor (FET) is a device that uses an electric field (or gate voltage) to control the shape—and hence the conductivity—of a channel of one type of charge carrier (electrons or holes) in a semiconductor material. It is used to either amplify or switch off electric currents. Due to their technological impact, FET devices are often regarded as one of the key inventions in electronics from the last century. Spin-FET devices would add the possibility to turn on/off or amplify spin polarized currents (i.e., where all the flowing electrons possess the same spin sign, up or down). Although envisaged in 1990 (Datta and Das 1990), the technological

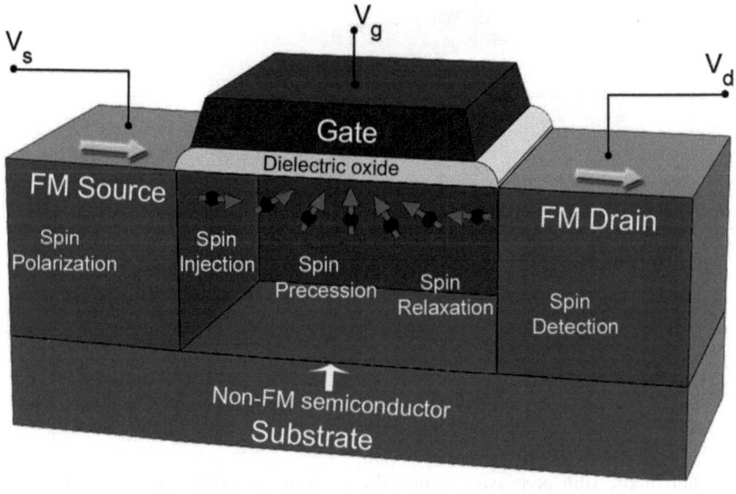

Fig. 7 Schematic diagram of a pin field-effect transistor (spin-FET)

realization of spin-FETs at RT remains elusive (see: "International Technology Roadmap for semiconductors", 2013 Edition). In a spin-FET device, the spins polarized in the source gate are injected into a narrow channel of a two-dimensional electron gas, made of a non-magnetic semiconductor (Fig. 7).

Due to the spin-orbit coupling, the spin orientation is progressively twisted as the electrons traverse the channel. The precession depends on the applied voltage at the gate (V_g) and the distance between the source and the drain. The electron precession within a semiconductor (spin orbit coupling) is a well-known relativistic effect (Bychkov and Rashba 1984). The gate voltage induces an electric field, $\vec{E}_g$, which brings about a net spin-orbit magnetic field, $\vec{B}_{S-O}$, given by: $\vec{B}_{S-O} \approx \alpha\left(\vec{E}_g \times \vec{k}\right)$, where $\vec{k}$ is the propagation unit vector of the electron and α is the so-called Rashba coefficient (see Fig. 8) (Bychkov and Rashba 1984). Due to the occurrence of $\vec{B}_{S-O}$ the electrons will experience a progressive rotation within the channel (Larmor effect). Hence, depending on the value of V_g these spins can reach the drain either in-phase or out-of-phase with respect to the drain magnetic orientation, so that either an increase or a blockage of the electrons with a given spin polarization is effectively accomplished.

Several physical phenomena converge together in a spin-FET and would need to be precisely controlled in order to make a real device: (i) spin polarization, which takes place as the electrons traverse the ferromagnetic source electrode; (ii) injection of spin-polarized currents from the source into the non-ferromagnetic channel; (iii) spin precession within the channel; (iv) spin relaxation and (v) spin detection in the ferromagnetic drain. Progress has been made in spin injection in semiconductors (Koo et al. 2007; Lou et al. 2007) and control of spin precession in the channel (Koo et al. 2009), although the later only in devices operating at very low

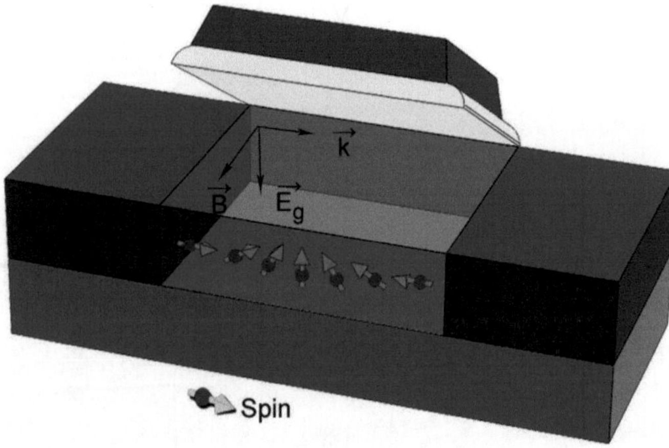

Fig. 8 Detail of the spin precession within the channel caused by the gate electric field ($\vec{E}_g$), which in turn generates the magnetic field $\vec{B}_{S-O}$. Note that $\vec{k}$ denotes the propagation direction of the electrons

temperatures. Spin injection requires of some procedures to adjust the conductivity of the source to match that of the channel. Inserting a tunnel or Schottky barrier between the source and the channel is necessary if electrons are to be injected from a ferromagnetic metal into the semiconductor (Lou et al. 2007). Alternatively, the source and channel can be both made of semiconductors with similar conductivity but, unfortunately, most ferromagnetic semiconductors (except ODMS) exhibit T_C below RT. In addition, a high quality semiconductor (i.e., single crystal or epitaxial layer) is necessary so that the electrons can adequately propagate along the channel without too much scattering. Because of their ability to either switch off or amplify electric currents, conventional transistors are often regarded as one of the key inventions in electronics. Spin-FET devices would add the possibility to turn on/off or amplify spin polarized currents (where all electrons possess the same spin, up or down). However, although envisaged in 1990 (Datta and Das 1990), the technological realization of spin-FETs at RT remains mainly elusive. Two of the main problems that hinder the implementation of spin-FETs are: (i) the lack of efficient injection of spin polarized currents from the metallic source to the semiconducting channel (due to their difference in conductivity) and (ii) the need of external magnetic fields to set the source and drain magnetic orientation.

7 Conclusions

During the last few years, the advent of nanotechnology has prompted a trend towards miniaturization in devices in diverse technological areas. Significant advances have been reached in nanoelectronics, with the occurrence of transistors with ever decreasing sizes (attaining lateral dimensions of only a few tens of nm). In turn, spintronics is currently revolutionizing information and communication technologies (ICT), offering non-volatile magnetic storage and new ways to process information, benefiting not only from the charge of electrons but also from their spin degree of freedom. The combination of nanotechnology with spintronics has led to the emerging field of nano-spintronics. In this review, particular attention has been given to spintronic effects that rely on the influence of applied voltage (rather than applied magnetic field or electric current) on the magnetic properties. Both fundamental aspects as well as possible applications of these effects have been overviewed. Voltage-driven nano-spintronic devices are highly appealing to enhance the energy efficiency in the ICT area, standing out as suitable systems to circumvent the problems associated with Joule heating dissipation and concomitant loss of electric power.

Acknowledgements J.Z. is grateful to the China Scholarship Council (CSC) for the PhD grant. E.P. acknowledges the Spanish Ministerio de Economía y Competitividad (MINECO) for the Ramon y Cajal contract (RYC-2012-10839). Partial financial support from the 2014-SGR-1015 project from D.G.U. Catalunya, SELECTA H2020-MSCA-ITN-2014 no. 642642 project and the SPIN-PORICS Consolidator Grant—2014 no. 648454 from the European Research Council is also acknowledged.

References

Albert FJ, Katine JA, Buhrman RA et al (2000) Spin-polarized current switching of a Co thin film nanomagnet. Appl Phys Lett 77:3809–3811

Astrov DN (1960) The magnetoelectric effect in antiferromagnetics. J Exp Theor Phys 38:708–709

Baibich MN, Broto JM, Fert A et al (1998) Giant magnetoresitance of (001)Fe/(001)Cr magnetic super lattice. Phys Rev Lett 61:2472–2475

Barnes SE, Leda J, Maekawa S (2014) Rashba spin-orbit anisotropy and the electric field control of magnetism. Sci Rep 4:4105

Binasch G, Grünberg P, Saurenbach F et al (1989) Enhanced magnetoresistance in layered magnetic structures with antiferromagnetic interlayer exchange. Phys Rev B 39:4828–4830

Bonell F, Murakami S, Shiota Y et al (2011) Large change in perpendicular magnetic anisotropy induced by an electric field in FePd ultrathin films. Appl Phys Lett 98:232510

Bychkov YA, Rashba ÉI (1984) Properties of a 2D electron gas with lifted spectral degeneracy. JETP Lett 39:78–81

Cherifi RO, Ivanovskaya V, Philips LC et al (2014) Electric-field control of magnetic order above room temperature. Nat Mater 13:345–351

Chiba D, Yamanouchi M, Matsukura F et al (2003) Electrical manipulation of magnetization reversal in a ferromagnetic semiconductor. Science 301:943–945

Chiba D, Sawichi M, Nishitani Y et al (2008) Magnetization vector manipulation by electric field. Nature 455:515–518

Chiba D, Fukami S, Shimamura K et al (2011) Electrical control of the ferromagnetic phase transition in cobalt at room temperature. Nat Mater 10:853–856

Coey JMD, Venkatesan M, Fitzgerald CB (2005) Donor impurity band exchange in dilute ferromagnetic oxides. Nat Mater 4:173–179

Cormier M, Ferré J, Mougin A et al (2008) High resolution polar Kerr magnetometer for nanomagnetism and nanospintronics. Rev Sci Instrum 79:033706

Curie P (1894) Sur la symétrie dans les phénomènes physiques, symétrie d'un champ électrique et d'un champ magnétique. J Phys Theor Appl 3:393–415
Datta S, Das B (1990) Electronic analog of the electro-optic modulator. Appl Phys Lett 56:664
Dieny B, Speriosu VS, Metin S et al (1991) Magnetotransport properties of magnetically soft spin-valve structures. J Appl Phys 69:4774–4779
Dieny B, Sousa RC, Hérault J et al (2010) Spin-transfer effect and its use in spintronic components. Int J Nanotechnol 7:591–614
Fiederling R, Keim M, Reuscher G et al (1999) Injection and detection of a spin-polarized current in a light-emitting diode. Nature 402:787–790
Hoffmann A (2013) Spin hall effects in metal. IEEE Trans Magn 49:5172–5193
Hu JM, Li Z, Chen LQ et al (2011) High-density magnetoresistive random access memory operating at ultralow voltage at room temperature. Nat Commun 2:553
Hu J, Haratipour N, Koester SJ (2015) The effect of output-input isolation on the scaling and energy consumption of all-spin logic devices. J Appl Phys 117:17B524
Huai Y (2008) Spin-transfer torque MRAM (STT-MRAM): challenges and prospects. AAPPS Bulletin 18:33–40
Huang YW, Lo CK, Yao YD et al (2004) Giant magnetocurrent in silicon-base magnetic tunneling transistor. J Magn Magn Mater 282:279–282
Kim HKD, Schelhas LT, Keller S et al (2013) Magnetoelectric control of superparamagnetism. Nano lett 13:884–88
Koo HC, Yi H, Ko JB et al (2007) Electrical spin injection and detection in an InAs quantum well. Appl Phys Lett 90:022101
Koo HC, Kwonj H, Eom J et al (2009) Control of spin precession in a spin-injected field effect transistor. Science 325:1515–1518
Lou X, Adelmann C, Crooker SA, Crowell PA et al (2007) Electrical detection of spin transport in lateral ferromagnet–semiconductor devices. Nat Phys 3:197–202
Maruyama T, Shiota Y, Nozaki T et al (2009) Large voltage-induced magnetic anisotropy change in a few atomic layers of iron. Nat Nanotech 4:158–161
Matsumoto Y, Murakami M, Shono T et al (2001) Room-temperature ferromagnetism in transparent transition metal-doped titanium dioxide. Science 291:854–856
Moodera JS, Kinder LR, Wong TM et al (1995) Large magnetoresistance at room temperature in ferromagnetic thin film tunnel junctions. Phys Rev Lett 74:3273–3276
Nakamura K, Shimabukura R, Fujiwara Y et al (2009) Giant modification of the magnetocrystalline anisotropy in transition-metal monolayers by an external electric field. Phys Rev Lett 102:187201
Ohno Y, Young DK, Beschoten B et al (1999) Electrical spin injection in a ferromagnetic semiconductor heterostructure. Nature 402:790–792
Ohno H, Chiba D, Matsukura F et al (2000) Electric-field control of ferromagnetism. Nature 408:944–946
Ovchinnikov IV, Wang KL (2009) Theory of electric-field-controlled surface ferromagnetic transition in metals. Phys Rev B 79:020402(R)
Pippard AB (2009) Magnetoresistance in metals. Cambridge University Press, New York
Prejbeanu IL, Kerekes M, Sousa RC et al (2007) Thermally assisted MRAM. J Phys: Condens Matter 19:165218
Prenat G, Jabeur K, Pendina GD et al (2015) Spintronics-based computing. In: Zhao W, Prenat G (eds) Spin orbit torque RAM SOT-MRAM for high speed and high reliability applications. Springer International Publishing, Switzerland, pp 145–157
Ralph DC, Cui YT, Liu LQ et al (2011) Spin-transfer torque in nanoscale magnetic devices. Philos Trans R Soc A 369:3617–3630
Ramesh R, Spaldin NA (2007) Multiferroics: progress and prospects in thin films. Nat Mater 6:21–29
Saitoh E, Ueda M, Miyajima H et al (2006) Conversion of spin current into charge current at room temperature: inverse spin-Hall effect. Appl Phys Lett 88:182509

Scheid M, Bercioux D, Richter K (2007) Zeeman ratchets: pure spin current generation in mesoscopic conductors with non-uniform magnetic fields. New J Phys 9:401
Schmid H (1994) Multi-ferroic magnetoelectrics. Ferroelectrics 162:317–338
Scogland T, Subramaniam B, Feng W (2013) The Green500 list: escapades to exascale. Comput Sci Res Dev 28:109–117
Shiota Y, Nozaki T, Bonell F et al (2011) Induction of coherent magnetization switching in a few atomic layer of FeCo using voltage pulses. Nat Mater 11:39–43
Slonczewski JC (1996) Current-driven excitation of magnetic multilayers. J Magn Magn Mater 159:L1–L7
Sort J, Baltz V, Garcia F et al (2005) Tailoring perpendicular exchange bias in [Pt/Co]-IrMn multilayers. Phys Rev B 71:054411
Stolichnov I, Riester SWE, Trodahl HJ et al (2008) Non-volatile ferroelectric control of ferromagnetism in (Ga, Mn)As. Nat Mater 7:464–467
Teyssedre G, Laurent C (2013) Advances in high-field insulating polymeric materials over the past 50 years. IEEE Electr Insul Mag 29:26–36
Valenzuela SO, Tinkham M (2006) Direct electronic measurement of the spin Hall effect. Nature 442:176–179
Wang Y, Hu J, Lin Y et al (2010) Multiferroic magnetoelectric composite nanostructures. NPG Asia mater 2:61–68
Wang W, Li M, Hageman S et al (2012) Electric-field-assisted switching in magnetic tunnel junctions. Nat Mater 11:64–68
Weisheit M, Fähler S, Marty A et al (2007) Electric field-induced modification of magnetism in thin-film ferromagnets. Science 315:349–351
Wiesendanger R, Güntherodt HJ (1990) Observation of vacuum tunneling of spin-polarized electrons with the scanning tunneling microscope. Phys Rev Lett 65:247–250
Wolf SA, Awschalom DD, Buhrman RA et al (2001) Spintronics: a spin-based electronics vision for the future. Science 294:1488–1495
Yamada Y, Ueno K, Fukumura T et al (2011) Electrically induced ferromagnetism at room temperature in cobalt-doped titanium dioxide. Science 332:1065–1067
Zhernenkov M, Fitzsimmons MR, Chlistunoff J, Majewshi J (2010) Electric-field modification of magnetism in a thin CoPd film. Phys Rev B 82:024420
Žutić I, Fabian J, Sarma SD (2004) Spintronics: fundamentals and applications. Rev Mod Phys 76:323–410

Ferroelectric and Piezoelectric Nanomaterials—Basic Properties, Characterization and Applications

Vesna Paunović, Zoran Prijić and Dragan Antić

Abstract Electroceramics have a wide range of applications and are a subject of extensive research activities in the fields of ion-conductivity (such as batteries and sensors), electrical insulators (substrates and multilayer integrated circuit packages), semiconductors (sensors), and superconductors. Among the electroceramic materials, ferroelectric and piezoelectric nanomaterials are technically the most challenging. They are known for their unique properties, such as high dielectric constant, as well as high piezoelectric constants, and are used in multilayer capacitors or as microwave devices within wireless communication systems. In addition, perovskite ferroelectric nanomaterials show potential for applications related to solar energy conversion and the production of storage memory devices. Ferroelectric films as functional materials are being extensively explored for various microsensor and microactuator applications, some of them being suitable for biomedical engineering. The most widely investigated perovskite ferroelectric and piezoelectric nanomaterials include $BaTiO_3$, $SrTiO_3$, $PbZrTiO_3$ along with suitable dopants and multiferroic oxides, and ZnO. In this review, the main concept of ferroelectricity in perovskite oxides and related nanomaterials is discussed. Fundamentals of ferroelectric nanomaterials, including size effects of ferroelectric nanomaterials properties and phase transitions are summarized. A detailed discussion on the synthesis, fabrication, nanostructure characterization of ferroelectric and piezoelectric nanomaterials such as $BaTiO_3$, $SrTiO_3$, and ZnO is presented. Progress in the research of ferroelectric perovskite oxide in nanometers scale is also highlighted.

Keywords Ferroelectric · Piezoelectric · Nanomaterials · $BaTiO_3$ · ZnO

V. Paunović (✉) · Z. Prijić · D. Antić
Faculty of Electronic Engineering, University of Niš, A. Medvedeva 14,
18000 Niš, Serbia
e-mail: vesna.paunovic@elfak.ni.ac.rs

D. Brabazon et al. (eds.), *Commercialization of Nanotechnologies–A Case Study Approach*, DOI 10.1007/978-3-319-56979-6_6

1 Introduction

Ferroelectricity as an effect is connected to the process of the spontaneous electric polarization in a material. Without the exterior influences, such as the electric field or pressure, positive and negative ions are moving, in the way resulting from the relation of local dipole moments. The appearance of spontaneous polarization is discovered in the relatively small group of materials, whose crystal structures are characterized by the absence of centers of symmetry. When the local electric dipole moments are turned towards the real crystallographic directions using the electric field, the material shows effects typical for ferroelectrics. The typical ferroelectric properties are spontaneous electric polarization, reversal polarization as well as the disappearance of the polarization above a phase transformation temperature. Ferroelectric materials have been a fertile field for the study of phase transitions, electron–phonon interactions, polaritons, and other optical and electrical phenomena in condensed materials.

Ferroelectricity is a phenomenon of crystalline materials that was discovered 1921 by Valasek in Rochelle salt (sodium potassium tartrate tetrahydrate, NaK-$C_4H4O_6 \cdot 4H_2O$). Pioneering studies on ferroelectricity in perovskite oxides occurred in the period of 1940–960s, mainly on barium titanate ($BaTiO_3$), lead titanate ($PbTiO_3$), and strontium titanate ($SrTiO_3$). Since then, many other ferroelectric oxides have been discovered, rendering the theoretical and application studies on ferroelectrics as one of the most active research areas in the field of solid state materials.

The phenomenon of piezoelectricity was first observed by brothers Curie in 1880. When a mechanical pressure is applied to piezoelectric crystals, positive and negative centers of the molecule are separated. Small dipoles are generated resulting in an electric field, proportional to the pressure. In contrast, the application of an external electric field leads to changes in the structure and size of the material.

In the first part of this review, the basic concept of ferroelectricity and piezoelectricity in perovskite oxide at nanometer scale is discussed. Discussion of the fundamentals of ferroelectrics nanomaterials, including size effects of ferroelectrics nanomaterials properties, and phase transition is presented. In the second part of the review an overview of synthesis of nanopowders and fabrication (sintering) of nanoceramics is given. In the next part of the review important methods employed for characterization of nanostructures, including the determination of the size and shape, atomic and electronic structures and electrical properties are described. At the end of the review, some important applications of ferroelectric and piezoelectric materials are highlighted.

2 Ferroelectricity and Piezoelectricity

2.1 Ferroelectricity in Perovskite Materials

Among ceramic materials, ferroelectricity was first observed in barium titanate. It is the ideal model for a discussion of the phenomenon from the point of view of a crystal structure and microstructure. As shown in Fig. 1, $BaTiO_3$ has typical ABX_3 perovskite structure, where A is mono, divalent or trivalent cation, B is corresponding tri, tetra or pentavalent element, and X is oxygen.

The concept of a phase transition is crucial for explaining ferroelectricity. The transition temperature is defined as the temperature at which the crystal is transformed from one phase to another. Upon decreasing the temperature, only a transition temperature at which a ferroelectric crystal transforms from a parental paraelectric phase to the first ferroelectric phase is called Curie temperature (T_c). Above its Curie temperature the unit cell $BaTiO_3$ is cubic with barium ions at the corner positions, titanium ions at the body central position and oxygen atoms at the face centered positions (Fig. 2a). Based on the perovskite structure, it can be concluded that $BaTiO_3$ belongs to the group of ferroelectrics with characteristic movement of the central Ti^{4+} ion in the crystal unit cell. At a temperature below the Curie temperature, due to displacement of titanium ion respect to the oxygen ions, there is a change of unit cell, causing a situation of asymmetry of positive and negative ions and the formation of electric dipoles (Moulson and Herbert 1995). Below the Curie temperature the crustal structure is slightly distorted to the tetragonal form with a dipole moment along the c direction. Other transformations occur at temperatures close to 0 °C and −80 °C. Below 0 °C the unit cell is orthorhombic and below −80 °C unit cell is rhombohedral (Tura et al. 1998).

The barium titanate unit cell in cubic and tetragonal phases is shown in Fig. 2. The cubic phase, Fig. 2a, exhibits an induced polarization, which returns to zero when the external field is removed; while the tetragonal phases retain a spontaneous

Fig. 1 Perovskite ABX_3 structure

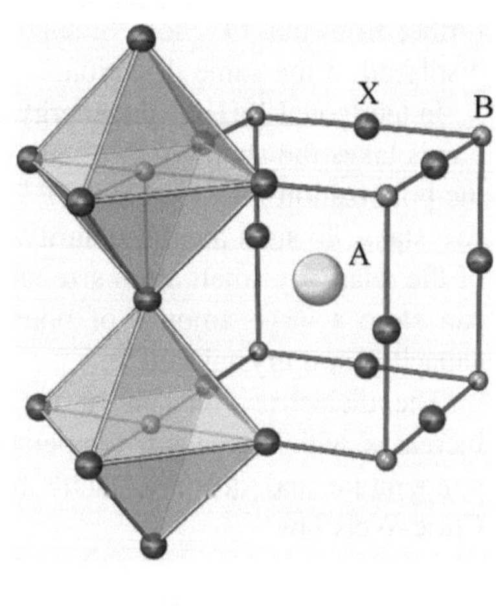

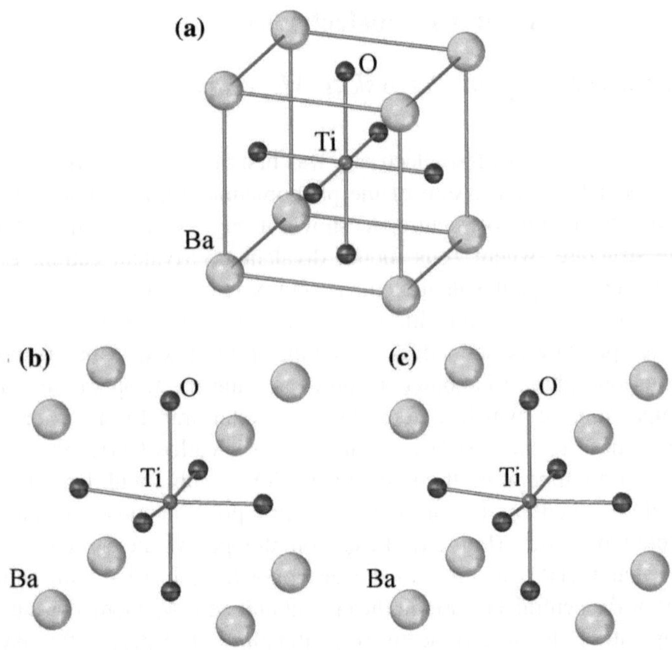

Fig. 2 Unit cell of $BaTiO_3$, **a** cubic, paraelectric phase, **b** and **c** tetragonal, ferroelectric phase

polarization. The two variations Fig. 2b, c, have a 180° difference in direction due to the off-center position of the titanium ion (Nuraje and Su 2013).

A consideration of the ion displacements accompanying the cubic-tetragonal transformation can give insight into how the spontaneous polarization might be coupled from unit cell to unit cell. X-ray studies have established that in the tetragonal form, taking the plane with four central oxygen ions in the cubic phase as a reference, titanium ions are slightly displaced. If the central T^{4+} ion is closer to one of the O^{2-}, it will be energetically favorable for the T^{4+} ion above to be located further from that O^{2-} ion. In such a way, all the T^{4+} ions in a lattice column are displaced in the same direction.

In tetragonal $BaTiO_3$ the energy of the Ti^{4+} ion in terms of its position along the c axis takes the shape of two valleys. An applied field in the opposite direction of the polarization may enable a Ti^{4+} ion to pass over the energy barrier between the two states. In such a case, polarity direction at that point will be reversed. Because of the relatively small atom size and high charge of Ti^{4+} the atom off-center effect can store a large amount of potential energy. This is a microscopic view of a capacitor in a crystal lattice.

The dielectric constant of the ferroelectrics depends on the temperature. It increases with increasing temperature, goes through a sharp maximum at Curie temperature and decreases with further increasing temperature according to the Curie-Weis law:

$$\varepsilon_r = \frac{C}{T - T_0}, \tag{1}$$

where C is the Curie constant, and T_0 the Curie-Weiss temperature, which is close to the Curie temperature.

Temperature of paraelectric to ferroelectric phase transition is of great importance for investigation and practical application of barium titanate ceramics (Paunovic et al. 2010a, b; Lobo et al. 1995). In addition, another interesting characteristic of ferroelectrics is the very high value of the dielectric constant on the Curie temperature. The multi-phase transition feature of $BaTiO_3$ results in exceptionally high dielectric constant, ε_r greater than 1000 (Zivkovic et al. 2006a), which makes $BaTiO_3$ attractive for many applications.

The effect of additives on the ferroelectric-paraelectric phase transformation temperature, as well as on Curie constant can be traced through the temperature dependence of the permittivity (Paunovic et al. 2012). The transition from the ferroelectric to the paraelectric phase can be with a sharp dielectric maximum (sharp phase transition) or with diffuse dielectric maximum which is a characteristic for "relaxor" ceramics (Paunovic et al. 2010b; Zhang et al. 2004). In general, modified $BaTiO_3$ exhibits a shift of Curie temperature towards lower temperature (Fang et al. 1933; Paunovic et al. 2014). One of the reasons they used a modified $BaTiO_3$ is that the additives have the effect of moving the Curie temperature in the temperature field that can be used effectively, significantly below 130 °C.

A large number of substitutions result in a decrease of the Curie point. $Ba_{1-x}Sr_xTiO_3$ has been one of the most extensively studied systems among these solid solutions. By adjusting the substitution level, the Curie point can be shifted to room temperature with a broad permittivity peak ($\varepsilon_r > 10000$) (Cross 1993). Broadening of the permittivity peak is attributed to inhomogeneity in the composition that led to the distribution of the Curie temperature. Furthermore, by tuning the ratio between barium titanate and strontium titanate, the Curie temperature can be changed to room temperature, which produces the material with a high dielectric constant at room temperature.

The transition from the paraelectric to ferroelectric phase of $BaTiO_3$ and $SrTiO_3$ is illustrated in Fig. 3.

The Curie temperature for barium titanate is 130 °C and for strontium titanate is −168 °C. Since the tetragonal phase is ferroelectric below the Curie temperature, barium titanate is ferroelectric at room temperature, while strontium titanate is paraelectric at room temperature.

Besides barium titanate and strontium titanate, other perovskite oxides used in electronic applications include $PbTiO_3$, $PbZrO_3$, $Pb(Zr_xTi_{1-x})O_3$, etc. (Parashar et al. 2004; Shukla and Choudhary 2010; Ahmad and Ganguli 2006).

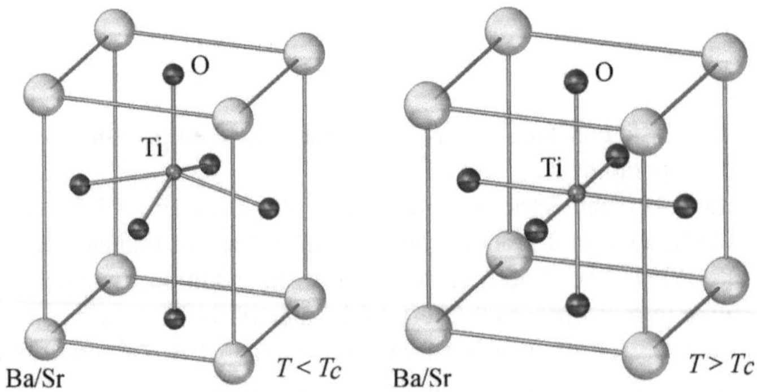

Fig. 3 Transition from paraelectric to ferroelectric phase for $BaTiO_3$ and $SrTiO_3$

2.2 *Piezoelectric Effect*

Piezoelectric ceramics are ferroelectric materials. Above the Curie temperature, these ceramics have a simple cubic symmetry without a dipole moment and do not show piezoelectric characteristics. Below the Curie point, crystal lattice becomes distorted and asymmetric. This brings about the formation of dipoles and the rhombohedral and tetragonal crystallite phases.

Adjoining dipoles form regions of local alignment are called domains. The alignment produces a net dipole moment, and thus a polarization of the domain. The polarization direction between neighbouring domains is random, so the ceramic materials have no overall polarization. Domain alignment is achieved by exposing the material to a strong DC electric field, slightly below the Curie temperature. During polarization domains that have an orientation closest to the filed direction are expanded and the material is lengthened. When the field is removed, the orientation of the domains is roughly preserved. The material has remanent polarization, and is permanently elongated.

2.2.1 Direct and Indirect Piezoelectric Effect

Direct piezoelectric effect is observed in certain materials, which respond to the external mechanical stress by producing a charge, thus converting mechanical into electrical energy. Vice versa, indirect piezoelectric effect is observed in materials that convert electrical into mechanical energy (Vijaya 2012; Tichý et al. 2010; Vives 2008).

When the piezoelectric material is subjected to mechanical stress, the internal structure is deformed by separation of positive and negative charge centers, thus producing an electrical field. By exploiting this direct piezoelectric effect, material can be used as a generator or a sensor, converting mechanical into electrical energy, as illustrated in Fig. 4. When a poled piezoelectric material is subjected to

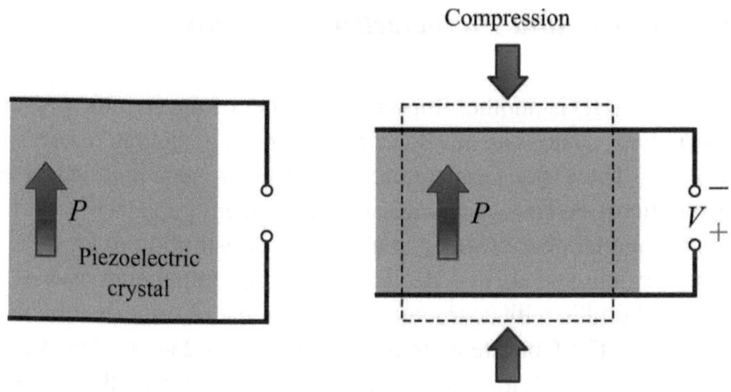

Fig. 4 Illustration of direct piezoelectric effect. *P* denotes poling direction

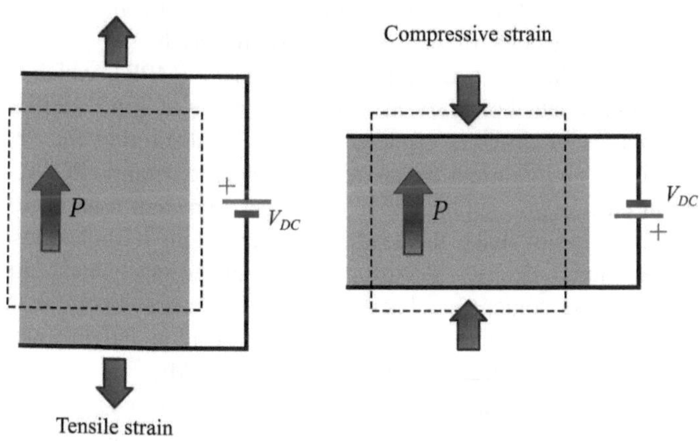

Fig. 5 Illustration of indirect piezoelectric effect. *P* denotes poling direction

compression in the direction parallel to the poling direction, a voltage is generated across the faces.

Conversely, an applied external electric field either stretches or compresses the piezoelectric material. The material can be used as an actuator converting electrical into mechanical energy. Indirect piezoelectric effect is illustrated in Fig. 5.

The material is extended if the polarity of the external voltage is the same as that of the field applied during poling and, when the voltage is applied in the reverse direction, the material is compressed.

The piezoelectric effect occurs both in monocrystalline materials and in polycrystalline ferroelectric ceramics. The first material developed as a piezoceramics was $BaTiO_3$. Polycrystalline ferroelectric ceramics such as barium titanate and lead zirconate titanate are known to exhibit significant displacements or induce larger electric voltages. Lead zirconate titanate is the most common piezoelectric material and is widely used for actuator or sensor applications. Special doping of the PZT ceramics with e.g. Ni, Bi, La, Nd, Nb ions make it possible to specifically optimize piezoelectric and dielectric parameters.

2.3 Ferroelectric and Piezoelectric Materials

Lead titanate ($PbTiO_3$) is another important member of the ferroelectric perovskites having a structure analogous to $BaTiO_3$. At room temperature, $PbTiO_3$ has a tetragonal phase. The Curie temperature of $PbTiO_3$ is very high and it is 490 °C. The characteristic of $PbTiO_3$ is a single transition from paraelectric cubic phase to ferroelectric tetragonal phase (Sheng et al. 2008). Substitution of Pb or Ti ions with ions of a similar radius leads to a decrease in Curie temperature (Sen et al. 2006). $PbZrO_3$ is orthorhombic with a structure similar to that of orthorhombic $BaTiO_3$ but is antiferroelectric. The Curie temperature of $PbZrO_3$ is 234 °C. The dipoles due to a displacement of the Zr^{4+}ions from the geometric center of the surrounding six O^{2-} ions are alternately directed in opposite senses so the spontaneous polarization is zero.

The solid solution of $PbTiO_3$-$PbZrO_3$ (PZT ceramics), shown in Fig. 6, is one of the most important ferroelectric and piezoelectric materials.

In this material, there is a morphotropic boundary (MPB), which is a nearly temperature independent boundary in a phase diagram of solid solution between ferroelectric phases of different composition. La^{3+} substitution for Pb^{2+} in perovskite PZT forms an important transparent ferroelectric ceramic PLZT, which has been found in electrooptic applications. The larger remanent polarization and relatively high Curie point make the lead based perovskite ferroelectrics favorable candidates for future electric, electrooptic, and electromechanical applications. Even though the lead content has been considered as a drawback for these ferroelectrics, the main shortage may be the high polarization fatigue, which causes reliability problems in applications like nonvolatile RAMs.

Zinc oxide (ZnO) is an important semiconducting compound of the group II–VI. ZnO normally forms in the hexagonal (wurtzite) crystal structure, shown in Fig. 7.

Wurtzite hexagonal structure of zinc oxide is the most stable and most commonly used. This is a hexagonal lattice, belonging to the space group $P6_3mc$ with

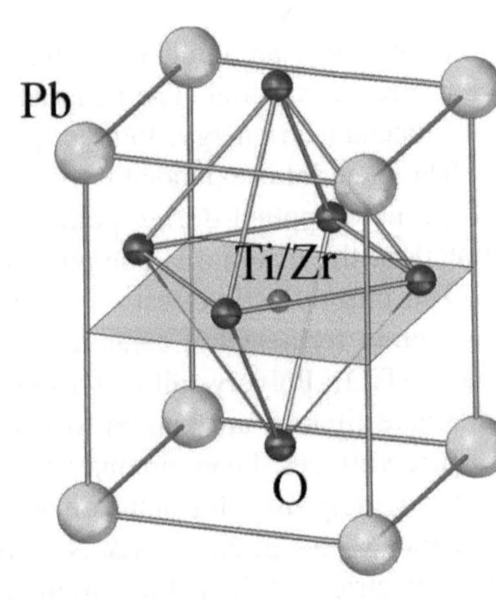

Fig. 6 Crystal lattice of $PbTiO_3/PbZrO_3$ (PZT ceramics)

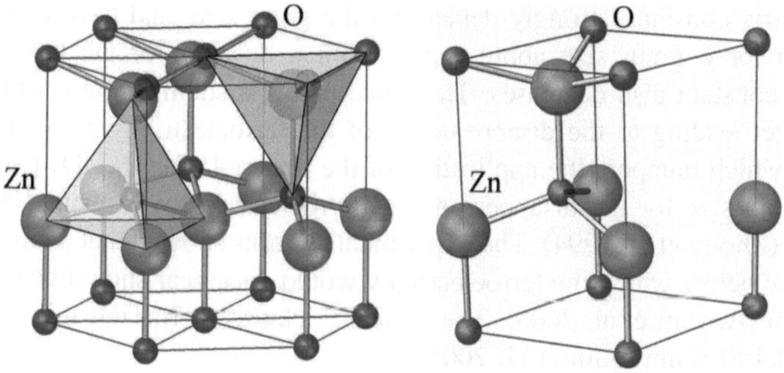

Fig. 7 Wurtzite crystal structure (*left*) and Wurtzite unit cell of ZnO (*right*)

two interconnecting sublattices of Zn^{2+} and O^{2-}. Each Zn ion is surrounded by tetrahedral of O ions. The hexagonal unit cell lattice parameters of wurtzite ZnO is a = 0.32495 nm (base) and c = 0.52069 nm (height). This tetrahedral coordination improves polar symmetry along the hexagonal axis. This polarity governs significant properties of ZnO like piezoelectricity and spontaneous polarization, and is a key factor in crystal growth, etching, and defect generation.

In addition to the wurtzite phase, ZnO is also known to crystallize in the cubic zinc blende and rocksalt or Rochelle salt (NaCl) structure that are less stable. The zinc blende ZnO structure is metastable and can be stabilized only by heteroepitaxial growth on cubic substrates, whilst the rocksalt structure is high pressure metastable phase forming at 10 GPa and cannot be epitaxially stabilized (Moezzi et al. 2012).

Hexagonal and zinc blende polymorphs have a center of symmetry. This and other lattice symmetry properties result in piezoelectricity of the hexagonal and zincblende ZnO, and piroelectricity of hexagonal ZnO. Real wurtzite structure deviates from the ideal, by changing the c/a ratio or the u value. Experimentally observed c/a ratios are smaller than ideal, where it has been postulated that not being so would lead to zinc blende phase. An inverse proportionality exists between the c/a ratio and the u parameter, i.e. when the c/a ratio decreases, the u parameter increases in such a way that those four tetrahedral distances remain nearly constant through a distortion of tetrahedral angles due to long-range polar interactions.

2.4 Grain Size Effects in Ferroelectric Nanomaterials

Due to potential applications in modern microelectronic devices, grain size effects in ferroelectric materials have attracted great attention. The effects of grain size on the ferroelectric properties and phase transitions of $BaTiO_3$ crystals have been studied both theoretically (Zhong et al. 1994) and experimentally (Kuroiwa et al. 2002; Niesz et al. 2011). The ferroelectric properties of $BaTiO_3$ ceramics, especially the dielectric constant, strongly depend on the grain size, and have a pronounced maximum for a grain size about 1 μm. With a decrease of the grain size, the dielectric constant also decreases. The cubic phase is stabilized at a wide temperature range, leading to the disappearance of the ferroelectricity above room temperature, which hampers the application of the material. For $BaTiO_3$ particles, the critical grain size for the disappearance of ferroelectricity was about 100 nm theoretically (Zhong et al. 1994). The experimental works showed that a critical size in the particle below which the ferroelectricity would disappear should be between 16 and 30 nm (Aoyagi et al. 2005). The critical thickness of $BaTiO_3$ thin film should be from 2.4 to 5 nm (Kim et al. 2005).

Ferroelectricity can been regarded because of balance between long-range Coulomb interaction and short range covalent rejections. As the grain size decreases into nano scale, the long-range Coulomb interaction is altered due to a lack of periodicity. On the other hand, a decrease of grain size will cause an increase of the fraction of ions in the surface layer. Ions in a surface layer may have a different number of nearest neighbors from those in the bulk, thus the short-range covalent interaction is modified. Therefore, in nano ferroelectrics, the balance is distorted with respect to the bulk.

In ceramics materials, individual crystallites are surrounded by grain boundaries, which greatly affect their overall performance. The internal stress caused by the phase transition can be relieved by the formation of 90° domains, so the contribution of the 90° domain walls to the dielectric constant is dominant in coarse-grained ceramics. With further reduction of grain size, the single domain becomes energetically favorable. Consequently, in fine-grained ceramics, the contribution of the stress should be taken into account due to the absence of 90° domains. It is for this reason that the dielectric constant of fine-grained $BaTiO_3$ ceramics has a maximum at about 0.7–1.0 μm and then decreases strongly with further reducing grain size.

If it is assumed that the grain is a single domain then the dielectric constant is affected only by grain boundary and stress contributes. The grain boundary has a lower dielectric constant than the ferroelectric grain, since with the decreasing of grain size, the fraction of grain boundary volume increases and the corresponding ceramics dielectric constant is reduced (Emelyanov et al. 2002; Frey et al. 1998). By measuring the dielectric constant temperature dependence for fine grain ceramics (nanoceramics), it was found that the phase transition temperature of cubic-tetragonal phase decreased (is carried out at a lower temperature) while those of tetragonal orthorhombic and orthorhombic rhombohedral increased with reducing grain size (Deng et al. 2006). The coexistence of the tetragonal and orthorhombic phases was observed in fine grained ceramics, at room temperature (Li et al. 2004) and the population of the orthorhombic phase increased with further decreasing grain size (Buscaglia et al. 2005).

Some problems arise during the preparation of ferroelectric nanocrystals, primarily due to their very high surface to volume ratio. The presence of ions of impurities on the surface of nanocrystals can lead to the formation of a depolarization field, thus preventing the growth of tetragonal phases (Dawber et al. 2003;

Ahn et al. 2004). After sintering at 1000 °C the tetragonal phase is obtained upon cooling the ceramics to room temperature. Although approaches with the trend towards low temperature synthesis of perovskite nanocrystals have been developed, most of the synthetic temperatures were higher than the Curie temperature (Niederberger et al. 2004).

Important issues for the synthesis of nanoceramics with tetragonal phase were lower synthesis temperature. At room temperature, tetragonal phase is thermodynamically more stable in $BaTiO_3$. However, the synthesis at low temperature usually results in the formation of the cubic phases. Sintering at around 1000 °C, followed by annealing through the Curie temperature, was requisite for obtaining the tetragonal polymorph. The high temperature sintering process results in agglomerations, which deteriorate the size distribution of the final products. For crystals in nanometer scale, the agglomeration is a more serious concern with regard to the particle distribution and size level. Therefore for synthesis of nanopowders various synthesis methods have been developed, including co-precipitation, hydrothermal and solvothermal methods, sol-gel process. In comparison with low-pressure sintering, high-pressure sintering could significantly increase the densification and dramatically reduce the growth rate. Thus, nanoceramics with an extremely small grain size and high density can be obtained by high pressure sintering.

Deng et al. (2006) and Buscaglia et al. (2006) have obtained dense $BaTiO_3$ ceramics with grain sizes of 20 and 30 nm, using spark plasma sintering. Wang et al. (2006) used two-step sintering techniques for obtaining high density $BaTiO_3$ ceramics with grain sizes of 30–35 nm. Their results point out that the ferroelectricity remains in barium titanate nanoceramics.

Starting from 10 nm $BaTiO_3$ raw powder and using a three-step high-pressure sintering method, Xiao et al. (2007) obtained dense $BaTiO_3$ ceramics with uniform microstructure and grain size about 30 nm. This $BaTiO_3$ ceramics displayed a similar successive phase transition to normal bulk samples as detected using variable-temperature X-ray diffraction. The ferroelectricity was characterized from dielectric measurements.

Significant physical properties of discussed materials are shown in Table 1.

3 Synthesis Methods of Nanomaterials

Over the last few years, nanomaterials have received extensive studies from scientific and engineering viewpoints. Properties of nanomaterials are different from those of bulk materials due to small-grained structure and large surface volume ratios. For the preparation of ferroelectric nanomaterials (nanoparticles), various synthesis methods were used. The techniques include high-temperature solid-state reaction, sol-gel technique, citrate-gel technique, hydrothermal synthesis, emulsion precipitation method, microemulsion, electrospinning method, etc.

Table 1 Summary of significant physical properties of perovskite materials

Material	Structure	Lattice param. (nm)	Density (g/cm^3)	ε_r	T_C (°C)	ρ (Ωcm)	Piezo coef.
$BaTiO_3$	Perovskite (cubic and tetragonal)	a = 0.399 b = 0.404	6.02	>1000	130	10^{10}	190
$PbTiO_3$	Perovskite tetragonal	a = 0.39 b = 0.41	7.52	150–400	490	10^7–10^8	20
PZT	Perovskite	a = 0.403 b = 0.419	7.6	800–2000	320	10^7	190–250
ZnO	Wurtzite	a = 0.325 b = 0.51	5.606	10–11	430	10^4	10.5–11.5
$SrTiO_3$	Perovskite (cubic)	a = 0.391	5.12	300	−168	10^{-2}	

3.1 *High Temperature Solid State Reaction—Ceramic Method*

In this method, to synthesize perovskite materials ABO_3 type, near stoichiometric ratios of A- and B-containing precursors are sintered above the eutectic temperature. The final products of this method are subject to parameters such as particle size, morphology, purity, and chemical composition of the precursors since the parameters affect the atomic diffusion during the solid-state reaction. The ceramic method cannot provide control over the oxide grain size at the nanometer scale, with grains typically >1 μm, due to the absence of control of nucleation and growth.

The solid-state reaction (Fig. 8) is usually applied to the synthesis $BaTiO_3$, $SrTiO_3$ and $PbZrTiO_3$ powders. Singh and Nath (2011) investigated the obtaining of $BaTiO_3$ nanoceramics. The starting raw materials $BaCO_3$ and TiO_2 (purity ≥ 99%) were weighed according to the stoichiometric formula to yield $BaTiO_3$ and homogeneously mixed in isopropanol. The powder was dried and calcined for 4 hours at 1050 °C. The calcined powder was then high-energy milled in the isopropyl alcohol using a planetary ball mill for 1, 10, 15, 20 and 30 h. During high-energy milling, a mass ratio of 1:5 for powder and balls was constant. The average grain size ranges from 105 to 25 nm and decrease with increase in milling time.

Preparation of $BaTiO_3$ nanopowders in the presence of nonionic surfactant and NaCl was investigated by Mao et al. (2003). $BaTiO_3$ nanopowder obtained by this method is characterized by a grain size from 50 to 80 nm and an aspect ratio higher than 25.

The $BaSrTiO_3$ powder was synthesized starting from pure powders $BaCO_3$, TiO_2 and $SrCO_3$ according to the stoichiometric proportion. The powder was

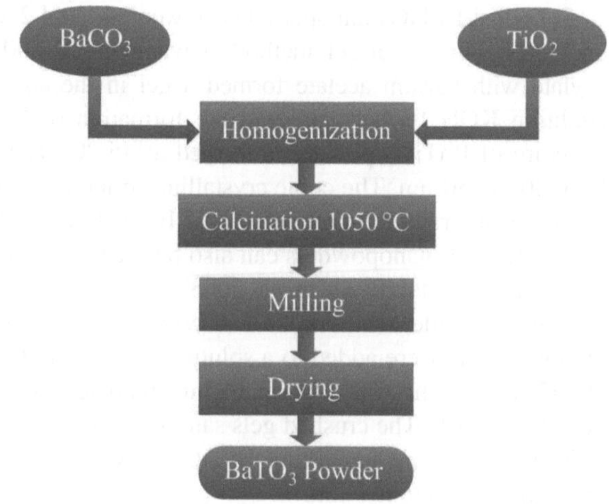

Fig. 8 High temperature solid state reaction—Ceramic method

ball-milled in a high energy vibratory mill for 6 h using zirconia balls as the milling media. The ball-to-powder weight-ratio was 10:1. The powders were uniaxially pressed in 10 mm diameter cylindrical disks using 1 GPa pressure. The compacts were reactively sintered using a two-step heating program. A final particle size obtained by this method is less than 50 nm (Vargas-Ortiz et al. 2012).

PZT nanoceramic was synthesized from high purity powders, PbO, ZrO_2, and TiO_2 by a high-energy ball milling technique (Parashar et al. 2004). The powders were milled in a planetary ball mill, with a different milling time, at room temperature. The milling speed was 300 rpm. The ball-to-powder weight ratio was 10:1. The powders were uniaxially compacted at 100 MPa in the pellets of 5 mm in diameter and 2 mm in thickness. The grain size of PZT powder is in the range from 25 to 93 nm.

3.2 *Sol-Gel Method*

Very fine initial powders with precise stoichiometry and better control of nanoparticle size can be obtained by sol-gel method and citrate-gel method. However, powders obtained in this way can have great agglomeration and intensive grains growth during sintering.

Therefore, for synthesis of barium titanate nanoparticles modified sol-gel techniques were developed (Wang et al. 2005). In these techniques, hydrolysis of metal alkoxide precursors was performed with the aid of hydrogen peroxide at a temperature of 100 °C. The oleic acid was applied as a passivating ligand. Crystalline nanoparticles between 4 and 12 nm with the cubic phase were obtained by adjusting experimental parameters.

Spherical $BaTiO_3$ nanoparticles between 40 and 250 nm were obtained through another modified sol-gel method (Hwang et al. 2004). The mixtures of titanyl acylate with barium acetate formed a gel in the aqueous solution of acetic acid. Addition KOH further catalyzed the formation of $BaTiO_3$. The cubic crystalline structure of $BaTiO_3$ powder is formed at 45 °C. The average particle size ranges from 40 to 250 nm. The cubic crystalline structure of $BaTiO_3$ particles transform to a tetragonal structure upon heating to 1000 °C for 1 h.

The $BaTiO_3$ nanopowders can also be prepared starting from barium acetate and titanium butoxide (Pazik et al. 2008). Acetil acetone was used as a solvent for titanium butoxide, while acetil acid was utilized to solve barium acetate. Drops of barium acetate were added to a solution of titanium butoxide while being stirred at 50 °C for 2 h. In aim to form barium titanate gel, the obtained sol was kept at 100 °C for 24 h. The crushed gels samples were heated at 700–1100 °C to form the $BaTiO_3$ powders. The starting $BaTiO_3$ powders are characterized by the smallest crystallite size about 7 nm.

In order to eliminate nanograin agglomeration, a three-step method involving high pressure was adopted to prepare dense uniform $BaTiO_3$ nanoceramics with grain of 10 nm (Xiao et al. 2007). The raw powder was uniaxially pressed at 8 MPa into a pellet at room temperature. In the second step, the pellet was cold pressed under higher pressure, such as 3GPa, and then it was milled into powder. The processed powder was pressed again as in the first step. The second step is crucial to crack the agglomeration and avoid exaggerated grain growth in the subsequent high-pressure sintering. The third step was to process the sample by high-pressure sintering.

Zinc chloride, zinc nitrate, and NaOH, were used as precursors for ZnO synthesis. ZnO nanocrystals are produced using a reactor with heating. A solution of $ZnCl_2$ and another of $Zn(NO_3)_2$ were slowly added (dripped) into the reactor. The dripping of the $ZnCl_2$ and $Zn(NO_3)_2$ solutions in an aqueous alkaline solution results in immediate precipitation of ZnO. After dripping, each solution remained agitated for a period of 2 h, maintaining the desired temperature. The obtained ZnO was filtered, washed with deionized water and dried several hours at 70 °C. The ZnO nanoparticles were obtained with sizes close to 30 nm (Gusatti et al. 2010).

3.3 *Citrate-Gel Method*

The citrate-gel method is used for synthesis of a number of multicomponent ceramics and films. The aqueous citrate-gel method uses citric acid and metal salts to form aqueous soluble complexes (Sale 1998). Dehydration of the transparent metal complexes solution yields an amorphous gel that can be crystallized to the target compound after high temperature calcination. The citric acid can be substituted with other polyfunctional hydroxy acids, among which EDTA. The high chelating ability of EDTA can minimize the segregation of precipitates resulting

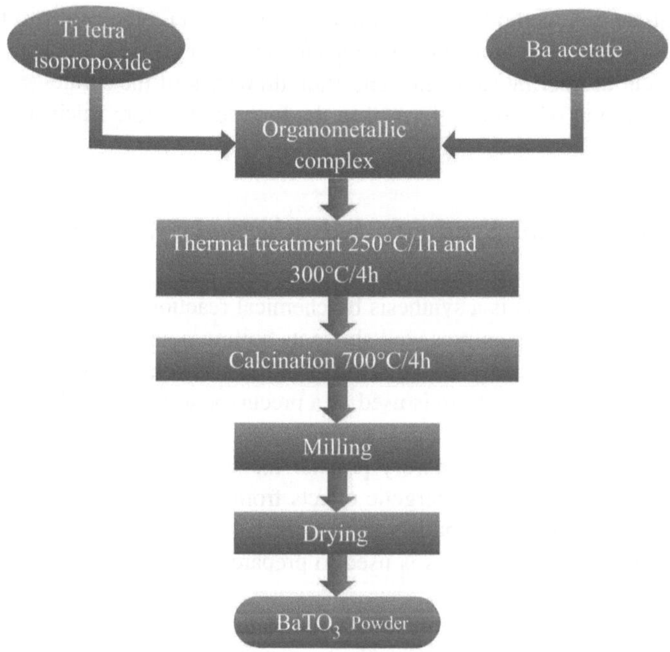

Fig. 9 Polymeric precursor method—Pechini method

from the low solubility of the starting chemicals or any other intermediate compounds.

Pure and doped $BaTiO_3$ ceramics can be prepared using a polymeric precursor method (Fig. 9) so called Pechini method (Pechini 1967). Titanium tetra-isopropoxide $Ti[OCH(CH_3)_2]_4$ and barium acetate $Ba((CH_3)COO)_2$ were used as starting materials for preparing the titanium citrate and barium citrate solutions. The molar ration of citrate acid to ethylene glycol for both citrate solutions was 4:16. Titanium citrate and barium citrate solutions were mixed and heated at 90 °C. To promote polymerization and remove the solvents, temperature was raised to 120–140 °C. Decomposition of organic resin was performed in the oven at 250 °C for 1 h and at 300 °C for 4 h. Subsequent thermal treatment of precursor was carried out at 500 °C for 4 h, 700 °C for 4 h and 800 °C for 2 h. After drying at room temperature the $BaTiO_3$ powder was obtained (Vijatović et al. 2011). The XRD analysis of powders indicate the formation of well-crystallized cubic phase of BT with crystallite size of about 20–30 nm.

The citrate-gel method was used to prepare lead zirconate titanate (PZT), with nitrates instead of chlorides as the starting materials. The pure PZT was obtained at a low calcination temperature (Sale 1998). The nitrate solutions of Zr and Ti were prepared separately by dissolving zirconium and titanium hydroxides in nitric acid. Using nitrates as starting materials, the XRD peaks belonging to the PZT phase were clearly observed after calcination at a temperature as low as 320 °C. Pure PZT was obtained after calcination at 600 °C.

The citrate-gel method has several advantages over other chemical solution techniques. It provides more homogeneous mixing, and giving better stoichiometric control (without filtering step), etc. The main drawback of the citrate-gel method is the agglomeration of particles formed in the high temperature calcination.

3.4 Hydrothermal Synthesis, Solvothermal Synthesis

Hydrothermal synthesis is a synthesis by chemical reactions in an autoclave, using water as solvent at temperatures well above its boiling points (Cushing et al. 2004). In the case of non-aqueous solvents, this synthesis is referred to as solvothermal synthesis. The organic medium is used as a precursor solution for the solvothermal method.

Hydrothermal syntheses are very popular methods for synthesizing perovskite nanomaterials, because the synergetic effects from solvent, temperature, and pressure can offer stable final products and prevent the formation of unwanted compounds. Hydrothermal synthesis is used to prepare $BaTiO_3$, $SrTiO_3$, $PbTiO_3$, ZnO nanomaterials.

For hydrothermal synthesis of $BaTiO_3$ the aqueous solutions of $Ba(OH)_2 \cdot 8H_2O$ (Dutta and Gregg 1992) or $BaCl_2$ (Testino et al. 2004) are used as Ba-source. Ti-source is typically TiO_2. An aqueous solution of NaOH are frequently used as mineralizers (Pithan et al. 2006). The ferroelectric tetragonal $BaTiO_3$ particles with size about 1 μm were hydrothermal synthesized from mixtures of $BaCl_2$, NaOH and TiO_2 in water at 200 °C. The absence of Cl ions resulted in only the cubic $BaTiO_3$. The typical crystallite size of synthesized $BaTiO_3$ is from 20 to 100 nm. The crystallite size can be decreased by using coordinating solvents or surfactants as soft templates to confine the particles growth.

The $PbTiO_3$ and $PbZrTiO_3$ powders with acicular morphology were synthesized using hydrothermal method (Hu et al. 2006; Gu et al. 2007). The phase composition and morphology of PZT particles in the method were dependent on the tetramethylammonium hydroxide pentahydrate (TMAH) concentration. At high TMAH concentration, cubic phase PZT were formed mainly through a dissolution precipitation process. At lower TMAH concentration, the PZT powder with acicular morphology was obtained.

Single crystalline tetragonal PZT nanopowders were fabricated in the addition of polymers such as polyvinyl alcohol (PVA) and polyacrylic acid (PAA). A hydrothermal reaction was conducted at 200 °C for 12 h. The nanopowder obtained by this method exhibited diameters from 150 to 175 nm and lengths of 1–1.6 mm.

Baruwati et al. (2006) have used zinc nitrate hexahydrate for aqueous synthesis of ZnO nanoparticles. The process was carried out in an autoclave at 120 °C after the pH has been set to a value of 7.5 by using ammonium hydroxide. After washing, the particles were dried at 80 °C overnight to obtain the powder form. The grain size of ZnO powder is around 20 nm.

The advantage of hydrothermal synthesis is that desired crystalline materials can be prepared at a low crystallization temperature without significant agglomeration. The drawbacks may be the difficulty of monitoring the reaction during the synthesis and the formation of unpredictable products.

3.5 Microemulsion Method

Among all the chemical techniques, the microemulsion is the most promising method to prepare nanoscale perovskite particles. Microemulsions are liquid, transparent, optically isotropic, and thermodynamically stable dispersions of a polar and a non-polar solvent. The solvents were water or an aqueous solution and a liquid hydrocarbon. The drops of 1–100 nm are stabilized by the addition of surfactants. Particles up to a few nanometers can be obtained by limiting crystal growth during particle formation.

The microemulsion technique is usable since the nanosized aqueous droplets act as nano-reactors for the synthesis of nanoparticles. In the reverse microemulsion system, water phase is a dispersed phase which is surrounded by the continuous phase which is oil. The size of the nano-reactor (water phase) is controlled by the ratio between water and the surfactant. The chemical reaction occurs inside the water phase, which prevents impurities from approaching.

Herrig and Hempelmann (1997) have developed a micro emulsion to prepare ternary metal oxide ($BaTiO_3$, $BaZrO_3$, $SrTiO_3$, $SrZrO_3$,) nanoparticle. The average nanocrystal size is about 9 and 18 nm for $BaTiO_3$ and $SrTiO_3$, respectively. The $BaTiO_3$ and $SrTiO_3$ nanoparticles were crystallized without calcination. That may be the first low temperature preparation of well-crystallized $BaTiO_3$ and $SrTiO_3$ nanoparticles using the microemulsion method. This method has been recently well repeated by Sakabe et al. (2005) and Pithan et al. (2006) to prepare $BaTiO_3$ nanoparticles with an average size of 8–10 nm using other microemulsion systems. Chen and Zhu (2007) also reported a fabrication of nanorods of Ba(Sr)TiO_3 with a diameter of 60–100 nm and a length of 450–1200 nm using a similar method at room temperature, which shows a single-crystalline tetragonal structure.

The microemulsion precipitation method provides a particular advantage of preventing agglomeration of the particles and simply controlling the particle size by variation of water content. The disadvantage of the microemulsion precipitation method may be the low yield with a sacrifice of a huge quantity of surfactants and solvents.

3.6 Sintering Process

The nanocrystalline powders consolidation can be a challenging process. A peculiarity of sintering nanocrystalline powders is a competition between the processes

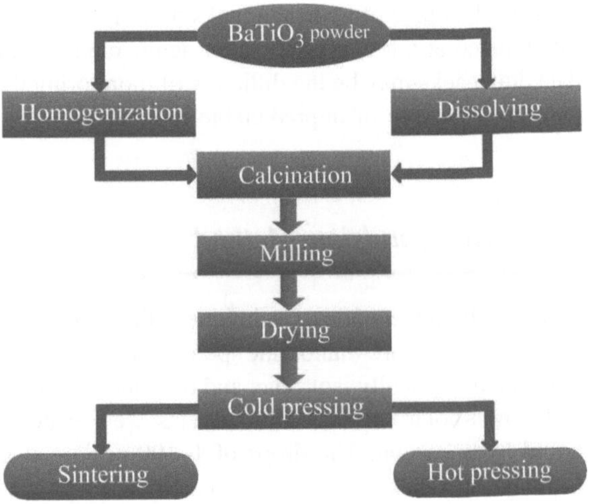

Fig. 10 Conventional solid-state sintering process

of densification and microstructure coarsening, which occur in parallel. High-density nanostructured ceramic materials are obtained mainly by conventional sintering of nanopowders using hot pressure assisted methods.

High sintering temperatures and long sintering times required for the consolidation of ceramic powders often result in grain coarsening and decomposition of the ceramic. This may lead to the deterioration of the mechanical properties of ceramics (Mitic et al. 2012). Also, high sintering temperature which is above eutectic point (1332 °C), apart from the fine grained matrix, leads to the formation of extreme grains and nonuniform nanostructure (Mitic et al. 2011). Therefore, it is difficult to obtain fine-grained and homogeneous $BaTiO_3$ ceramics with grain sizes less than 50 nm by a conventional sintering method.

In aim to prepare a material with high density, close to the theoretical and a grain size as small as possible, various techniques have been developed. Of these, methods using fast heating, high pressure, and addition of various agents that accelerate shrinkage and inhibit grain growth are the best known.

During conventional solid-state reaction (Fig. 10) the starting oxide powders are first milled and calcined. Raw materials of ceramic ($BaTiO_3$, $SrTiO_3$, etc.) powders were homogenized and ball milled in ethyl alcohol for 24 h. After homogenization, the starting powders were calcined. Calcination is a reaction in the solid phase and is carried out at a temperature of 900–1000 °C. The reacted powders (having the desired stoichiometry) are milled once again to reduce the particle size by using a classical ball mill with alumina or zirconia balls (10 mm balls diameter) as the milling media. Milling in the classical ball mill can be wet and dry. For the wet milling, water, acetone or alcohol are used as liquids. After milling, the slurries are dried in oven at 200 °C for few hours.

The dried powders were then pressed under a uniaxial pressure of 90-150 MPa into disk of 3–6 mm in diameter and 2 mm of thickness (Haertling 1999; Zivkovic et al. 2006b). Optimal density compacts is 55–65% of TD. The density of compacted powders exponentially increases with increasing pressure pressing until the critical pressure. Finally, the samples were sintered.

At the sintering temperature, the ceramic particles coalesce with each other to form grains, the material shrinks and the pores are eliminated. The ceramics are treated at high temperature (1000 °C) to obtain the perovskite type structure. A second heating step is subsequently applied to sinter the ceramic. In this way, dense ceramics (>90% of the TD) are obtained at 1250 °C for pure $BaTiO_3$ and 1300 °C for pure $SrTiO_3$. Such reduced sintering temperatures can only be applied to powders with particle sizes smaller than 50 nm, which is uniquely attained, thanks to the high-energy milling. Whenever the $BaTiO_3$ samples were sintered at temperatures higher than 1250 °C, severe strain was observed as the melting temperature of the material (1652 °C) was approached.

Hot pressing of some ceramic materials was discovered to allow the occurrence of densification at temperatures much lower than the ones during conventional sintering. Advantages are increase of densification and limiting of grain growth. Disadvantages are limited geometry of the final product, and the expensive equipment required.

To overcome the problem of grain growth, advanced sintering and densification techniques were used. Advanced sintering techniques include: the grain growth inhibitors in solid solution, high-pressure densification, spark plasma sintering and similar techniques (Kim and Khalil 2006; Vijayan and Varma 2002; Allen et al. 1996; Veljovic et al. 2009).

4 Characterization of Ferroelectric and Piezoelectric Nanomaterials

Structural characterization is essential for nanomaterials research. The control of particle size, distribution, and dispersion at nanolevel needs various characterization techniques:

- Scanning probe microscopy (SPM),
- Scanning tunneling microscopy (STM),
- Atomic force microscopy (AFM),
- Scanning electron microscopy (SEM),
- Transmission electron microscopy (TEM),
- X-ray diffraction (XRD),
- X-ray absorption spectroscopy (XAS),
- Energy dispersive X-ray spectroscopy (EDS).

Some of these techniques are more surface sensitive than others. Some techniques are directly element-dependent while others are not. The choice of techniques depends strongly on the information being sought about the material (Zhang et al. 2003; Wang 2000; Wang et al. 2000).

4.1 Scanning Probe Microscopy (SPM)

Scanning probe microscopy has become a revolutionary method in probing non-destructive domain structures of ferroelectric thin films and testing the ferroelectric structures at a nanometer scale. Scanning probe microscopy is a set of techniques, which includes atomic force microscopy (AFM) and scanning tunneling microscopy (STM). These techniques have been extensively applied to characterize nanostructures with atomic or subatomic spatial resolution (Bonnell 2000). A common characteristic of STM and AFM is a sharp tip used to scan the sample surface. The images are formed either by measuring the force acting on the tip, or by measuring the current flowing through it. In this way, it allows direct imaging of inorganic surfaces and organic molecules.

4.1.1 Atomic Force Microscopy (AFM)

AFM is based on measuring the force between the tip and the solid surface and is used for measuring surface roughness on the nanometer scale (Meyer 2007). The force acting on the tip reflects the distance from the tip atom to the surface atom.

AFM construction includes a sharp tip set to a microscale cantilever, laser, position-sensitive detector, PZT scanner and electronic control. The cantilever is made of silicon or silicon nitride with a tip radius of the order of a nanometer. The repulsive force on the tip is set by pushing the cantilever against the surface using a piezoelectric positioning element. There are three scan modes for AFM: contact mode, non-contact mode, and tapping mode. In the contact mode, the tip scans the specimen in close contact with the surface of the materials. In the non-contact mode, the scanning tip hovers about few nm above the specimen's surfaces. The deflection of the cantilever is measured and an image is generated through a feedback loop between the optical detection system and piezoelectric scanners. In the tapping mode, the cantilever oscillates at its resonant frequency using a piezoelectric crystal, and both the amplitude of oscillation and frequency are changed near the specimen.

AFM is nondestructive technique, used in nonconductive material and it has a very high three dimensional spatial resolution. Atomic force microscopy is able to provide very detailed information on the atomic scale, which is important for understanding the bonds within molecules and the electronic structure of atoms.

4.1.2 Scanning Tunneling Microscopy (STM)

Measurement of the tunneling current between a metallic tip and the sample is used as an operating principle of Scanning tunneling microscopy (STM). When a metal tip is brought very close to conducting or semiconducting surface, a bias voltage applied between them can lead electrons to tunnel through (Bai 2007).

A small voltage, between a few mV and a few V, is applied between the tip and the specimen. When the tip is away from the sample surface, the circuit is considered as open and the current is zero. However, at very small distances a weak tunneling current flows between the specimen and the tip. For STM, good resolution is considered to be 0.1 nm laterally and 0.01 nm vertically.

Images are generated by keeping the current between the tip and the specimen at some constant value and by using a piezoelectric crystal to adjust the distance between the tip and specimen surface. This distance control is achieved using feedback electronics and the tip is moved in a direction parallel to the surface. Measured piezo voltage in the feedback loop contains information about the surface topology of the specimen. The STM is applicable mainly for conductive specimens. STM is used in ultra-high vacuum, but also in air atmosphere and various gas or liquid and wide temperature range.

4.2 *Electron Microscopy*

Microstructure with grain size smaller than 1 μm cannot be observed with optical microscopy. If imaging at considerably higher resolution is required, electromagnetic radiation of shorter wavelengths must be used. Electron microscopes are able to achieve magnifications of the order of 1 million and disclose details with a resolution of about 0.1 nm. The image techniques such as scanning electron microscopy (SEM) and transmission electron microscopy (TEM) are used to observe the sub-micron size materials. When an electron beam interacts with a sample, electrons can be transmitted, backscattered, and diffracted, as illustrated in Fig. 11.

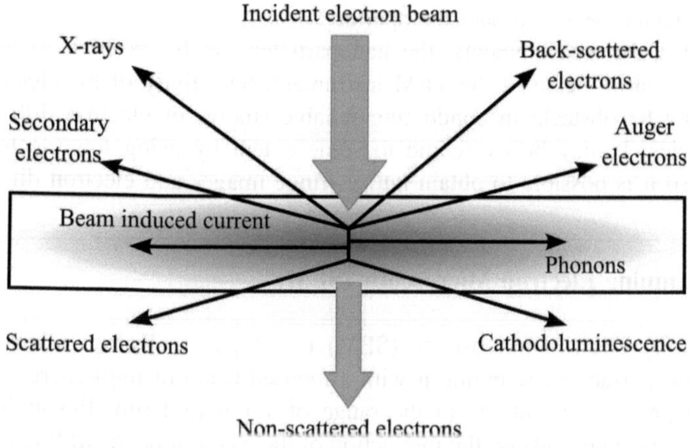

Fig. 11 Illustration of interaction between incident electron beam and specimen

TEM uses the transmitted electrons to form a sample image, while SEM uses backscattered electrons and secondary electrons emitted from the sample. Electrons can also be obtained by diffraction on the particles, which provide crystallographic information.

4.2.1 Transmission Electron Microscopy (TEM)

Transmission electron microscopy (TEM) is one of the most exploited techniques for characterizing nanomaterials (Wang 2003). It allows obtaining of structural information down to atomic resolutions. It also allows direct correlation to chemical and physical properties.

A modern TEM is capable of direct imaging of atoms in crystalline samples at resolutions close to 0.1 nm. An electron beam can also be focused to a diameter smaller than 0.3 nm, allowing quantitative chemical analysis from a single nanocrystal. TEM can be used to characterize nanomaterials to gain information about particle size, shape, crystallinity, and antiparticle interaction.

A stream of monochromatic electrons produced by an electron gun is focused into a coherent beam by two condenser lenses. The beam is restricted by the condenser aperture to remove high angle electrons before it reaches the specimen. To enable some of the electrons to be transmitted through the sample, a very thin specimen is required. Interaction between the beam and the specimen generates elastically and inelastically scattered electrons, along with some unscattered electrons, in the forward direction after the sample has been detected. The detected signal contains information about the sample.

To achieve a high contrast image, it is necessary to block electrons deflected away from the optical axis of the microscope by setting the aperture to allow only unscattered electrons trough. This produces a variation in the electron intensity that reveals information on the crystal structure. This technique is known as bright field or light field. It is also possible to produce dark field image, by allowing only some of the diffracted beams to pass the aperture.

During TEM measurements, the nanoparticles can be exposed to the highly energetic electron beam of the TEM instrument. Sensitivity of the electron beam makes may be obstacle in conducting reliable studies of electron diffraction on nanoparticles. It may be overcome to some extent by using low electron beam currents, so it is possible to obtain lattice fringe images and electron diffraction.

4.2.2 Scanning Electron Microscopy (SEM)

The scanning electron microscope (SEM) is a type of electron microscope for imaging the surface by scanning it with a focused beam of high-energy electrons. SEM can provide resolutions in the range of 1 μm to 1 nm. Resolution of the generated image depends on the interaction of the electron beam with the specimen and on the probe itself. The interaction of an incident electron beam with the

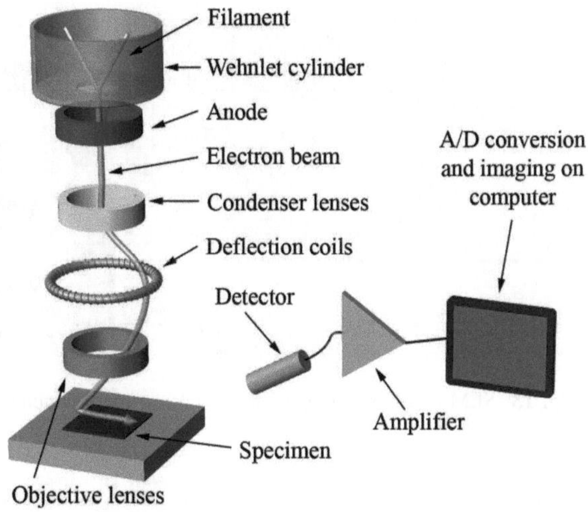

Fig. 12 Illustration of the scanning electron microscope (SEM)

specimen produces secondary electrons, with energies typically smaller than 50 eV. The emission efficiency depends on surface geometry, surface chemical characteristics and bulk chemical composition (Hanada et al. 2008; Goldstein et al. 1992). SEM can get information on surface topography, crystalline structure, microstructure morphology, and chemical composition. The high-resolution capability of SEM makes it convenient for probing nanomaterials of which the structural features on the nanoscale are critical to their properties and functionalities.

A stream of monochromatic electrons is generated by an electron gun, as illustrated in Fig. 12. The condenser lenses and a condenser aperture were used for a focused electron beam. It passes through pairs of scanning coils in the objective lens and pairs of deflector plates, which deflect the beam to produce a raster over a rectangular area of the specimen surface. After the electron beam interacts with the surface, the primary electrons are inelastically scattered by atoms.

Consequently, the primary electron beam effectively spreads and fills a teardrop-shaped volume that extends less than 100–5000 nm into the surface (Ma et al. 2006). Interactions in this region lead to the subsequent emission of secondary electrons, backscattered electrons, Auger electrons, and characteristic X-ray which are then detected for SEM imaging, and for EDS, WDS (Wavelength dispersive spectrometry), or SAM (Scanning Auger microscopy) microanalysis.

SEM requires that the specimens should be conductive for the electron beam to scan the surface and that the electrons have a path to ground for conventional imaging. Non–conductive solid specimens are generally coated with a layer of conductive material by low vacuum sputter coating.

SEM microphotographs of $BaTiO_3$ powders obtained by the high temperature solid-state process are shown in Fig. 13. The powders of $BaTiO_3$ were first dissolved in ethanol and then homogenized. Gold is deposited over the surface in order to insure electrical conductivity.

Fig. 13 SEM microphotographs of $BaTiO_3$ powders obtained by solid-state reaction

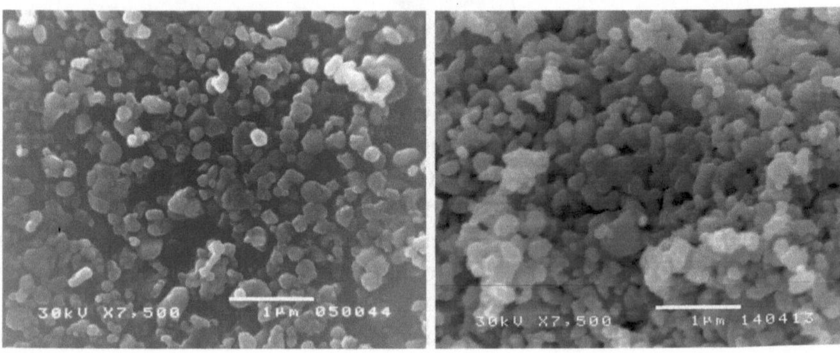

Fig. 14 SEM microphotographs of $BaTiO_3$ ceramics sintered at 1290 °C for 2 h (*left*) and 4 h (*right*)

In Fig. 14 SEM micrograph of $BaTiO_3$ ceramics obtained by conventional solid-state process and sintered at 1290 °C, with average grain size ranged from 80–120 nm, are shown. The density of $BaTiO_3$ specimens obtained by conventional method is 91% of theoretical density.

While both TEM and SEM techniques can provide topological, morphological and compositional information about the sample, TEM can also provide crystallographic information. In addition, TEM allows diffraction patterns to be detected thus giving useful crystallographic information about the sample.

4.3 X-Ray

X rays are a form of electromagnetic radiation similar to light, but with a much shorter wavelength and a much higher energy. They are produced when there are decelerated electrically charged particles of sufficient energy. X rays are produced at the point of impact, and radiate in all directions. For the characterization of nanomaterials a few X-ray techniques can be used:

- X-ray diffraction (XRD),
- X-ray absorption spectroscopy (XAS),
- Energy dispersive X-ray spectroscopy (EDS),
- X-ray photoelectron spectroscopy (XPS).

X-ray spectroscopy is used for determination of the materials chemical composition (Barr and Modern 2008; Zubavichus et al. 2002). The main principle is based on the detection and analysis of radiation that is absorbed or emitted from the sample after excitation with X-rays. The spectroscopic features are characteristic of specific elements and thereby can be used for sample elemental analysis.

4.3.1 X-Ray Diffraction (XRD)

X-ray diffraction (XRD) is used to determine the crystal structure. By examining the diffraction pattern, one can identify the crystalline phase of the material. Phase identification using XRD is mainly based on the diffraction peak positions according to the well-known Bragg's law, and to some extent on the relative peak intensities. XRD analysis uses data obtained by interacting an incident beam of monochromatic X-rays with a crystalline sample. X-rays are generated by a vacuum cathode ray tube. They are then filtered and focused, producing a monochromatic radiation which is directed towards the sample. The interaction of the incident rays with the specimens produces constructive interference and a diffracted ray when conditions satisfy Bragg's Law. The characteristic X-ray diffraction pattern (Fig. 15) generated in a typical XRD analysis provides a unique "fingerprint" of the crystals present in the sample. XRD results on sintered $BaTiO_3$ indicate formation of the perovskite type tetragonal structure, which is confirmed by X-ray reflection at $2\theta = 22.36$, 31.82, 39.24, 45.62, 51.38, 56.70, 66.5 and 75.62. Appearance of the two splitting peaks, (200) and (002) is perceived as a typical signature of tetragonal $BaTiO_3$ phase.

By comparison with standard reference patterns and measurements, this fingerprint allows the identification of the crystalline form.

4.3.2 Energy Dispersive X-Ray Spectroscopy (EDS)

EDS is a technique used for the elemental composition analysis of materials, for analysis of near surface elements and estimating their proportion at different position, thus giving an overall mapping of the sample. The EDS technique is used in conjunction with SEM. The electron beam energy used at EDS analysis is in the range of 10 to 20 keV. Interaction of the primary beam with atoms in the sample causes shell transitions, which result in the emission of an X-ray from the material.

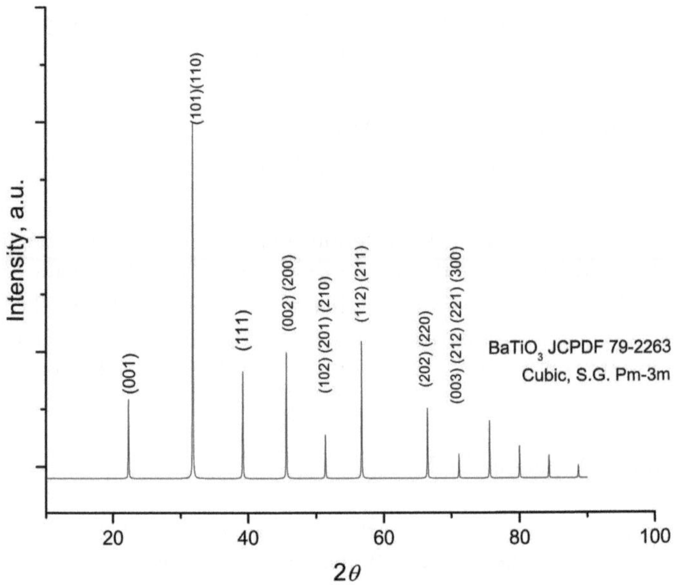

Fig. 15 XRD diffraction spectra

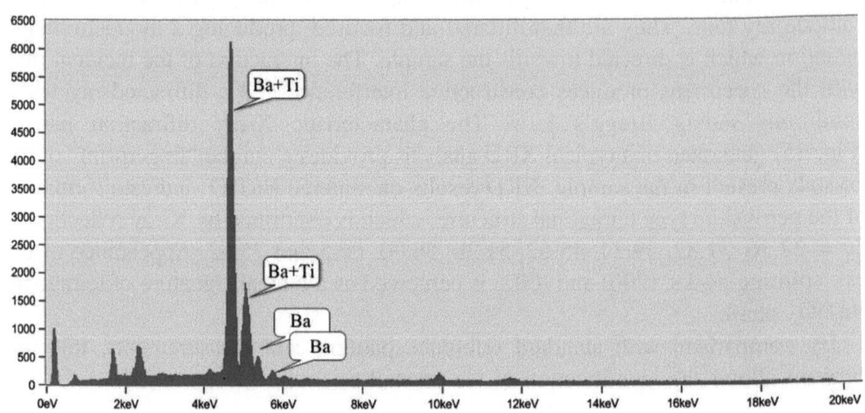

Fig. 16 EDS spectra of $BaTiO_3$ ceramics

The energy of the emitted X-rays depends on the material under examination. EDS spectra can provide rapid qualitative (Fig. 16) or quantitative composition analysis with a sampling depth of 1–2 μm (Mitic et al. 2014).

X-rays may also be used to form maps or line profiles, showing the elemental distribution in a sample surface (Mitic et al. 2010). By moving the electron beam over the material, the image of elements in the investigated specimen can be obtained.

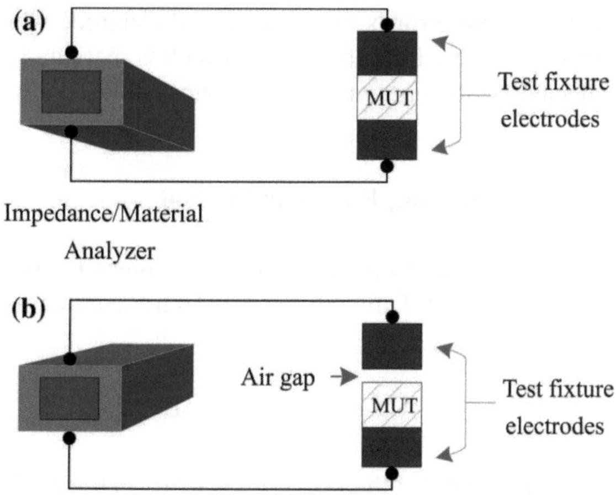

Fig. 17 Parallel plate measurement method: **a** contacting electrode method; **b** non-contacting electrode method

4.4 Electrical Characterization

In order to characterize nanomaterials, various measurements, such as frequency and temperature response, must be performed. A measurement system is comprised of three main pieces: precise measurement instruments, test fixtures that hold the material under test, and software that can model, calculate and display basic material parameters.

4.4.1 Parallel Plate Measurement Method

For accurate measurement of impedance of nanomaterials the parallel plate method is usually employed. This technique requires an LCR meter or impedance analyzer. The parallel plate method, involves sandwiching a thin sheet of material under test (MUT) with known dimensions between two electrodes to form a capacitor. AC bias is applied to the MUT. Developed voltage across MUT is measured, and capacitance is derived, from which the dielectric constant can be calculated.

There are the two types of parallel plate measurement methods: contacting and non-contacting electrode method, as illustrated in Fig. 17.

In the contacting electrode method, a thin film electrode is applied onto surfaces of the MUT, which achieves greater measurement accuracy. The main advantage of this method is that only one measurement provides a value of dielectric constant. On the other side, the non-contacting electrode method has medium measurement accuracy and involves very simple sample preparation and setup. The MUT should be a solid material with a flat and smooth surface. The dielectric constant is derived by using the results of two capacitance measurements.

The edge capacitance on the electrodes causes a measurement error, since the current flows through the dielectric material and edge capacitor. A solution for the measurement error is to use the guard electrode which absorbs the electric field at the edge. Then, the capacitance, which is measured between the electrodes, is only affected by the current that flows through the dielectric material.

4.4.2 Contacting Electrode Method

The contacting electrode method determines permittivity by measuring the capacitance of the MUT directly. The dielectric constant is calculated using the equation:

$$\varepsilon_r = \frac{tC_p}{S\varepsilon_0} = \frac{tC_p}{\pi\left(\frac{d}{2}\right)^2 \varepsilon_0}, \tag{2}$$

where t is material thickness, C_p is equivalent parallel capacitance, S is the guarded electrode's surface area, d is diameter, and ε_0 is the dielectric constant of the vacuum. The contacting electrode method does not require material preparation and the operation involved when measuring is simple. Air gaps may appear between the MUT and the electrodes. This could lead to a measurement error, because a parasitic capacitance will be in series with MUT capacitance. The error may have serious effect on the measurements of thin samples. Apart from applying larger pressure by text fixture (which could deform the sample), another way to reduce the error is to use thin film electrodes deposited directly on MUT. This technique involves the application of a thin film electrode onto the surfaces of the dielectric material in order to increase the contacting area and to eliminate the air gap. The electrode is produced onto the dielectric material using high-conductivity silver paste. The electrode film must be thinner than the MUT. In this case, it is important to position the electrode onto the MUT appropriately, to contact precisely the electrodes of test fixture. The measurement error is determined by:

$$\delta_{r\varepsilon} = \frac{\varepsilon_r - 1}{\varepsilon_r + \frac{t_m}{t_a}} \cdot 100(\%), \tag{3}$$

where ε_r and t_m are permittivity and thickness of the MUT, respectively, and t_a is the thickness of the air gap.

4.4.3 Non-contacting Method

The non-contacting method does not use thin film electrodes. The method was developed to unite the advantages and eliminate the disadvantages of the contacting electrode method (Agilent Technologies 2013). Two capacitance measurements, obtained with the MUT and without it, are necessary and the dielectric constant is determined by:

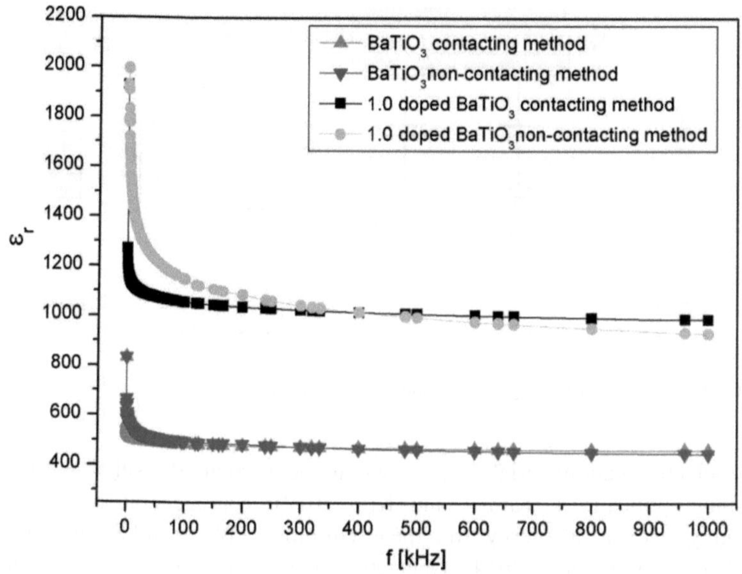

Fig. 18 Dielectric constant versus frequency for $BaTiO_3$ samples measured with contacting and non-contacting electrode methods

$$\varepsilon_r = \frac{1}{1 - \left(1 - \frac{C_1}{C_2}\right) \cdot \frac{t_g}{t_m}}, \tag{4}$$

where C_1 and C_2 are capacitances without and with the MUT inserted, respectively, t_g is an electrode gap, and t_m is the average thickness of the MUT. It is necessary that the condition $t_g - t_m \ll t_m$ should be met, i.e. the gap between the electrode and the MUT should be much smaller than the thickness of the MUT.

The dielectric constant-frequency dependence, measured with two methods, is shown in Fig. 18 (Marjanović et al. 2014).

5 Applications

The ferroelectric and piezoelectric effects are used in many electronic applications. Ferroelectric nanomaterials are promising for production of various electronic devices like miniaturized capacitors, sensors and memories (Mao et al. 2010). Ferroelectric capacitors are used in medical ultrasound machines, infrared cameras, acoustic and vibration sensors, etc. Nanowire and thin film Field-Effect-Transistors (FETs) based on ZnO and $BaTiO_3$ structures are realized and their usage as non-volatile memory devices is subject of extensive research (Brandt et al. 2009; Liu et al. 2012). Apart from the dimension scaling, main potential benefits are non-destructive readout and increased overall reliability.

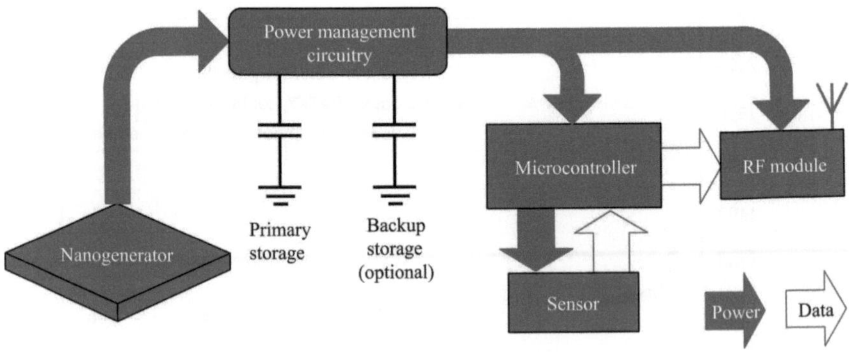

Fig. 19 Block diagram of energy harvesting wireless sensor node

Piezoelectric nanomaterials are shown to be particularly suitable for energy harvesting applications (Kumar and Kim 2012). Nanogenerators are used for scavenging mechanical energy from the environment (Wang and Song 2006; Wang 2012; Vocca and Cottone 2014). Generators may be formed by sandwiching ZnO nanorods or nanowires between electrodes (Nour et al. 2014), or using PZT nanofibers on flexible substrates (Chang et al. 2012). Various autonomous micro-electromechanical sensors and systems (MEMS) are realized (Murillo et al. 2011; Wang and Wu 2012). Systems are designed as self-powered nodes in wireless sensor networks (WSN). A typical node consists of a piezoelectric nanogenerator, power management circuitry, one or more sensors, ultralow power microcontroller and RF module, as illustrated in Fig. 19.

Data from the sensors are acquired by the microcontroller and transmitted wirelessly using the RF module. It is to be noted that nanogenerators, delivering power of the order of tenths to hundreds of microwatts, still are not efficient enough to enable continuous supply of the system. Rather, nodes are designed to operate using duty cycling, spending a part of the operating cycle in sleep/idle mode (Nechibvute et al. 2012). Although many MEMS based nodes are realized using system on board integration, practical realization of the node with nanogenerator implies system on chip approach.

Applications are in telemetry, covering environmental monitoring, automotive industry, wearable devices, preventive maintenance of the structures exposed to vibrations, etc. (Lu 2014; Calio et al. 2014). Special attention is given to the health monitoring and biomedical applications through construction of implantable devices (Wang and Shi 2012; Diao et al. 2014), where the human body is used as a source of mechanical energy which is converted to electrical energy by piezoelectric nanogenerators. There are also applications in the fields of pressure sensors, transducers, and ultrasonic waves. Ultrasonic is used in engineering diagnostics as well as in medical applications.

It is notable that piezoelectric nanomaterials are found to be suitable for development of novel electronic devices such as magnetic field sensors and magnetoelectric random access memories. Structure of these devices includes both magnetic and ferroelectric materials. Working principle is based on the coupled effect between the voltage induced piezo strain and magnetostriction (Tiercelin et al. 2016). Ultra-high density memories released as nanostructures have been reported (Hu et al. 2015).

6 Conclusion

Ferroelectric and piezoelectric nanomaterials are an attractive choice for the development of many electronic devices. Some of them exhibit unique properties that come from both semiconductor and piezoelectric nature. A variety of technologies have emerged in the recent years, demonstrating advanced design and production of shapes like nanorods, nanofibers, nanobelts, and other more complex structures. Nanowire memory cells and nanogenerators are representative examples of technological potential in nanoelectronics. Extensive research and development is being carried out, with the goal to improve their performance, particularly in the fields of reliability and efficiency.

Acknowledgements The printing of this work is financed by EU project 543898-TEMPUS-1-2013-1-ES-TEMPUS-JPHES. Parts of the scientific works are supported by Ministry of Education, Science and Technological Development, Serbia, projects OI172057 and TR32026.

References

Agilent Technologies (2013) Agilent impedance measurement handbook 4th edn. http://cp.literature.agilent.com/litweb/pdf/5950-3000.pdf. Accessed 30 Jan 2015

Ahmad T, Ganguli AK (2006) Reverse micellar route to nanocrystalline titanates ($SrTiO_3$, Sr_2TiO_4, and $PbTiO_3$): structural aspects and dielectric properties. J Am Ceram Soc 89: 1326–1332

Ahn CH, Rabe KM, Triscone JM (2004) Ferroelectricity at the nanoscale: local polarization in oxide thin films and heterostructures. Science 303:488–491

Allen AJ, Kruegger S, Skandan G et al (1996) Microstructural evolution during the sintering of nanostructured ceramic oxides. J Am Ceram Soc 79:1201–1212

Aoyagi S, Kuroiwa Y, Sawada A et al (2005) Size Effect on Crystal Structure and Chemical Bonding Nature in $BaTiO_3$ Nanopowder. J Therm Anal Calorim 81:627

Bai C (2007) Scanning tunneling microscopy and its applications. Springer-Verlag Telos, New York, p 366

Barr TL, Modern ESCA (2008) The principles and practice of x-ray photoelectron spectroscopy. CRC Press, Boca Raton, FL, p 376

Baruwati B, Kumar DK, Manorama SV (2006) Hydrothermal synthesis of highly crystaline ZnO nanoparticles. Sens Actuators B 119:676

Bonnell D (2000) Scanning probe microscopy and spectroscopy. In: Theory, techniques, and applications. New York, Wiley-VCH

Brandt M, Frenzel H, Hochmuth H et al (2009) Ferroelectric thin film field-effect transistors based on ZnO/$BaTiO_3$ heterostructures. J Vac Sci Technol, B 27:1789
Buscaglia V, Buscaglia MT, Viviani M et al (2005) Raman and AFM piezoresponse of dense $BaTiO_3$ nanocrystalline ceramics. J Eur Ceram Soc 25:3059–3062
Buscaglia MT, Viviani M, Buscaglia V et al (2006) High dielectric constant and frozen macroscopic polarization in dense nanocrystalline $BaTiO_3$ ceramic. Phys Rev B 73: 064114-1/064114-10
Caliò R, Rongala BU, Domenico C et al (2014) Piezoelectric energy harvesting solutions. Sensors 14:4755–4790
Chang J, Dommer M, Chang C et al (2012) Piezoelectric nanofibers for energy scavenging applications. Nano Energy 1:356–371
Chen W, Zhu Q (2007) Synthesis of barium strontium titanate nanorods in revers microemulsion. Mater Lett 61:3378–3380
Cross LE (1993) Ferroelectric ceramics: tailoring properties for specific applications. Birkhauser Verlag, Basel Switzerland; Boston, MA, USA, pp 1–85
Cushing BL, Kolesnichenko VL, O'Connor C (2004) Recent advances in the liquid phase syntheses of inorganic nanoparticles. J Chem Rev 104:3893–3946
Dawber M, Chandra P, Littlewood PB et al (2003) Depolarization corrections to the coercive field in thin film ferroelectrics. J Phys: Condens Matter 15:393–398
Deng X, Wang ZX, Wen H et al (2006) Phase transitions in nanocrystalline barium titanate ceramics prepared by spark plasma sintering. J Am Ceram Soc 89:1059–1064
Diao LW, Zheng J, Pan XD et al (2014) Application of piezoelectric nanogenerator in medicine: bio-experiment and theoretical exploration. J Thorac Dis 6:1300–1306
Dutta PK, Gregg JR (1992) Hydrothermal synthesis of tetragonal barium titanate ($BaTiO_3$). Chem Mater 4:843–846
Emelyanov AYu, Pertsev NA, Hoffmann-Eifert S et al (2002) Grain boundary effect on the Curie Weiss law of ferroelectric ceramics and polycrystalline thin films: calculation by the method of effective medium. J Electroceram 9:5–16
Fang TT, Hsich HL, Shiau FS (1933) Effect of pore morphology and grain size on the dielectric properties and tetragonal-cubic phase transition of high purity barium titanate. J Am Ceram Soc 76(5):1205–1211
Frey MH, Xu Z, Han P et al (1998) The role of interfaces on an apparent grain size effect on the dielectric properties for ferroelectric barium titanate ceramics. Ferroelectrics 206:337–353
Goldstein JI, Newbury DE, Echlin P et al (1992) Scanning electron microscopy and x-ray microanalysis, 2nd edn. Plenum Press, New York, p 820
Gu H, Hu Y, Wang H et al (2007) Fabrication of lead titanate single crystalline nanowires by hydrothermal method and their characterization. J Sol Gel Sci Technol 42:293–297
Gusatti M, Rosário AJ, Campos EMC et al (2010) Production and characterization of ZnO nanocrystals obtained by solochemical processing at different temperatures. J Nanosci Nanotechnol 10:1–4
Haertling GH (1999) Ferroelectric ceramics: history and technology. J Am Ceram Soc 82(4): 797–818
Hanada N, Hirotoshi E, Chikawa T et al (2008) SEM and TEM characterization of magnesium hydride catalyzed with Ni nano-particle or Nb_2O_5. J Alloy Compd 450:395–399
Herrig H, Hempelmann R (1997) Microemulsion mediated synthesis of ternary and quaternary nanoscale mixed oxide ceramics powders. Nanostruct Mater 9:241–244
Hu Y, Gu H, Sun X, You J et al (2006) Photoluminescence and Raman scattering studies on PbTiO3 nanowires fabricated by hydrothermal method at low temperature. Appl Phys Lett 88:193120
Hu J, Nan T, Sun N et al (2015) Multiferroic magnetoelectric nanostructures for novel device applications. MRS Bulletin 40:728–735
Hwang UY, Park HS, Koo KK (2004) Low temperature synthesis of fully crystallized spherical $BaTiO_3$ particles by the Gel-sol method. J Am Ceram Soc 87:2168–2174

Kim SW, Khalil KAR (2006) High-frequency induction heat sintering of mechanically alloyed alumina–yttria-stabilized zirconia nano-bioceramics. J Am Ceram Soc 89:1280–1285
KimYS Kim DH, Kim JD et al (2005) Critical thickness of ultrathin ferroelectric $BaTiO_3$ films. Appl Phys Lett 86:102907
Kumar B, Kim SW (2012) Energy harvesting based on semiconducting piezoelectric ZnO nanostructures. Nano Energy 1:342–355
Kuroiwa Y, Aoyagi S, Sawada A et al (2002) Structural study of perovskite-type fine particles by synchrotron radiation powder diffraction. J Therm Anal Cal 69:933–938
Li B, Wang X, Li L et al (2004) Dielectric properties of fine grained $BaTiO_3$ prepared by spark plasma sintering. Mater Chem Phys 83:23–28
Liu X, Liu Y, Chen W et al (2012) Ferroelectric memory based on nanostructures. Nanoscale Res Lett 7:285
Lobo RPSM, Mohallem NDS, Moreira RL (1995) Grain-size effects on diffuse phase transitions of sol-gel prepared barium titanate ceramics. J Am Ceram Soc 78(5):1343–1346
Lu K (2014) Nanomaterials: bringing new excitements to the energy world. Ann J Mater Sci Eng 1(1):2
Ma H, Shieh KJ, Qiao TX (2006) Stady of transmission electron microscopy and scanning electron microscopy (SEM). Nat Sci 4:14–22
Mao Y, Banerjee S, Wong SS (2003) Large-scale synthesis of single-crystalline perovskite nanostructures. J Am Chem Soc 125:15718–15719
Mao Y, Zhou H, Wong SS (2010) Synthesis, properties, and applications of Perovskite-Phase metal oxide nanostructures. Mater Matters 5(2):50
Marjanović M, Paunović V, Prijić Z et al (2014) On the measurement methods for dielectric constant determination in Nb/$BaTiO_3$ ceramics. INDEL, pp 38–41
Meyer E (2007) Atomic force microscopy: fundamentals to most advanced applications, vol 1. New York, Springer-Verlag TELOS, p 250
Mitić V, Nikolić Z, Pavlović V et al (2010) Influence of rare-earth dopants on barium titanate ceramics microstructure and corresponding electrical properties. J Am Ceram Soc 93(1): 132–137
Mitic VV, Paunovic V, Pavlovic V, Lj Zivkovic (2011) Sintering process influence on microstructure and intergranular impedance of rare-earth modified $BaTiO_3$-ceramics. Sci Sinter 43(3):277–287
Mitic VV, Paunovic V, Purenovic J, Lj Kocic, Pavlovic V (2012) The processing parameters influence on $BaTiO_3$-ceramics fractal microstructure and dielectric characteristics. Adv Appl Ceram: Struct, Funct Bioceramics 111(5&6):360–366
Mitic VV, Paunovic V, Pavlovic V (2014) Microstructure and dielectric properties of rare-earth doped $BaTiO_3$ ceramics. Ferroelectrics 470(1):159–167
Moezzi A, McDonagh A, Cortie B (2012) Zinc oxide particles: synthesis, properties and applications. Chem Eng J 185–186:1–22
Moulson AJ, Herbert JM (1995) Electroceramics. Chapman & Hall, London
Murillo G, Lee M, Xu C et al (2011) Hybrid resonant energy harvester integrating ZnO NWs with MEMS for enabling zero-power wireless sensor nodes. Nano Commun Netw 2:235–241
Nechibvute A, Chawanda C, Luhanga P (2012) Piezoelectric energy harvesting devices: an alternative energy source for wireless sensors. Smart Mater Res. doi:10.1155/2012/853481
Niederberger M, Garnweitner G, Pinna N et al (2004) Nonaqueous and halide free route to crystaline BaTiO3, SrTiO3 and (Ba, Sr) TiO3 and nanoparticles via a mechanism involving c-c nond formation. J Am Chem Soc 126:9120–9126
Niesz K, Ould-Ely T, Tsukamoto H et al (2011) Engineering grain size and electrical properties of donor-doped barium titanate ceramics. Ceram Int 37:303–311
Nour SE, Khan A, Nur O et al (2014) A flexible sandwich nanogenerator for harvesting piezoelectric potential from single crystalline zinc oxide nanowires. Nanomaterials Nanotechnol 4:24. doi:10.5772/59068
Nuraje N, Su K (2013) Perovskite ferroelectric nanomaterials. Nanoscale 5:8752–8780

Parashar SKS, Choudhary RNP, Murty BS (2004) Electrical properties of Gd-doped PZT nanoceramic synthesized by high-energy ball milling. Mater Sci Eng, B 110:58–63
Paunovic V, Lj Zivkovic, Mitic V (2010a) Influence of rare-earth additives (La, Sm and Dy) on the microstructure and dielectric properties of doped $BaTiO_3$ ceramics. Sci Sinter 42:69–79
Paunovic V, Mitic V, Pavlovic V et al (2010b) Microstructure evolution and phase transition in La/Mn doped barium titanate ceramics. Process Appl Ceram 4(4):253–258
Paunovic V, Mitic VV, Miljkovic M, Lj Zivkovic (2012) Ho_2O_3 Additive effects on $BaTiO_3$ ceramics microstructure and dielectric properties. Sci Sinter 44(2):223–233
Paunovic V, Mitic V, Prijic Z et al (2014) Microstructure and dielectric properties of Dy/Mn doped $BaTiO_3$ ceramics. Ceram Int 40:4277–4284
Pazik R, Kaczorowski D, Hreniak D, Stre W, Lojkowski W (2008) Synthesis, structure and magnetic properties of $BaTiO_3$ nanoceramics. Chem Phys Lett 452:144–147
Pechini MP (1967) US Patent No. 3,330,697
Pithan C, Shiratori Y, Waser R et al (2006) Preparation, processing and characterization of nano crystalline $BaTiO_3$ powders and ceramics derived from microemulsion mediated synthesis. J Am Ceram Soc 89:2908–2916
Sakabe Y, Yamashita Y, Yamamoto H (2005) Dielectric properties of nanocrystalline $BaTiO_3$ synthesis by microemulsion method. J Eur Ceram Soc 25:2739–2742
Sale FR (1998) The citrate-gel processing of electronic and magnetic ceramics. In: IEE colloquium on Sol-gel materials for device applications, vol 412. London, UK, IEE, p 4/1–4/7
Sen S, Choudhary RNP, Tarafdar A et al (2006) Impedance spectroscopy study of strontium modified lead zirconate titanate ceramics. J Appl Phys 99(12):124114/1–124114/8
Sheng G, Zhang JX, Li YL et al (2008) Domain stability of $PbTiO_3$ thin film under anisotropic misfit strains: phase field simulations. J Appl Phys 104:054105/1
Shukla A, Choudhary RNP (2010) Ferroelectric phase-transition and conductivity analysis of La^{3+}/Mn^{4+} modified $PbTiO_3$ nanoceramics. Phys B 405:2508–2515
Singh KC, Nath AK (2011) Barium titanate nanoparticles produced by planetary ball milling and piezoelectric properties of corresponding ceramics. Mater Lett 65:970–973
Testino A, Buscaglia MT, Buscaglia V et al (2004) Kinetics and mechanism of aqueous chemical synthesis of $BaTiO_3$ particles. Chem Mater 16:1536–1543
Tichý J, Erhart J, Kittinger E, Prívratská J (2010) Fundamentals of piezoelectric sensorics. Springer
Tiercelin N, Dusch Y, Giordano S et al (2016) Strain mediated magnetoelectric memory. In: Atulasimha J (ed) Nanomagnetic and spintronic devices for energy-efficient memory and computing. Wiley, p 221
Tura V, Mitoseriu L, Papusoi C et al (1998) Ferroelectric to paraelectric phase transition in barium titanate ceramic investigated by pyrocharge measurements. J Electroceram 2(3):163–169
Valasek J (1921) Piezoelectric and allied phenomena in Rochelle. Phys Rev 17:475–481
Vargas-Ortiz RA, Espinoza-Beltran FJ, Munoz-Saldana J (2012) $Ba_{1-X}Sr_XTiO_3$ ceramics synthesized by an alternative solid-state reaction route. In: Sintering of ceramics—new emerging techniques, pp 437–466
Veljovic DJ, Jokic B, Petrovic R et al (2009) Processing of dense nanostructured HAP ceramics by sintering and hot pressing. Ceram Int 35:1407–1413
Vijatović Petrović MM, Bobić JD, Ramoška T et al (2011) Electrical properties of lanthanum doped barium titanate ceramics. Mater Charact 62:1000–1006
Vijaya MS (2012) Piezoelectric materials and devices: applications in engineering and medical science. CRC Press, London
Vijayan S, Varma H (2002) Microwave sintering of nanosized hydroxyapatite powder compacts. Mater Lett 56:827–831
Vives AA (ed) (2008) Piezoelectric transducers and applications. Springer, Berlin Heidelberg
Vocca H, Cottone F (2014) Kinetic energy harvesting. In: Fagas G et al (eds) ICT—energy—concepts towards zero—power information and communication technology. InTech doi:10.5772/57091
Wang ZL (2000) Transmission electron microscopy of shape-controlled nanocrystals and their assemblies. J Phys Chem B 104:1153–1175

Wang ZL (2003) New developments in transmission electron microscopy for nanotechnology. Adv Mater 15:1497–1514
Wang X (2012) Piezoelectric nanogenerators—harvesting ambient mechanical energy at the nanometer scale. Nano Energy 1:13–24
Wang LZ, Song HJ (2006) Piezoelectric nanogenerators based on zincoxide nanowire arrays. Science 312:242–246
Wang X, Shi J (2012) Piezoelectric nanomaterials for biomedical applications. In: Ciofani G, Menciassi A (eds) Piezoelectric nanogenerators for self-powered nanodevices. Nanomed Nanotoxicology, Springer, Berlin, pp 135–172
Wang LZ, Wu W (2012) Nanotechnology-enabled energy harvesting for self-powered micro-/ nanosystems. Angew Chem Int Ed 51:2–24
Wang XH, Deng XY, Bai HL et al (2006) Two step sintering of ceramics with constante grain size II: BaTiO3 and Ni-Cu-Zn ferite. J Am Ceram Soc 89:438–443
Wang ZL, Poncharal P, de Heer WA (2000) Proceedings of the carbon nanotubes by in situ TEM. Microsc Microanal 6:224–230
Wang X, Zhuang J, Peng Q et al (2005) A general strategy for nanocrystal synthesis. Nature 437:121–124
Xiao CJ, Zhang WW, Chi ZH et al (2007) Ferroelectric $BaTiO_3$ nanoceramics prepared by a three-step high-pressure sintering method. Phys Status Solidi A 204(3):874–880
Zhang JZ, Wang ZL, Liu J et al (2003) Self-assembled nanostructures. In: Nanoscale science and technology. Kluwer Academic/Plenum Publishers, New York, p 316
Zhang R, Li JF, Viehland D (2004) Effect of aliovalent substituents on the ferroelectric properties of modified barium titanate ceramics: relaxor ferroelectric behavior. J Am Ceram Soc 87: 864–870
Zhong WL, Wang YG, Zhang PL et al (1994) Phenomenological study of the size effect on phase transition in ferroelectric particles. Phys Rev B 50:698
Zivkovic L, Paunovic V, Miljkovic M, Ristic MM (2006a) Microstructure evolution and dielectric properties of Nb/Mn and Dy/Mn doped barium titanate ceramics, in recent developments in advanced materials and processes. Mater Sci Forum 518:229–234
Zivković L, Paunović V, Stamenkov N et al (2006b) The effect of secondary abnormal grain growth on the dielectric properties of La/Mn co-doped $BaTiO_3$ ceramics. Sci Sinter 38(3): 273–281
Zubavichus YV, Slovokhotov YL, Nazeeruddin MK et al (2002) Structural characterization of solar cell prototypes based on nanocrystalline TiO_2 anatase sensitized with Ru complexes. X-ray diffraction, XPS, and XAFS spectroscopy study. Chem Mater 14:3556–3583

Friction at Nanoscale—Self-assembled Monolayers

Jelena Manojlović

Abstract In many technical fields a contact between two surfaces is very important and often the subject of research. The numerous physical phenomena that occur at the contact between two materials indicate the complexity of the processes that take place at the macro, micro or nanoscale. Therefore, friction, lubrication and wear are the subjects that have been attracting attention for many years, especially as part of tribological investigations. The research has shown that these three components are of fundamental importance for surfaces in contact. The aim of this chapter is to primarily describe friction as a tribological component and lubrication as a process to control friction, at scales of various lengths, especially at the atomic level. At the atomic and molecular scale there are materials with the property to spontaneously assemble themselves into ordered structures and many surface properties are influenced by the formation of such a film. One of the procedures to make these ultrathin organic films of controlled thickness is to prepare self-assembled monolayers. These monolayers are described as a model system to study boundary lubrication.

Keywords Friction · Lubrication · Tribology · Nanotribology · Boundary lubrication

1 Introduction

When two surfaces move relative to each other, many phenomena occur and they can be described and discussed. The total resistance of the surface of two bodies in their relative motion is defined as friction. These interactions of two solid surfaces may result in slight or significant removals of material from the surfaces, the process known as "wear". Usually, there is a need to minimize wear by using

J. Manojlović (✉)
Faculty of Mechanical Engineering, University of Niš, A. Medvedeva 14, 18000 Niš, Serbia
e-mail: jmanojlo@gmail.com

D. Brabazon et al. (eds.), *Commercialization of Nanotechnologies–A Case Study Approach*, DOI 10.1007/978-3-319-56979-6_7

lubricants. This method modifies surface properties, resulting in wear minimization (Bhushan 1994; Singer 1994; Hsu et al. 2014). The science that investigates friction, wear, and lubrication is known as tribology. The source of the word tribology is the Greek word "tribos" which means "rubbing". The study of friction, wear, and lubrication is of great practical importance for many systems, for example mechanical systems, because the work of those systems often depends on the values of parameters such as friction. Therefore, tribology is a branch of many engineering areas, mechanical engineering or materials science, and many other sciences such as biology (Gebeshuber et al. 2008), or medicine (Ramanpreet and Sreedhar 2014).

The interaction between two moving surfaces and the related processes at the moving interface are the research subjects of the science of tribology. The nature of interactions of the surfaces in contact impacts friction or lubrication. In order to understand tribological processes and to solve technical problems related to them, it is important to clarify these interactions with special emphasis on physical and chemical actions at the interface. The objective of this paper is to primarily describe friction as a tribological component and lubrication as a process to control friction at the atomic level. For this purpose, it is important to explain these concepts and indicate their presence in many technical processes (Spikes 2001).

In the last several decades, the process of miniaturization has been increasingly present in technology (Singh and Suh 2013). Thus, it has been possible to manufacture very small machines such as microscopic devices micro-electro-mechanical systems (MEMS) and, as the next step of the miniaturization process, nano-electro-mechanical systems (NEMS). For such small devices, primarily those with moving parts and a very small distance between them, it is highly important to investigate their behavior, including friction and lubrication at the atomic scale. These are just some of the reasons why today more attention is directed towards nanotechnology (Hsu 2004a, b; Crone 2008).

The study of phenomena, structures and materials on the scale of a nanometer is usually described as nanoscience. Accordingly, the subject of interest in nanotechnology is the study of very small parts (about 1–100 nm) and the application of nanostructures in small devices (nanoscale devices) (Ratner and Ratner 2006). Having in mind the size of things that we know, for example, the thickness of a sheet of paper is about 100.000 nm, a human hair is about 50.000–100.000 nm wide, DNA is just 2 nm wide, it is obvious why nanoscience is often defined as "the science of small" (Williams 2007). Since the properties of materials depend on the atomic and molecular species, their size and their arrangement, nanoscience needs to be involved in scientific research in engineering, materials science, chemistry, biology, physics and many other fields. Nanoscience and nanotechnology also include the ability to see and to control or manipulate individual atoms and molecules.

Friction is perhaps one of the oldest studied scientific subjects, but the fundamental comprehension of friction remains incomplete (Zhang 2012). In the last few decades it has been possible to study the nature of friction forces at the atomic level, with the help of new devices and methods (Hsu et al. 2014). The rapid developments in this fundamental research have led to the emergence of a new scientific field described as nanotribology, the term first used by Krim in a publication 25 years ago (Krim 1991). The object of study of this science is exploring adhesion, friction, wear and lubrication and the phenomena related to them at the nanoscale (Dedkov 2000; Manojlovic 2010; Zhang 2012). What is the source of friction, does friction appear due to adhesion or not, and many other questions are still open for scientists (Quing et al. 2007; Gawali and Kumawat 2011; Mathew 2008).

As a research subject, friction has been examined and described by many authors, usually considered as an engineering problem and rarely as a problem of physics (Krilov and Frenken 2014). Since the contact interface is an interaction of molecules and atoms of both contact surfaces, the study of physical and chemical actions between them would allow that. Thus, it is expected that scientists will be able to better understand these phenomena, but also more easily resolve new tribological problems by applying analytical methods or by developing more successful empirical approaches. The task of this chapter is to elucidate certain important tribological facts, discovered by studying the phenomena and mechanisms in the processes of friction or adhesion at the scale of atoms and molecules (Adams et al. 2001).

2 Fundamental Aspects of Friction and Lubrication

We acquire the principles of tribology early in our childhood, without even realizing that they have anything to do with it, because friction is needed in many of our everyday activities. We could not walk without friction between our shoes and the ground, the friction is present when we write with a pencil, and without friction a car would not start moving etc. These are just several of the many cases when friction is helpful (Zimmerli et al. 2005).

In contrast to the above examples, there are cases when friction is not desirable, primarily in various branches of engineering. The reason for that lies in the fact that the force of friction generally causes energy losses (Taylor 2012). It is well known that the risk of overheating between two parts rubbing against each other due to friction is present and also undesirable. Overheating leads to an unwanted damage of machine parts and it should be prevented. As a matter of fact, many industries, such as mechanical, automotive, and microelectronics, are affected by certain tribological phenomena (friction and wear). Thus, their understanding is of great relevance.

According to the examination results from the US industry, for example, wear represents a great problem, especially considering the fact that it affects the lifetime of individual parts of a machine (e.g. parts in an automobile or an airplane). Losses due to wear are estimated at about \$8 billion annually (Adams et al. 2001) and losses due to friction account for around 70% of all energy, which is very significant (Zimmerli et al. 2005). A detailed study of the phenomena that hide behind the word tribology can have a positive impact on the very important issues in the industry—reducing frictional losses and damage of machines (Widmer 2002). By solving problems or finding better solutions for tribological problems, it is possible to realize large savings (Hironaka 1984).

As can be inferred from the above, one of the main points in tribological investigations is to find good technology solutions to the rubbing of two solid surfaces. The requirements usually have to do with less friction, which implies appropriate material selection for the surfaces and the corresponding lubricant between them. In accordance with that, the lubrication regime is defined as the process with the ability to control friction and wear between two surfaces in relative motion. Between two surfaces in contact, a lubricant can usually be found in the form of a lubricating oil or a chemically formed layer from one or both of the contacting bodies. This research is a very complex task because of many parameters involved in such an examination (Matsumoto 2003).

The study of lubrication has thus far been mainly empirical in nature, and the mechanism of their functioning is still insufficiently known (Hsu 2004a, b). In this century, the great technical discoveries spurred the development of modern complex lubricants made of oils and chemical additives (Somers et al. 2013). New methods in the study of nanotechnology have considerably facilitated the examination and obtaining of new knowledge and better understanding of lubrication systems.

3 Basics of Friction at the Macro and Micro Scale

Friction plays an important role in everyday life, technology, medicine and many other fields, and it has been investigated since ancient times. It has been determined that there are many complex phenomena at the interface of two contacting bodies. However, the influence and contributions of adhesion, roughness and asperities, elastic and plastic deformation and the interatomic forces on friction are still not entirely clear (Widmer 2002). Therefore, tribology is everywhere around us, in simple and complex applications and in scales from large to small.

Modern tribology is divided into three fields according to the scale of the friction study. Macrotribology (tribology) deals with large objects and loads, which is opposite to nanotribology that studies friction at the scale of individual atoms and their interactions. An intermediate field between these two is microtribology (Broitman 2014). In order to reduce frictional forces between contacting objects, which is most frequently required to be achieved, the topics of interest have become both macrotribology and nanotribology, because they deal with origins of friction and wear, and they study the role of lubricants (Widmer 2002).

3.1 Friction at the Macro and Micro Scale

Amontons' laws of friction from the 17th century indicate that the history of research in this area is very long (Mattias 2008). The basic knowledge about friction usually comes from the observations of many scientists, such as Leonardo da Vinci, Guillaume Amontons, Leonard Euler und Charles Coulomb (Zimmerli et al. 2005; Popova and Popov 2015; Hyun-Joon and Dae-Eun 2009).

The friction force F_f which appears between two macroscopic bodies, made of different materials, when they touch as they move against each other, is usually described by the equation known as the 'law of friction' or Amontons' law, $F_f = \mu F_N$, where F_N denotes the external normal force and μ the friction coefficient (depends on the type of material of which the surfaces in contact are made) (Broitman 2014). The force of friction F_f is always in the direction that opposes movement or potential movement between two surfaces. Equation (Amontons' law) provides a few important pieces of information about friction: the friction force F_f is proportional to the normal force F_N, and friction does not depend on two parameters—the surface area in contact and the sliding velocity (Krilov and Frenken 2014). When we consider the relative motion of two bodies, due to their different physical origins, we can distinguish between static and dynamic friction, very often described in detail in the literature (Charitidis et al. 2013).

Macrotribology investigates friction, wear and lubrication at the macroscopic scale, along with the reasons for possible deformations at the contact surfaces (Broitman 2014). Measurements show the difficulties which one faces trying to predict how friction changes under various experimental conditions. Many researchers have therefore pointed to the possibility of defining friction at the macro, micro and nanoscale (Broitman 2014). The interesting fact is that the coefficient of friction in these cases can be different at the macro, micro and nanoscale, which of course indicates the complexity of friction analysis as a tribological phenomenon (Riedoa and Brune 2003). Despite these findings, the above factors are not included in the simple empirical relation defined as the Amontons' law of friction, and this equation is still used in the design and manufacture of machines and devices.

Significant improvement in the tribological research occurred in the middle of the 20th century when Bowden and Tabor introduced the physical understanding of the laws of friction at the atomic scale. Since they considered the surfaces in a tribological contact as the objects made of molecules and atoms, it was pointed out that different friction strongly depends on the physical and chemical properties of molecules and atoms on both surfaces. Therefore, there is a need to conduct research at the atomic scale and use the findings to clarify friction at the macroscopic level.

The first simplification in the macroscopic friction described above was that the surfaces in contact are clean and flat, and very smooth at the macro level. But if we look through a microscope, these surfaces are rough, with many asperities, Fig. 1a. The real contact takes place only at certain points as shown in Fig. 1b, indicating the fact that surface roughness limits the contact between surfaces (Yang 2008).

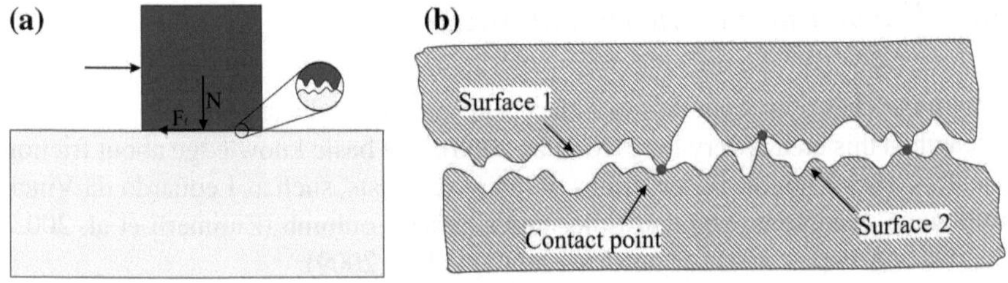

Fig. 1 Contact between surfaces **a** and asperities **b**

In order to determine friction at the micro or molecular level, it is important to determine the nature of the contact between surfaces (Broitman 2014; Blau 2008). Surface roughness plays an important role in the contact between surfaces. It influences the dissipation of frictional energy and is considered very often as the "fundamental" in tribology (Krim 2012). Microtribology studies such individual contacts (Fig. 1b).

The understanding of friction is important for scales of different length. There are two reasons why there is a need to study the microscopic origin of friction. One is due to the development of nanodevices and the second one is that the findings can be often helpful to understand the complex friction process at a higher scale. In microtribological research it is important to define factors that affect friction. For example, it has been observed by the experiments that the adsorbed films of few nm in thickness on many surfaces can influence friction and wear at this scale (Bushan and Kulkarni 1996).

The investigation of the microscopic origins of friction is not a new scientific field. If the contact is examined at the microscopic level, when we bring two surfaces into contact, the real contact will be between asperities (Fig. 1b), and this contact is much smaller than the apparent one (Zimmerli et al. 2005). This means that at the microscale the contact between asperities and their deformation is dominant, with a significant influence on the friction force (Adams et al. 2001; Braun and Naumovets 2006). The influence of the real contact area on friction due to micro/nano scale asperities was introduced by Bowden and Tabor in 1939 (Bowden and Tabor 1938; Shivaprakash 2013).

The relation between frictional force and loading force in microtribology is more complex than on the macroscale. There is a significant difference in the working conditions in the tribological systems at the macro scale (relatively large contact stresses and speeds) and the micro/nano scale (low loads and speeds), and the dominant mechanisms of friction in macro-scale systems may be different from those in micro/nano-scale systems. It has been observed that the tribological interactions at small scales are often influenced by surface forces. Numerous phenomena such as adhesion, surface tension, and stiction are critical to micro/nano-scale systems (Sung et al. 2003). However, the study and the analysis of the mechanism of friction and wear at the nano scale is a very complex task.

4 Tribology at the Nanoscale

When the friction coefficient and wear for the materials of interest have to be defined, we usually talk about macroscopic tribology (Hu and Granick 1998). Nevertheless, the conclusions reached regarding the basic macroscopic friction cannot be accepted as a valid general theory (Holscher et al. 2008). Recent studies and results in the area of microscopic friction imply a significant difference between friction at the atomic scale and macroscopic friction, and this can be helpful in considering the origin of friction at the molecular level (Holscher et al. 2008). When the contact between two surfaces is very small, compared to a macro contact, many experiments have shown that the empirical laws of friction do not always hold (Gnecco and Meyer 2007). These are just some of the reasons why studying friction at the micro or nanoscale is still necessary (Szlufarska et al. 2008).

The study of friction, adhesion, lubrication and wear at contacts measured in nanometers is known as nanotribology (Szlufarska et al. 2008). The aim of this discipline is the characterization and the modification of the surface, for both scientific and technological needs. Nanotribology is important for scientists in many fields such as engineering or natural sciences. Studying the phenomenon when two surfaces are in a sliding contact can help in uncovering the causes of fundamental phenomena responsible for friction or energy dissipation. Due to the minimization of technical devices (MEMS and NEMS), surface phenomena and surface forces have become important and they can provide or prevent the operation of a nanodevice (Szlufarska et al. 2008; Manojlovic et al. 2015).

Friction and occurrences connected with it, such as lubrication and wear, are the topics which have implications at the nanoscale. The reason why we should deal with nanotechnology can be described as a need to better understand the basis of friction at the nanoscale. Indeed, even though tribology has been known for many centuries, as described above, it is still complex at any scale and even for many macroscopic problems, which indicates that tribology appears more complex the more we learn about it (Urbakh et al. 2004). The challenging task in nanotribology research is to define and better understand three important subjects (adhesion, lubrication and wear) connected to this problem, as described in a review paper by Adams et al. (2001).

Nanotribology deals with the interaction between individual atoms. In order to understand the phenomena at the atomic level, it is essential to introduce several concepts at the nanoscale, such as the material properties at the nanoscale, the research techniques applicable at such a small scale, and many others.

4.1 Nanomaterials

Numerous experimental results show that "conventional" bulk materials may have significantly different properties from nanosize materials (Lue 2007). It has been observed that some materials can have properties at the nanoscale different than they do at larger sizes (become chemically active, conduct electricity better, have different magnetic properties) (Filipponi and Sutherland 2010). For example, bulk gold (Au) has a melting point of 1064 °C, but melting point of 5 nm size gold nanoparticles is reduced to around 600 °C; the color of gold particles depends on their size etc. (Schmid and Corain 2003).

Surface testing indicates that properties of a surface are determined by the chemical groups found on it. This becomes especially important for nanomaterials, since a big part of the total sample molecules of a very small particle is on its surface. This has a big effect on reactions that occur at the surface and at the contact between two moving surfaces. These descriptions related to the interactions at the nanoscale can indicate a variability of properties at the atomic and molecular level (Filipponi and Sutherland 2010).

Nanotechnology has pointed out that many materials can spontaneously assemble themselves into ordered structures at the nanoscale, affecting the specific material properties. The big challenge in nanotechnology is to arrange molecules in a controlled way, with the aim of creating new substances with new, very special properties. This is one of the reasons why nanotechnology can exert influence on manufacturing new kinds of materials, which makes nanotechnology a promising field for many technical applications (Theodore 2006).

The behavior at the nanoscale can be explained by quantum mechanics, developed to describe the motion and energy of atoms and electrons. A brief overview of the most important quantum effects, such as tunneling, quantization of energy, increased surface-to-volume ratio, have been described in the literature (Filipponi and Sutherland 2010), and they are essential to understanding the exceptional properties of nanomaterials and phenomena at such a small scale.

4.2 *Experimental Techniques and Methods Applied in Nanotribology*

In order to understand friction between two moving surfaces and how two materials in contact interact, it is very important to know the properties of the surfaces and perform physical and chemical analysis of the surfaces in contact. On the other hand, nanodevices are the products very sensitive to nanoparticles contamination in the working process, which is an additional reason for detailed surface analysis. Numerous surface analytical techniques have been developed to analyze the studied surfaces, and they provide the tools to explore the surface and subsurface chemistry of solid materials, or to characterize new materials (Mowat et al. 2007).

The modern spectroscopic methods of surface analysis were first applied several decades ago (around 1960). They are able to provide a lot of information about surfaces such as elemental or quantitative analysis, and by these techniques it is possible to investigate specific compounds present on a surface.

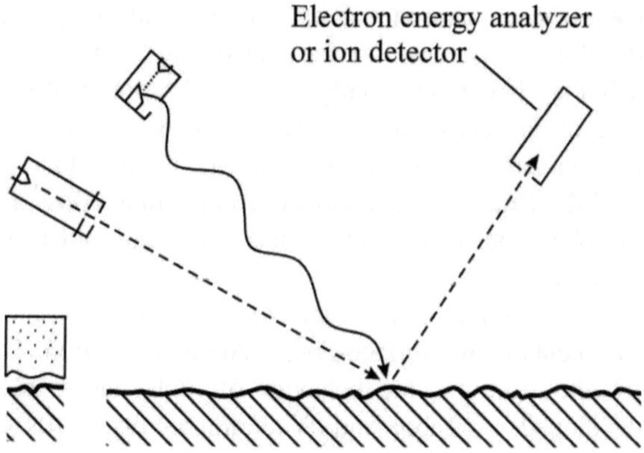

Fig. 2 "Beam in, beam out" mechanism

The spectroscopic techniques are based on a "beam in, beam out" mechanism. The sample is exposed to a beam of electrons, ions or photons of certain energy, which penetrates to the depth that is dependent on the beam characteristics (Fig. 2). A second beam, created as a result of the interactions between the first beam and solid, exits from the surface and is analyzed by a spectrometer. The energy of the ejected and scattered species has been recorded so as to get a picture of the surface under research.

The fact that it is necessary to point out concerning surface analysis experiments is that they are performed under vacuum conditions. One of the important reasons for that is to keep the studied surfaces clean and without contamination during the experiment, which will provide more accurate information on the studied sample. The change of energy of the entering beam and the type of the particles in the entering beam, resulted in the formation of many of surface analytical techniques, such as X-ray photoelectron spectroscopy (XPS, or electron spectroscopy for chemical analysis—ESCA), secondary ion mass spectroscopy (SIMS), Auger electron spectroscopy (AES), and many others, described in details in the literature (Matsumoto 2003). The X-ray Photoelectron Spectroscopy is a very often used technique to investigate the chemical composition of surfaces. It is important to underline that not all surface analytical techniques are suitable for all applications. Usually, it is necessary to apply more than one technique to provide sufficient information on the surface of the analyzed sample.

For friction measurements it is very important to have information about both surfaces in contact, their structure, chemistry and physics, and special attention is needed if we have to measure friction at the nanoscale. For that aim, and to understand the molecular surface structure, we need appropriate equipment and particular procedures. Depending on the engineering field and the size of system components, micro contacts can be the smallest level of interest. Due to the size of surface asperities, a contacting body on the interface can be a micrometer or less in dimension (the size of individual atoms, corresponding to the size of a nanometer). In order to study tribological phenomena at the very small scales, there is a need to have appropriate technical equipment. Numerous nanotribological studies have usually been carried out by experimental analysis and computer simulations of tribological systems. At the end of the 20th century a new important period in tribology began, especially with the development of new experimental techniques such as the surface-force apparatus (SFA), the atomic-force microscope (AFM), the friction-force microscope (FFM), and the quartz-crystal microbalance (QCM) (Braun and Naumovets 2006; Braun 2005; Yang et al. 2014). The study of the phenomena in tribology has been achieved at the microscopic and even the nanoscale.

One example of the very complex and very precise instrument for force measurements is the Surface Forces Apparatus (SFA). This instrument is very popular for the study of the behavior of molecularly thin films. The instrument was developed by Tabor and Winterton in the 1970s (Zhang 2012). Thanks to the devices such as the atomic force microscope (AFM) or the scanning force microscope (SFM) (Bhushan et al. 1994; Rymuza 2010) and many others, in the last few decades it has been possible to carry out friction measurements and to make single-asperity contact measurements. For tribological tests these experimental techniques allow a better definition of surfaces that come into contact, as well as a better understanding of the consequences of their contacts.

In addition to techniques applied in the direct measurement in tribological process, there are other research options. Due to the progress in computer science, researchers have been able to perform simulations of friction processes at the atomic scale. Tribological systems can be investigated by simulation with molecular dynamics (MD) (Braun and Naumovets 2006). Molecular dynamics (MD) is defined as a computer simulation method applied to study the physical movements of atoms and molecules that govern behaviors of physical systems (Allen 2004; Xuan 2014). Atoms and molecules are allowed to interact, change the arrangement and a simulation can give a picture of the way in which a molecular system changes from one configuration to another. Today, atomistic MD simulations are very accurate and helpful to understand and predict the system behavior in such cases as liquids under nano confinement, friction between surfaces with asperities, etc. (Gnecco and Meyer 2007). A significant progress in nanotribology has been achieved by computer simulations (Szlufarska et al. 2008).

4.3 Nanometer Scale Friction

For technological and fundamental research the understanding of friction at the nanoscale is very important. The findings about friction at the nanoscale are also relevant for the manipulation of small devices, such as MEMS and NEMS. With the development of new experimental techniques described above, experimental and theoretical studies of friction at the atomic scale have received additional attention (Berman and Krim 2013). Actually, the nanometer scale tribology implies the study of many subjects, such as adhesion, friction and wear at the molecular level (Weymouth et al. 2013).

The history of atomic friction research is several decades long. Some authors believe that the first atomic friction measurements (Mate et al. 1987) were the beginning of nanotribology. Almost a decade ago, the European Science Foundation Program "Nanotribology" (NATRIBO) was formed with the aim of connecting researchers in theory and practical investigation of nanotechnology to better explain surface contact at the nanoscale. The connecting knowledge and findings can give a deeper insight into the complex phenomena of nanotribology (Gnecco and Meyer 2007). Since nanotechnology is becoming predominant in both the industry and everyday life (cosmetics, computing and healthcare), it is understandable why these studies deserve more attention.

The results of research in nanotribology indicate that this force has the origin in many unexpected sources (Krim 1996). The tasks that nanotribologists have at the nanoscale are to understand the origin of friction at the micro and nanoscale and find the key underlying friction (Mo et al. 2009). Achieving these objectives would contribute to controlling surfaces more easily, which is important for very small machines that are already being produced (nanomachines).

Another very important cause of the resistive force of friction is molecular adhesion. Adhesion is defined as the molecular force obtained when two bodies are brought into contact, and these forces between molecules have to be broken if objects are trying to slide against each other. In macroscopic experiments, for example, in several studies we can find the data that larger adhesion leads to higher friction. However, there are cases where high friction is associated with low adhesion forces (Zhang et al. 2007). This indicates that the relationship between friction and adhesion is probably much more complex. The influence of adhesion on friction was first pointed out several centuries ago by John Theophilus Desaguliers (at the end of the 17th and the first half of the 18th century) (Shivaprakash 2013). At the nanoscale friction is significantly adhesion dependent (Egberts and Carpick 2013), which represents an example where Amontons' Law cannot predict the friction force. At the macroscopic scale such observations are impossible to record, which points to the necessity for detailed consideration to better understand the process of friction at smaller scales. From the physical point of view, and according to the results obtained by the molecular dynamics method, a review of the mechanism of friction was provided several years ago by Braun (2005). The results showed great progress in the field of friction research, pointing to the still major challenges in the field of lubrication at the molecular level.

The first nanotribological experimental results at the atomic level were obtained by using the surface force apparatus (Israelachvili and Tabor 1973). They were very useful in the development and manufacture of micro-and nanomachines (Bhushan et al. 1994; Krim 2002a, b).

The great progress in computer science has allowed the simulation of real tribosystems using the molecular dynamics method (Braun 2005). The friction force is very often calculated after the experimental measurements, or can be defined by molecular dynamics simulations (Beer et al. 2015). This way of studying is helpful for scientists to "see what is happening" at the atomic and molecular level at the interface. A very interesting review of friction at the atomic scale, especially from the standpoint of energy losses, can be found in the literature (Singer 1994). By applying molecular dynamics simulations, scientists are able to model friction at the atomic level and explain what takes place during the sliding contact. One simple model is usually sufficient not only to describe a phenomenon or process of friction, but even to predict the system behavior.

In the tribological research there are two very often used friction models, the Tomlinson model and the Frenkel-Kantorova model, but there are also many other friction models described in the literature (Braun 2005; Krim 2002a, b; Mattias 2008; Vanossi et al. 2013; Braun and Kivshar 2004). Many years ago (1929) Tomlinson reported an atomic friction model and in the literature this is usually identified as the first explanation of atomic-scale friction. This model was very much different compared with the friction description at that time (Wang et al. 2015; Vanossi et al. 2013). This model of friction starts from the assumption that a surface is without asperities, but due to the atoms at the surface, modeled as spheres, the surface is not perfectly flat. Frictional behavior is described by interference between atoms located on the two surfaces in contact as they slide against each other. This model can provide only a qualitative description of the problem, and it is very often used in the interpretation of tribological experiments. The first model is usually applied to tip motion in tip-based devices and not for contact between two macroscopic solid surfaces (Kim and Kim 2009; Biswas et al. 2015; Braun and Naumovets 2006).

Another theoretical model for atomic-scale friction widely used in tribology is the Frenkel-Kontorova (FK) model (1938). This model has been applied to describe dislocations in solids, to describe adsorbed films on a surface or for a contact of two solid surfaces (to describe the case of "dry" friction) (Kim and Kim 2009; Biswas et al. 2015; Braun and Naumovets 2006).

It is clear that the friction force cannot be said to depend on a defined number of parameters such as normal load, surface roughness, or material type of the tribological system. Tribological phenomena, especially nanotribological, can be significantly influenced by many other factors—temperature, humidity, which are very often neglected (Hyun and Dae-Eun 2009). The numerous results that scientists have obtained show the importance of a well-controlled specimen, and how experimental conditions should guarantee the reliability of the results at the nanoscale. However, to control friction is still a challenge because many fundamental facts about friction at the atomic level are not yet understood (Gnecco and Meyer 2007).

Experimental results and molecular dynamics have together provided some answers, but also opened up some new questions on the topic of friction and energy losses at the sliding interface (Bhushan 1994).

5 Lubrication

A very important process in tribology, usually inseparable from friction, is lubrication. When friction appears, lubricants play an important role, because they can decrease it. The role of a lubricant is to reduce friction and to prevent/minimize wear. This is carried out by elimination of contact between two solid moving surfaces. In the industry, lubricant application can prevent corrosion or can lower the operating temperature of some objects.

The distance between moving parts or the thickness of lubrication films can be very much different in practice in comparison with experimental measurements. The distance between moving surfaces can be significant or measure only a few molecular diameters. In some tribological systems, the thickness of one molecular layer can be sufficient to ensure moving the surfaces without contact and wear.

It is common to distinguish between a few types of lubrication.

Hydrodynamic lubrication or full film lubrication. In hydrodynamic lubrication we have a combination of both liquid and relative motion to reduce friction (hydro means liquid and dynamic means relative motion). When the operating conditions and the contact geometry are such that the surfaces are completely separated by the lubricated film, this can be defined as the hydrodynamic lubrication. The fluid film is thicker than the asperity heights on the surfaces.

Elastohydrodynamic lubrication: In this regime the load is high enough, and the elastic deformation of the surfaces is possible, such as an increase in the viscosity of the lubricant. Solid surfaces can be deformed due to high film pressure, but direct contact is still prevented by the film thickness of the order of 1 μm.

Partial or mixed lubrication: The regime between the full film elastohydrodynamic and boundary lubrication regimes is described as mixed lubrication, in which a lubricant film is not thick enough and contact between asperities can occur. Film thickness is reduced compared with the previous regimes and has the value of the size of surface asperity (of the order of 0.1 μm), and discontinuous contact is obtained.

Boundary lubrication. Boundary lubrication is lubrication by a liquid under conditions where the distance between solid surfaces is very small and the contact between asperities on surfaces occur. The sliding surfaces are separated only by adsorbed molecular films. Usually boundary lubrication occurs under specific conditions, such as high load and low speed, the film thickness is commonly a single adsorbed layer and intermittent solid contact may take place. Brief history of boundary lubrication is reported by Spickes (1996).

Friction behavior under the defined regimes can be described by the Stribeck curve (Fig. 3). The curve shows that the value of the friction coefficient is the smallest for the thinnest hydrodynamic film, increasing rapidly for thinner films. When the film of lubricant gets thicker, due to the viscosity effect, the friction coefficient becomes higher. The Stribeck curve has been very often described in more detail in the literature (e.g. Hahner and Spencer 1998).

5.1 Nanoscale Boundary Lubrication

The development of nanotechnology, new instrumentation and techniques designed for its needs, offer great opportunity to analyze the chemistry of the surfaces and

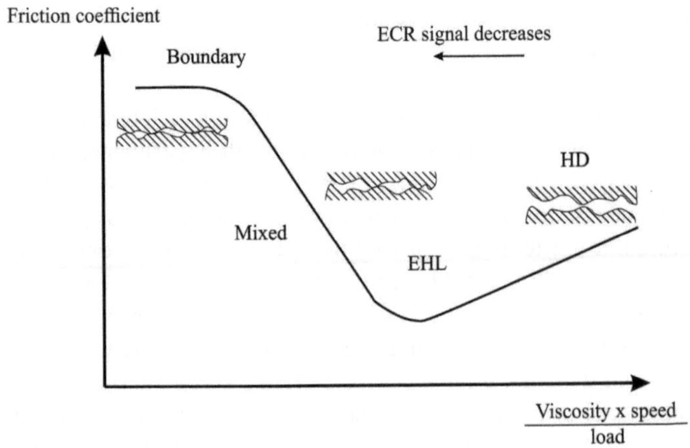

Fig. 3 Stribeck curve

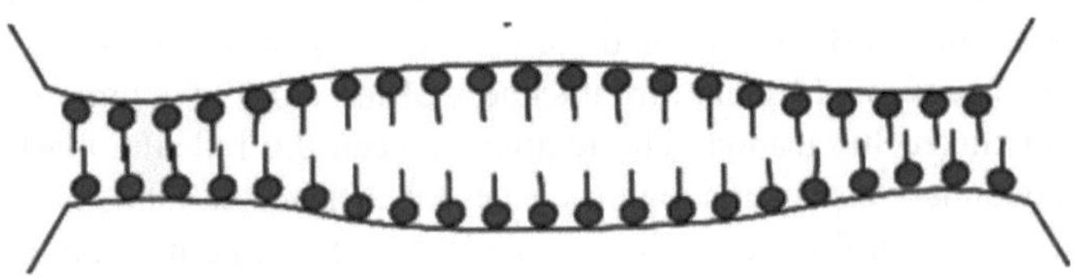

Fig. 4 Boundary lubrication (Manojlovic 2006)

phenomena and mechanisms in the boundary lubrication regime. The results provide the possibility for better understanding of even complex lubrication systems. Because surface forces become important in providing or preventing the operation of a nano-device, the behavior of lubricants at the nanoscale is significantly different than at the micro and macro scale (Krim 2002a, b; Hsu 2004a, b).

If the distance between contact surfaces is a few nanometers, which amounts to a few molecular layers, or where the working conditions do not allow the formation of a hydrodynamic film, we can define boundary lubrication (Fig. 4).

Between two surfaces in the boundary lubrication regime the primary interaction is between asperities that are touching each other, producing friction, which can sometimes lead to elastic or plastic deformation with wear. The thickness of the lubricant layer between two surfaces can be as small as a single molecule, indicating that even a single molecular layer of certain substances can have significant effects on a tribological system (Gnecco and Meyer 2007).

When liquids of nm thickness are confined between two atomically smooth solid surfaces, many experiments show that a certain new structural and dynamic phenomenon occurs. These observations, together with numerical and analytical results, have been important for tribological research. Indeed, many phenomena are still unexplained (Gnecco and Meyer 2007). Some of the experimental techniques applied in friction measurements can be applied in lubrication research too, such as the SFA. Developed more than 40 years ago, the SFA is usually applied to study properties of molecularly thin films, confined between two molecularly smooth macroscopic surfaces, with surface separations at the angstrom level (Bhushan et al. 1994). Using the SFA experiments surface forces can be measured.

A number of questions in tribology are connected to the boundary phenomena, while the answers are directly related to the physical and chemical properties of the surface. For the lubricant film, certain properties of adsorbed films have already been well described, and this knowledge can be used to get more details about the occurrences in friction contacts (Jun and Yonggang 2015).

In recent years there have been extensive studies of the behavior of liquids at an interface between two solid surfaces with the aim of understanding how boundary lubrication might occur. Some of the results (experimental or simulations) have led to a physical mechanism that deals with how such a thin layer of liquid is able to reduce friction, considering the surface forces that are normally dominant at the corresponding scale, micro or nano (Sokoloff 2003; Braun 2005). However, closer examination indicates that some films are protective and some are prowear. Despite all this, a complete theory of boundary lubrication is still missing. Even though there are many models which describe reactions at the contact between two surfaces at the atomic level, many physical and chemical processes at this level are still an enigma to scientists (Bhushan 2001).

Special attention in scientific research is paid to the chemical reaction between lubricant molecules and surfaces because of its influence on the friction force. This reaction is caused by the chemical nature of both components, the surfaces and the lubricant. The results demonstrate that only a clearer picture of these phenomena—fundamental, atomic-scale study of the critical processes, can be helpful and positively influence technical applications.

Boundary films have been studied for many decades where special attention has been paid to the dynamics of lubricants. The molecular behavior of lubricants in boundary lubrication and their dynamics during sliding become more complex due to a change in some of the experimental parameters, such as temperature. Computer simulation can be very helpful in describing the processes during sliding when multitude of atoms are involved in this regime. Those results indicate that the atomic processes cannot be neglected in a nanotribological experiment (Bhushan 1994; Bhushan 2001).

In the research at the molecular level, several available methods can be applied (Miyake et al. 2012). They are used to describe several topics under boundary regimes, such as atomic interactions between surfaces, lubricants and their reactions, formation of boundary lubricating film, and application of the appropriate model to predict boundary lubrication process (Bhushan 2001). However, there are still some uncertainties when it comes to physics and chemistry of contact between surfaces.

A molecular basis of how lubricants and additives operate in lubrication can provide results that can change the current lubricating technology (Hsu 2004a, b). Scientific research pays special attention to monomolecular thin lubrication films. Some of the films are only studied as models, but some practical solutions in this area can greatly contribute to the knowledge of lubrication (Hsu 2004a, b).

5.2 *Boundary Lubrication by Self-assembled Monolayers*

Numerous organic compounds have the capability to form thin organic films on solid surfaces. Those films can demonstrate special chemical, mechanical, optical or electrical properties. Thin organic films are an emerging area of materials chemistry and are applied in many areas. One of the procedures used to make ultrathin organic films of controlled thickness is the preparation of self-assembled monolayers (SAMs). Due to the possibility to use these films for surface modification and manufacturing molecular layers with the particular properties with numerous potential applications, the attention in SAM formation and structure has been increasing recently (Beaumont et al. 2011).

Depending on using different types of molecules and different substrates, numerous self-assembled monolayers can be prepared. The examples of SAM model systems very often studied and described in the literature are: adsorbed thiol molecules onto gold (Carpick and Salmeron 1997; Abe et al. 2000), silanes on an oxide surfaces (Carpick and Salmeron 1997), or alkane phosphate monolayers (Brovelli et al. 1999). The choice of the substrates used in the self-assembling process is dictated by the molecules and their interactions, as well as the final application, and a variety of systems has been reviewed in detail (Ulman 2013).

Organic self-assembled monolayers (SAMs) are model systems which are very attractive for boundary lubrication (Manojlovic 2013). The process of self-assembly is considered as a very important example of equilibrium in structural organization at the molecular scale, and a very simple way to make ultrathin organic films of controlled thickness. SAMs have many applications in organic chemistry (Doudevski and Schwartz 2001a, b), in controlling the surface properties in micro-electromechanical devices (Scherge and Schaefer 1998), controlling and limiting the corrosion of surfaces (Rajkumar and Sethuraman 2014) or production of nanoscale organic transistors (Kumar et al. 2014).

Special attention is paid to the possibility of producing molecular layers with particular, desired properties. Molecular self-assembly is recognized as a powerful strategy for the fabrication of nanoscale structures (Egberts and Carpick 2013; Manojlovic 2013), and for further investigations into understanding and solving friction, lubrication and related problems (Mattias 2008). For the very small, atomically structured devices, self-assembly is a promising approach. Recently, a great challenge in scientific research into lubrication in small-size systems, such as MEMS or NEMS (Liu et al. 2009), has been the study of new kind of lubricants. As a good candidate in MEMS lubrication, different types of monolayers attached to the contact surfaces have been proposed. The understanding of the behavior of those monolayer films of few nanometers in thickness is very important in nanotribological experiments. With the development of a number of powerful techniques in surface analysis, scientific interest in SAMs has increased, with the opportunities to investigate the growth and the structure of such layers at the nanometer scale (Kopta 2001).

5.2.1 Surfactant Self-assembly

When the surfactant solution is put in contact with a clean solid surface, a key factor for surfactant adsorption is the nature of the substrate. The preparation process of self-assembled monolayers happens spontaneously, when specified molecules (e.g. surfactants) are adsorbed from the solution onto a solid surface (Manojlovic 2013; Kopta 2001). Surfactant molecules have a property to adsorb onto solid surfaces, ideally forming a very ordered adsorbed layer, known as the self-assembled monolayer (Fig. 5). The hydrophilic headgroup of the surfactant molecule usually has an attractive interaction with the substrate surface. The hydrophobic tail of the molecule, which does not have a strong interaction with the surface, can be chemically tailored to exhibit specific chemical properties. Therefore, SAMs offer the capability to form ordered organic surface coatings, suitable for various applications such as wetting or corrosion protection.

The bonding of surfactant molecules to the substrate, and the interactions between molecules in the monolayer are two important forces involved in the process of self-assembly. Depending on the properties of the substrate and the surfactant, the balance between these two forces can be shifted. The monolayer, which forms as an equilibrium structure in solution, can be stable or metastable when it is removed from the solution. The stability of the formed layer is mainly determined by the strong inter-chain van der Waals interactions (Chaki and Vijayamohanan 2002).

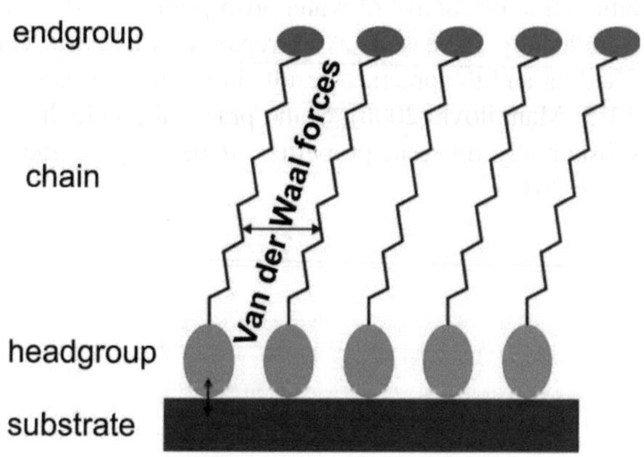

Fig. 5 An organized monolayer on a substrate (Manojlovic 2006)

5.2.2 Adsorption of Surfactants to Solid Surfaces

SAMs are formed onto solid surfaces by adsorption of organic molecules, which is in balance with desorption (Doudevski and Schwartz 2000). Molecules on a substrate can be in two different adsorption states, physisorbed or chemisorbed. Through a strong chemisorption it is possible to form a rather stable monolayer, e.g. covalently bound thiols on gold (Xue et al. 2014). Chemisorbed molecular layers have recently been proposed as good lubricants by various authors (Vakarelski et al. 2004). Alternatively, monolayers or multilayers can be formed in the physisorbed state, bound to a surface via electrostatic forces, hydrogen-bonding or van der Waals forces (Chen and Israelachvili 1992). Physisorbed monolayers are often adsorbed from solution or from the Langmuir-Blodget transfer technique. These self-assembling surfactant systems are usually used as wetting agents, dispersant or emulsifiers (Vakarelski et al. 2004).

The ability of surfactant molecules to adsorb at the solid-liquid interface depends on the properties of all components of the system such as the solid, the surfactant, and the solvent (Manojlovic 2012). The influence of the surfactant concentration is also very important in the adsorption process and it has been shown by numerous experiments that the concentration of surfactant molecules affects the structure of the adsorbed layers (Doudevski and Schwartz 2001a, b, Doudevski and Schwartz 2001a, b). A monolayer can be formed at a low surfactant concentration, while discrete aggregates can be formed at a higher concentration (Zana 2005).

Apart from the surfactant type and surface properties, numerous examinations indicate that a number of other factors can have an influence on monolayer or multilayer formation such as surfactant solution temperature (Mellott et al. 2004), or solvent quality (Clear and Nealey 2001), or environmental conditions, such as ambient temperature and humidity (Chen and Israelachvili 1992). Due to the wide variety of ambient laboratory conditions, the diversity of reported SAM morphologies is hardly surprising.

Such a monolayer film can affect certain surface properties, for example, chemical or physical hydrophobicity or electrostatics (Grosse and Estel 2000). Organic molecules have many interesting applications, such as biosensors, for lubrication or control of water adsorption. Although the examination of SAMs has significantly increased, a discrepancy still exists between the theoretical understanding and the practical results in the formation of such layers (Zhang and Meng 2015; Manojlovic 2006). Some practical results have shown that SAMs can have substantially different properties, depending on the variety of parameters (Manojlovic 2012).

6 Conclusions

The study of friction pays no special attention to certain particularly interesting phenomena. Usually, only static and kinetic friction are described, or sometimes rolling friction and friction in fluids. However, there are complex and interesting facts about the process of friction, such as suggested in 1998 by Hahner and Spencer: "Though simply expressed, the laws of friction encapsulate a host of microscopic and nanoscopic phenomena whose elucidation has become one of the most fascinating pursuits in applied physics" (Hahner and Spencer 1998). Tribology is becoming an area that attracts the attention of a great number of researchers in many fields, such as physics, chemistry, materials science, especially in theoretical studies and experimental measurements.

The clarification of the concept of friction requires the cooperation of many scientific disciplines, natural and technical (engineering) sciences, leading to an increasing progress in the field of tribology. As explained above, a contact between two surfaces is the main subject of tribology. However, the area of contact is often very inaccessible for analysis and can be highly complex.

Therefore, this research devotes more attention to the study of phenomena at the atomic level, resulting in the appearance of new scientific fields such as nanotribology. Many empirical results receive their explanations at the level of molecules and become clearer thanks to the application of physics. Bearing in mind the rapid development of science, these areas of research can soon provide new solutions to the old problems related to friction, wear and lubrication.

Scientists are trying to understand many phenomena at the nanoscale, and simultaneously develop nanoscale devices that can be controlled and manipulated at the atomic and molecular levels. Great things are expected of these small devices in the field of medicine, with the hope that nanorobots will soon be able to detect problems in the body and deliver the drug only to a diseased cell, keeping healthy cells safe from the medication. After several decades of research in nanotechnology, great potential has been observed, and we can expect that nanotechnology will begin to influence our everyday life more and more.

References

Abe K, Takiguchi H, Tamada K (2000) Dynamic contact angle measurement of Au (111)-thiol self-assembled monolayers by the Wilhelmy plate method. Langmuir 16(5):2394–2397

Adams BJ, Hector LG Jr, Siegel DJ et al (2001) Adhesion, lubrication and wear on the atomic scale. Surf Interface Anal 31:619–626

Allen M (2004) Introduction to molecular dynamics simulation. NIC Ser 23:1–28

Beaumont N, Hancox I, Sullivan P et al (2011) Increased efficiency in small molecule organic photovoltaic cells through electrode modification with self-assembled monolayers. Energy Environ Sci 4:1708–1711

Beer S, Kenmoe GD, Muser MH (2015) On the friction and adhesion hysteresis between polymer brushes attached to curved surfaces: rate and solvation effects. Friction 3(2):148–160

Berman D, Krim J (2013) Surface science, MEMS and NEMS: progress and opportunities for surface science research performed on, or by, microdevices. Prog Surf Sci 88:171–211
Bhushan B, Israelachvili JN, Landman U (1994) Nanotribology: friction, wear and lubrication at the atomic scale. Nature 374:607–616
Bhushan B, Kulkarni AV (1996) Effect of normal load on microscale friction measurements. Thin Solid Films 278:49–56
Bhushan B (2001) Modern tribology handbook. CRC Press, New York
Biswas S, Ray P, Chakrabarti BK (2015) Statistical physics of fracture, breakdown, and earthquake: effects of disorder and heterogeneity. Wiley-VCH Verlag GmbH & Co, KGaA
Blau PJ (2008) Friction science and technology: from concepts to applications. CRC Press
Bowden FP, Tabor D (1938) The area of contact between stationary and between moving surfaces. In: Proceedings of the royal society of London. Ser A, Math Phys Sci: 391–413
Braun OM, Kivshar YS (2004) The frenkel-kontorova model. Springer, Berlin
Braun O (2005) Nanotribology: mechanisms of friction on the atomic scale, printing material. Kiev, Ukraine
Braun OM, Naumovets AG (2006) Nanotribology: microscopic mechanisms of friction. Surf Sci Rep 60:79–158
Broitman E (2014) The nature of the frictional force at the macro-, micro-, and nano-scales. Friction 2(1):40–46
Brovelli D, Hahner G, Ruiz L et al (1999) Highly oriented, self-assembled alkane phosphate monolayers on tantalum (V) oxide surfaces. Langmuir 15(13):4324–4327
Carpick RW, Salmeron M (1997) Scratching the surface: fundamental investigations of tribology with atomic force microscopy. Chem Rev 97(4):1163–1194
Chaki NK, Vijayamohanan K (2002) Self-assembled monolayers as a tunable platform for biosensor applications. Biosens Bioelectron 17(1–2):1–12
Charitidis CA, Koumoulos EP, Dragatogiannis DA (2013) Nanotribological behavior of carbon based thin films: friction and lubricity mechanisms at the nanoscale (review). Lubricants 1: 22–47
Chen YL, Israelachvili JN (1992) Effects of ambient conditions on adsorbed surfactant and polymer monolayers. J Phys Chem 96(19):7752–7760
Clear SC, Nealey PF (2001) Lateral force microscopy study of the frictional behavior of self-assembled monolayers of octadecyltrichlorosilane on silicon/silicon dioxide immersed in n-alcohols. Langmuir 17(3):720–732
Crone W (2008) A brief introduction to MEMS and NEMS. Springer handbook of experimental solid mechanics. Springer Science + Business Media, New York, pp. 203–228
Dedkov GV (2000) Nanotribology: experimental facts and theoretical models. Physics Uspekhi 43 (6):541–572
Doudevski I, Schwartz DK (2000) Mechanisms of self-assembled monolayer desorption determined using in situ atomic force microscopy. Langmuir 16(24):9381–9384
Doudevski I, Schwartz DK (2001a) Self-assembled monolayers in the context of epitaxial film growth. Appl Surf Sci 175:17–26
Doudevski I, Schwartz DK (2001b) Concentration dependence of self-assembled monolayer island nucleation and growth. J Am Chem Soc 123(28):6867–6872
Egberts P, Carpick RW (2013) Friction at the Atomic Scale. Physics 6:102
Filipponi L, Sutherland D (2010) NANOYOU. International Nanoscience Center (iNANO), Aarhus University
Gawali AL, Kumawat SC (2011) Nanotribology. Int J Adv Eng Technol 2:300–310
Gebeshuber IC, Drack M, Scherge M (2008) Tribology in biology. Tribology 2(4):200–212
Gnecco E, Meyer E (2007) Fundamentals of friction and wear. Springer, Berlin
Grosse I, Estel K (2000) Thin surfactant layers at the solid interface. Colloid Polym Sci 278 (10):1000–1006
Hahner G, Spencer N (1998) Rubbing and scrubbing. Phys Today 22–27
Hironaka S (1984) Boundary lubrication and lubricants. Three Bond Technical News 1(9):1–8

Holscher H, Schirmeisen A, Schwarz U (2008) Principles of atomic friction: from sticking atoms to superlubric sliding. Phil Trans R Soc A 366:1383–1404
Hsu S (2004a) Molecular basis of lubrication. Tribol Int 37(7):553–559
Hsu S (2004b) Nano-lubrication: concept and design. Tribol Int 37:537–545
Hsu S, Ying C, Zhao F (2014) The nature of friction: a critical assessment. Friction 2(1):1–26
Hu YZ, Granick S (1998) Microscopic study on thin film lubrication and its contributions to macroscopic tribology. Tribol Lett 5:81–88
Hyun-Joon K, Dae-Eun K (2009) Nano-scale friction: a review. Int J Precisi Eng Manuf 10 (2):141–151
Israelachvili J, Tabor D (1973) Van der Waals forces: theory and experiment. Prog Surf Membr Sci 7:1–55
Jun Z, Yonggang M (2015) Boundary lubrication by adsorption film. Friction 3(2):115–147
Kim H, Kim E (2009) Nano-scale friction a review. Int J Precis Eng Manuf 10(2):141–151
Kopta S (2001) Nanometer scale friction and wear on self-assembled monolayers investigated by atomic force microscopy. Dissertation, University Basel
Krilov SY, Frenken WM (2014) The physics of atomic-scale friction: basic considerations and open questions (review article). Phys Status Solid B 251(4):711–736
Krim J (1996) Atomic-scale origins of friction. Langmuir 12:4564–4566
Krim J (2002a) Resource letter: FMMLS-1: friction at macroscopic and microscopic length scales. Am J Phys 70(9):890–897
Krim J (2002b) Surface science and the atomic-scale origins of friction: what once was old is new again. Surf Sci 500:741–758
Krim J (2012) Friction and energy dissipation mechanisms in adsorbed molecules and molecularly thin films (review). Adv Phys 61(3):155–323
Krim J, Solina DH, Chiarello R (1991) Nanotribology of a Kr monolayer: a quartz-crystal microbalance study of atomic-scale friction. Phys Rev Lett 66:181–184
Kumar B, Kumar KB, Negi YS (2014) Organic thin film transistors: structures, models, materials, fabrication, and applications: a review. Polym Rev 54:33–111
Liu M, Lu Y, Zhang J et al (2009) MEMS/microelectronics self-assembly based on analogy of Langmuir-Blodgett approach. Microelectron Eng 86(11):2279–2282
Lue JT (2007) Physical properties of nanomaterials. Encycl Nanosci Nanotechnol X:1–46
Manojlovic J (2006) Structure, morphology and history effects in surfactant self-assembly. Dissertation, ETH Zuerich
Manojlovic J (2010) Friction and lubrication at the atomic level. In: Paper presented at the international conference, mechanical engineering in XXI century, MASING 2010, University of Niš, Serbia, 25–26 Nov 2010
Manojlovic J (2012) The Krafft temperature of surfactant solutions. Thermal Sci 16(2):S633–S642
Manojlovic J (2013) Self-assembled monolayers in lubrication on atomic level. In: The 2nd international conference, mechanical engineering in XXI century, faculty of mechanical engineering, Nis, Serbia, 20–21 June 2013
Manojlovic J (2013) Preparation and characterization of quaternary ammonium surfactants on muscovite mica. SERBIATRIB 2013. In: 13th international conference on tribology faculty of engineering, Kragujevac, 15–17 May 2013
Manojlovic J, Milojevic A, Tomic M et al (2015) Tribology in micromechatronical devices. In: SERBIATRIB, 2015, 14th international conference on tribology, Belgrade, 13–15 May 2015
Mate MC, McClelland MG, Erlandsson et al (1987) Atomic-scale friction of a tungsten tip on a graphite surface. Phys Rev Lett 59(17):1942–1946
Mathew CM (2008) Tribology on the small scale. Oxford University Press Inc., New York
Mattias L (2008) Self lubrication on the atomic scale design. Dissertation, Uppsala University
Matsumoto K (2003) Surface chemical and tribological investigations of phosphorus-containing lubricant additives. Dissertation, ETH Zuerich
Mellott M, Hayes W, Schwartz DK (2004) Kinetics of octadecyltrimethylammonium bromide self-assembled monolayer growth at mica from an aqueous solution. Langmuir 20(6): 2341–2348

Miyake K, Kume T, Nakano M et al (2012) Effects of surface chemical properties on the frictional properties of self-assembled monolayers lubricated with oleic acid. Tribol Online 7(4):218–224
Mo Y, Turner KT, Szlufaraska I (2009) Friction laws at the nanoscale. Nature 457:1116–1119
Mowat I, Moskito J, Ward I et al (2007) Analytical methods for nanotechnology. NSTI-Nanotech 4:20–23
Popova E, Popov LV (2015) The research works of Coulomb and Amontons and generalized laws of friction. Friction 3(2):183–190
Quing Z, Qi Y, Hector LG et al (2007) Origin of static friction and its relationship to adhesion at the atomic scale. Phys Rev B 75(144114):1–7
Rajkumar G, Sethuraman MG (2014) Corrosion protection ability of self-assembled monolayer of 3-amino-5-mercapto-1, 2, 4-triazole on copper electrode. Thin Sold Films 562:32–36
Ramanpreet S, Sreedhar M (2014) Bio-tribology and its applications in medical sciences-a review. Int J Mech Eng Rob Res 3(3):141–144
Ratner M, Ratner D (2006) Nanotechnology. Pearson Education, Inc
Riedoa E, Brune H (2003) Young modulus dependence of nanoscopic friction coefficient in hard coatings. Appl Phys Lett 83(10):1986–1988
Rymuza Z (2010) Advanced techniques for nanotribological studies. Sci Probl Mach Oper Maint 1 (161):33–43
Scherge M, Schaefer JA (1998) Microtribological investigations of stick/slip phenomena using a novel oscillatory friction and adhesion tester. Tribol Lett 4:37–42
Schmid G, Corain B (2003) Nanoparticulated gold: syntheses, structures, electronics, and reactivities. Eur J Inorg Chem 17:3081–3098
Shivaprakash NR (2013) Adhesion and tribology on well-defined, nano-scale rough surfaces: a gradient approach. Dissertation, ETH Zuerich
Singer IL (1994) Friction and energy dissipation at the atomic scale: a review. J Vac Sci Technol A 12(5):2605–2616
Singh A, Suh K (2013) Biomimetic patterned surfaces for controllable friction in micro- and nanoscale devices. Micro Nano Syst Lett 1:6
Sokoloff J (2003) A mechanism for lubrication between surfaces with atomic level roughness. Physics Faculty Publications. Paper 4, Northeastern University, Boston
Somers EA, Howlett PC, MacFarlane DR et al (2013) A review of ionic liquid lubricants. Lubricants 1:3–21
Spikes H (1996) Direct observation of boundary layers. Langmuir 12:4567–4573
Spikes H (2001) Tribology research in the twenty-first century. Tribol Int 34:789–799
Sung IH, Lee HS, Kim DE (2003) Effect of surface topography on the frictional behavior at the micro/nano-scale. Wear 254:1019–1031
Szlufarska I, Chandross M, Carpick RW (2008) Recent advances in single-asperity nanotribology (topical review). J Phys D Appl Phys 41(123001):1–39
Taylor RI (2012) Tribology and energy efficiency: from molecules to lubricated contacts to complete machines. Royal Soc Chem Faraday Discuss 156:361–382
Theodore L (2006) Nanotechnology: basic calculations for engineers and scientists. Wiley, New York
Ulman A (2013) An introduction to ultrathin organic films: from Langmuir-Blodgett to self-assembly. Academic Press
Urbakh M, Krafter J, Gourdon D et al (2004) The nonlinear nature of friction. Nature 430:525–528
Vakarelski IU, Brown SC, Rabinovich YI et al (2004) Lateral force microscopy investigation of surfactant-mediated lubrication from aqueous solution. Langmuir 20(5):1724–1731
Vanossi A, Manini N, Urbakh M et al (2013) Modeling friction: from nano to meso scales. Rev Mod Phys 85(2):529–552
Wang Z-J, Ma T-B, Hu Y-Z et al (2015) Energy dissipation of atomic-scale friction based on onedimensional Prandtl-Tomlinson model. Friction 3(2):170–182
Weymouth AJ, Meuer D, Wutscher MT et al (2013) Atomic structure affects the directional dependence of friction. Phys Rev Lett 111(126103):1–4
Widmer M (2002) Modified molecular friction in artificial hip joints. Dissertation, ETH Zurich

Williams L (2007) Nanotechnology demystified. McGraw-Hill Companies
Xuan Z (2014) Molecular dynamics simulation of boundary lubricated contacts. Dissertation, University of Wollongong
Xue Y, Li X, Li H et al (2014) Quantifying thiol–gold interactions towards the efficient strength control. Nat Commun 5:4348. doi:10.1038/ncomms5348
Yang C (2008) Role of surface roughness in tribology: from atomic to macroscopic scale. Dissertation, Berlin
Yang Y, Singh J, Ruths M (2014) Friction of aromatic thiol monolayers on silver: SFA and AFM studies of adhesive and non-adhesive contacts. Royal Soc Chem 4:18801–18810
Zana R (2005) Dynamics of surfactant self-assemblies. CRC Press, Taylor & Francis
Zhang J (2012) Nanotribological and nanomechanical investigation of nanomaterials. Dissertation, Rice University
Zhang J, Meng Y (2015) Boundary lubrication by adsorption film. Friction 3(2):115–147
Zhang Q, Qi Y, Hector LG et al (2007) Origin of static friction and its relationship to adhesion at the atomic scale. Phys Rev B 75(144114):1–7
Zimmerli L, Nuller F, Hidber H-R et al (2005) Exploring the nano-world: friction and light at the nanometer scalenano3d visualization. http://nano-world2.cs.unibas.ch/Events/nmt-master/doc.pdf

Modeling of Self-healing Materials with Nanocontainers Using Discrete and Continuum Methods

Nenad Filipovic, Dejan Petrovic, Milica Nikolic, Svetlana Jovanovic, Aleksandar Jovanovic and Nebojsa Zdravkovic

Abstract Corrosion degradation of materials is an important issue that leads to depreciation of investment goods. In production of materials a huge challenge is to design 'smart' synthetic systems that can actively re-establish the continuity and integrity of a damaged area. Nanocontainers represent new technology for smart nanocoating interfaces. This chapter describes the solutions based on an innovative integrated modeling approach, including nano- and macro-scale in the automotive, aerospace and biomedical industry. Two different modeling approaches, discrete and continuum, are used to investigate coating substrates that contain nanoscale defects with healing agents. Dissipative Particle Dynamics (DPD) method uses three forces: repulsive, dissipative and random forces, as well as additional forces which bound healing agents to a metal substrate. Finite Element Method (FEM) is continuum modeling method with different diffusivity and fluxes. The chapter includes the real case examples from industry with different concentrations of inhibitors inside the primer layer. These findings could be used for guidelines for formulating nanocomposite coatings and healing effects of the surfaces through the self-assembly of the particles into the defects.

Keywords Modeling · Nanocontainers · Self-healing · DPD · FEM

N. Filipovic (✉) · D. Petrovic · M. Nikolic
Faculty of Engineering, University of Kragujevac, Kragujevac, Serbia
e-mail: fica@kg.ac.rs

N. Filipovic · D. Petrovic · M. Nikolic
Research and Development Center for Bioengineering, BioIRC, Kragujevac, Serbia

S. Jovanovic · A. Jovanovic
Risk-technologies, GmbH, Stuttgart, Germany

N. Zdravkovic
Faculty of Medical Science, University of Kragujevac, Kragujevac, Serbia

D. Brabazon et al. (eds.), *Commercialization of Nanotechnologies–A Case Study Approach*, DOI 10.1007/978-3-319-56979-6_8

1 Introduction

1.1 Corrosion Problem

The problem of corrosion is very important. As an example, in the USA it is estimated that direct costs of corrosion are US $18 billion annually, while the total national costs in all sectors of the economy are US $275 billion annually (Corrosion Cost and Preventive Strategies in the United States, NACE International, 2013). In certain industries, such as chemical, pulp and paper or oil refining industry, relative costs incurred due to corrosion damage can be much higher. The cost of wear can also be tremendous in some industries. In Germany alone, the costs caused by wear were estimated to be EUR 16–20 billion, while in Finland the costs of wear were EUR 2.7 billion (Survey on Tribology in Finland, VTT Manufacturing Technology, 1999). In industries such as mining, construction and recycling the costs of component wear can be significant. In typical mining and construction industry applications wear parts and consumables may represent 40% of life cycle costs, other main cost items being energy 25%, maintenance labor 5–10% and initial plant investment only 15–20%.

1.2 Bio-Inspired Self-healing Concept

Microcracking induced by thermal and mechanical fatigue is also a long-standing problem in polymer adhesives (Lee et al. 1991). During the last several decades the only successful crack-healing methods that have been reported so far required some form of manual intervention (Stavrinidis et al. 1983; Jud et al. 1981). It is always a challenge to automatically conduct healing of the cracks. We are using a biological analogy and approach for thrombus formation. When the balance between pro- and anti-coagulation factors present in the blood is not sustained, pathological platelet aggregation is triggered. This imbalance initiates the coagulation cascade which enhances platelet activation (Loscalzo and Schafer 2003; Kuharsky and Fogelson 2001). A scheme of platelet adhesion and aggregation is shown in Fig. 1.

Similarly, the non-live material incorporates a microencapsulated healing agent that is released upon crack intrusion. Polymerization of the healing agent is then triggered by the contact with an embedded catalyst, bonding the crack faces (White et al. 2001).

The definition of multiple nanocontainers is represented in Fig. 2. Yellow particles represent the healing agent and red particles denote the material which is located initially inside the crack space.

The definition of binding force for healing agents to crack and nanocontainers wall is presented in Fig. 3. Yellow particles represent the healing agent. The force of binding to the wall is 10 times larger than the repulsive force between the particles and it is denoted with $F' = Fe * 10$.

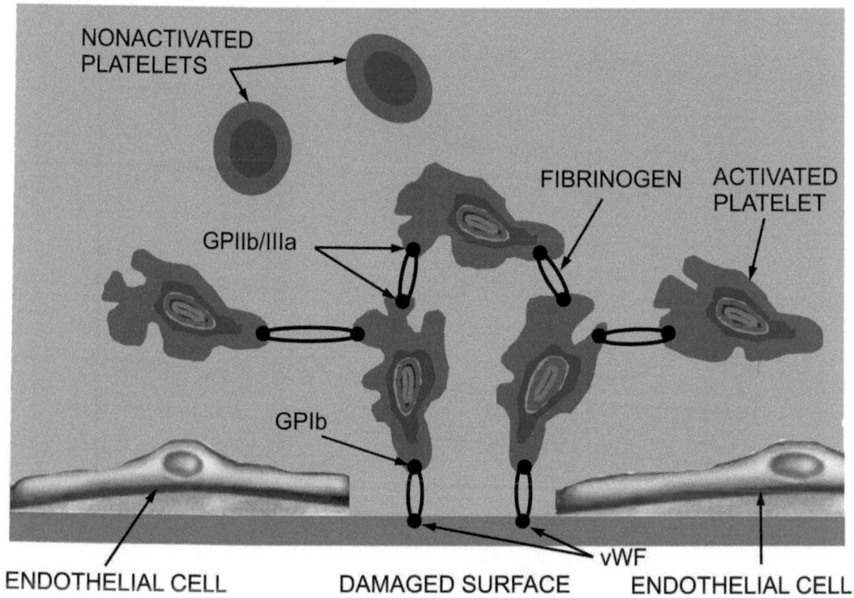

Fig. 1 Schematic representation of the mechanisms of platelet adhesion and aggregation in flowing blood, (*GP* Glycoprotein, *vWF* Von Willebrand Factor)

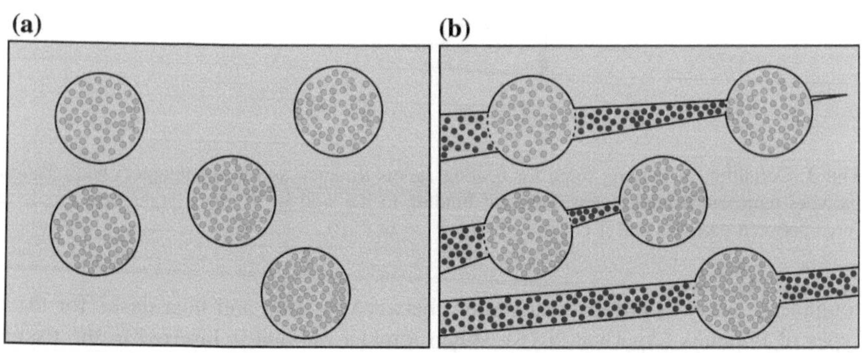

Fig. 2 Multiple nanocontainers. **a** Healing agents inside nanocontainers, **b** Breaking of nanocontainers with random crack positions. *Yellow* particles denote healing agents

1.3 Application in Biomedical Materials

Metals are used for manufacturing larger implants such as those used for knee and hip replacement (Nag 2012). Biomaterials made from metals are used for their strength, biocompatibility and durability (Kheirkhah et al. 2015), especially when mechanical support is important (long bones, vertebral fractures) (Agarwal 2015). Their advantages compared to ceramics are mechanical strength and fracture

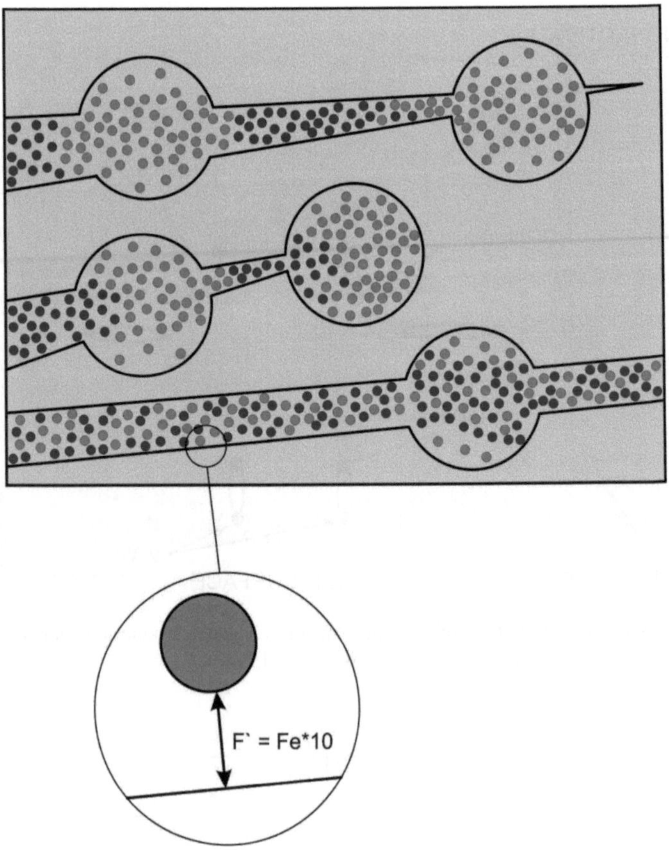

Fig. 3 Definition of binding force for healing agents to crack and nanocontainers wall. *Yellow* particles represent healing agent. Force of binding to the wall is 10 times larger than repulsive force between particles and it is denoted with $F' = Fe * 10$

toughness. There is minimal movement between implant and host tissue for these types of implants (Agarwal 2015). When a metal implant is inserted in the tissue, most metals become encapsulated by a thin layer. Adhesion between the metal implant and the surrounding tissue is comparatively weak. Bones contain hydroxyapatite (a form of calcium phosphate) as a structural matrix. If a metal implant is coated with hydroxyapatite, there is no intervening layer for bone cells which adhere onto the surface of the apatite coating. There is excellent adhesion of the coated implant to the bone because the hydroxyapatite matrix of the bone cells becomes integral with the hydroxyapatite coating (Batchelor 2004).

One of the first metal biomaterials used was stainless steel. It is used for bone fixation plates and screws. The problem with this material is that it corrodes in biofluids (Kheirkhah et al. 2015). In order to reduce corrosion, some surface modifications need to be performed. One of those modifications is the use of bioactive ceramics as a coating material. In this way, the implant keeps mechanical properties of the metal as well as biological properties of ceramics (Kheirkhah et al. 2015). Besides stainless steel, titanium and its alloys also add to the available materials for bone-replacing devices. They are mainly used for their strength but are lighter, although the thin oxide layer that naturally forms on their surface also acts as a passivation protective barrier, conferring on this metal its known corrosion resistance under physiological conditions. Calcium phosphates, however, are chosen for their unrivaled compatibility with human bone: they initiate a rapid biological response, improving adhesion between the bone and the implant and providing a scaffold for bone growth (Leon 2009). In order to eliminate this problem, implants could be improved by coating them with titanium or with calcium phosphate ceramics. This approach would combine the mechanical strength of titanium with the biological properties of calcium phosphate (Leon 2009).

For bone growth and regeneration polymeric implants are used. Commonly used polymers are silicone, polyethylene and hydrogels Ultra-high-molecular-weight polyethylene (UHMW PE) and polymethylmethacrylate (PMMA) polymers have found application in acetabular cups, patellar prostheses and in dentistry or vertebroplasty. Upon implantation in a bone, most of these biomaterials are tolerated by human tissues, leading to neither necrosis nor adverse inflammatory reactions. These artificial implants are simply encapsulated by fibrous tissue, not being in a direct contact with bone tissue. The materials are classified as bioinert (e.g. zirconia, alumina) and biotolerant (e.g. PMMA, titanium, Co-Cr) (Leon 2009).

Biocomposites, i.e. composite materials intended for use as biomaterials, also add to the range of potential materials (Batchelor 2004).

Previously the low E-modulus of nanostructured materials were often attributed to the unusual grain boundary structures present, but this is actually determined by the defect structure such as porosity. This leads to the concept of optimizing both hardness to a (sufficiently) high value and elastic modulus to an appropriate (relatively low) level, when designing tribological coatings (Cavaleiro 2006). Further, it was anticipated in previous research that control of the grain size dispersion is extremely important in the experimental design of these nanostructured coatings. A nanostructured material with a broad grain size dispersion will exhibit a lower overall flow stress compared to a material with the same average grain size (Cavaleiro 2006).

While in most cases the biomedical material is intended to last at least as long as the probable remaining life of a patient, some biomedical materials are designed to degrade quickly in applications requiring a temporary presence in the body. A typical example of an implant could be a bone fracture fixation device where the device is intended to provide a mechanical support to injured tissue until tissue repair is complete. After the healing process is completed, the fixation devices could cause discomfort to the patient and warrant removal of the device (Batchelor 2004).

For similar reasons, every drug delivery system needs to be biodegradable. The drugs are enclosed in a capsule usually made of a polymer and upon release of the drug at the desired location the capsule is expected to degrade so that it does not trigger a foreign object immune rejection response in the surrounding tissue membranes (Batchelor 2004). An antibiotic or pharmaceutical medicine can be coated with a degradable organic polymer, which contains a standard amount of the drug which prevents contamination. Some drugs can be destroyed by gastric acid inside the stomach, so the coating is made resistant to acidic but not to alkaline conditions. This allows the pill to be resistant through the stomach before the coating dissolves in the alkaline milieu of the small intestine to release the drug (Batchelor 2004).

In the case of nanocontainers, a possibility for generating polymer hollow spheres is to form a polymer shell around a preformed template particle, which can subsequently be removed. Despite these synthetic routes to such polymeric nanocontainers, their high stability, their major drawback is prevention of effective loading of preformed capsules or releasing the encapsulated material in a controlled way towards the desired target (Landfester 2001).

Although implants are used in order to provide a better life quality, there are some disadvantages such as development of a biomaterial-associated infection. The human body provides comparatively severe conditions for most biomedical materials. The health and, indeed, safety of a user may depend upon the reliability of the implant (Batchelor 2004). The implant not only has to continue to function within the human body, but there should also be an absence of side effects engendered by the implant's presence.

After inserting a foreign material into a body, there will be an interaction between that material and proteins. In a short period of time the surface of the inserted material will be coated by those proteins that have a crucial role in determining the body reaction to the foreign material. Being able to avoid immune reaction to biomaterials would reduce the healing time. Some of the basic problems, when it comes to interaction between an implant and the human body as well as immune reaction, are: inflammation, pain, destruction of surrounding tissue or, in the worst case, implant failure and rejection of inserted device (Nag and Banerjee 2012).

Current estimates suggest that up to 2.5% of primary hip and knee arthroplasties and up to 20% of revision arthroplasties are complicated by periprosthetic joint infection. According to some authors (Dale 2009), not only are these numbers underestimates but they are also on the rise. A major problem with orthopedic implants is the release of wear-particles from the moving surfaces (Batchelor 2004). Unless the implant is made from a truly inert material, some corrosion will probably occur. The body has only a limited tolerance for corrosion products and wear particles released by an implant. A deep infection leads to implant removal and ensuing increased morbidity (Zmistowski 2013). In some cases, cells that surround an implant will reject the foreign object and that leads to chronic inflammation and potential removal of the implant. There are many reasons for that such as wear or fracture of the implant, overgrowth of the tissue, degradation of material, leaching out of ions (implant made from metal), etc. (Nag 2012).

Infection is one of the possible complications that can lead to a prolonged recovery. Some research is headed toward developing antibacterial surfaces that would prevent bacterial adhesion as well as colonization and proliferation into the surrounding tissues (Gallo 2014). Many different bacteria can affect implants but almost 70% of orthopedic implant infections are caused by Staphylococcus bacteria (Raphel 2016). Although infection usually shows soon after the implant is inserted, it is important to think about a long-term prevention of infection. Some solutions are a long-term presentation of antibiotics and slow release of antimicrobial coatings (Raphel 2016). Orthopedic implants that have a controlled release of antibiotics in coatings have shown to be successful for prevention of bone and tissue infection (de Breij et al. 2016).

1.4 Surface Coating for Biomaterial

Surface coatings have been developed as sheathing of implant in a coating for metal implants, because they are more acceptable to the human body (Lentino 2003).

After an implant is positioned inside the body, there will be some gaps between the implant and the surrounding tissue. When considering orthopedic implants, it is important to reduce that gap as much as possible in order to limit micro motion that can lead to implant failure (Valstar et al. 2002). One of the possible solutions is a coating that would bridge gaps and reduce micro motion. Different types of coatings are used to reduce infection. Those are: adhesion resistant coating, coating with antibiotics, silver impregnated coatings, bioactive molecules and cytokines (Goodman et al. 2013). Hydrogels, foams or metallic structures could be used for this solution as well as nitinol (Fournier et al. 2014). For better bone grow on implants, calcium phosphate coating are used. A coating created from calcium phosphate and additives help with osteoconductivity and improve antibacterial properties (Raphel et al. 2016).

Previous research showed that coating implant with carbonated hydroxyapatite enhances new bone formation on the implant surface and they seal interface from wear particles and macrophage that are associated with periprosthetic osteolysis. For delivery of growth factors, DNA and bioactive molecules, this type of coating has been used (Goodman et al. 2013).

Covering implant with hydrogel coating is one of the most used methods. Hydrogel coatings are easy to coat and they are a very popular solution for orthopedic implants (Goodman et al. 2013). In order to reduce body reaction to a foreign material, implants can be coated with bisphosphonates that help with bone formation.

1.5 State of the Art in Modeling

Advances in technology allow simulation of a great number of different situations in order to learn about the process without the need for procedures on patients. A finite element (FE) study can help researchers learn about reasons for failure based on stress distribution. One of those studies was performed regarding a simulation of dental implant insertion process (Guan et al. 2011). They concluded that the best approach for implantation regarding levels of stress is to use a threaded implant. The theoretical interfacial shear strength can be expressed with the absolute mean slope for dental implants (Hansson et al. 2010). In order to simulate removal torque situations, Hansson et al. (2011) developed a mathematical model that can estimate interfacial shear stress between a bone and an implant in situation when the gap between the bone and the implant can arise during shear situations. Vanegas-Acosta et al. (2011) created a mathematical model of osseointegration at the bone-dental implant interface. Using the FEM, they showed that their model is able to simulate wound healing process that includes blood clotting, cell migration, new osteoid formation, etc.

Corrosion degradation of materials is an important issue that leads to depreciation of investment goods. Active and passive approaches exist for corrosion protection. The passive corrosion protection is explained by deposition of a barrier layer preventing contact of the material with the corrosive environment. When the main barrier is damaged, the active corrosion protection decreases the corrosion rate (Zheludkevich et al. 2007).

Several strategies exist which provide inhibitor encapsulation, solubility control and delivery. These include conventional microencapsulation (Agarwala et al. 1989; Enos et al. 1998; Bibbs et al. 1999), self-assembled silica nano-vesicles and hollow silica fibers formed by evaporation induced self-assembly of organic and inorganic precursors (Lu et al. 1999), plasma polymerization of chemical compounds with particulate materials (van Ooij 2002).

Self-healing materials are firstly mentioned with description(White et al. 2001). Computer simulation of the self-healing system is explained in (Balazs 2007). Hybrid approach with coupled lattice Boltzman model (LBM) and the lattice spring model (LSM) was used by Verberg et al. (2007) for simulation of the motion of microcapsules caused by an imposed flow to move on a substrate with an adhesive coating.

One of the very first methods for discrete modeling Molecular Dynamics (MD) was first introduced by Alder and Wainwright in the late 50s (Alder et al. 1959). Here authors describe interactions between the particles by using Newton's equations of motion on an atomistic scale. Certainly, MD has been used in materials science since the 1970s and the first MD simulation of a real system was the simulation of liquid water in 1974 (Stillinger et al. 1974). Internal molecular temperature was explained in (Srinivasa et al. 2004).

In MD it is important to have a small time step in order to avoid discretization errors. In order to capture a macroscopic effect, using MD will require a large number of time steps. Simulation time could be very long. To overcome this drawback, parallel processing has been used (Hendrickson et al. 1995; Plimpton 1995) for solving this problem. The philosophy of the parallel processing is split up tasks and execution on multiple processors.

Several studies using MD have been used for a wide variety of applications ranging from evaluating the liquid properties of Pd-Ni alloys (Kart et al. 2004) to mechanical response of high-performance polymers (Cagin 1993).

Another discrete method is the Discrete Element Method (DEM) which was introduced by Cundall in 1971 to solve problems in rock mechanics. Williams et al. in (1985) gave the theoretical basis for DEM (Williams et al. 1985). This method considers a large system of distinct interacting general shaped (deformable or rigid) bodies or particles. The DEM method can be used to model particles with a non-spherical shape. DEM method uses contact forces between any two interacting particles for the purpose of evolution of particle positions and velocities. It is widely used in problems connected to granular media. Typical industries using this method are Pharmaceutical, Mining, Oil and gas, Chemical and Agriculture and food handling.

The Lattice-Boltzmann method (LBM) (Chen et al. 1992; Qian et al. 1992; Wolf-Gladrow 2000) is a mesoscopic method for fluid domain description, based on a mechanical system of particles. It comes from Cellular Automata models (Frisch et al. 1986; Wolfram 1986), it has recently been shown (He and Luo 1997; Abe 1997; Shan and He 1998; Shan et al. 2006; Philipp et al. 2006) that the LBM is based on a special discrete form of the continuous Boltzmann equation. The LBM is built on the mesoscopic scale, with a single-particle distribution function, $fi(\mathbf{x}, t)$, representing the number of particles with velocity **ci** at the site **x** and time t, where $i = 0, \ldots, b$. In the LBM, the particles are restricted to a discrete lattice where each group of particles can only move along a finite number of directions and with a limited number of velocities. In this method both the physical and the velocity spaces are discretized.

In this chapter two different modeling approaches, DPD (discrete) and FEM (continuum) for self-healing materials, as well as the mesoscopic bridge method, are presented in detail. Different effect of the nanocontainers fillings, particle-particle interactions, diffusivity parameter are analyzed. This could be guide for formulating nanocomposite coatings that has effect on the surface healing through the self-assembly of the particles into the defects.

2 Methods

2.1 Dissipative Particle Dynamics (DPD) Modeling

The coating layer with nanoscopic scratch can be modeled using molecular dynamics (MD) (Tyagi et al. 2004). A mesoscopic modeling using the DPD method could be consider as another approach of this problem (Groot and Warren 1997; Filipovic et al. 2008a, b). The motion of each DPD particle is described by Newton law equation

$$m_i \dot{\mathbf{v}}_i = \sum_j \left(\mathbf{F}_{ij}^C + \mathbf{F}_{ij}^D + \mathbf{F}_{ij}^R \right) + \mathbf{F}_i^{ext} \quad (1)$$

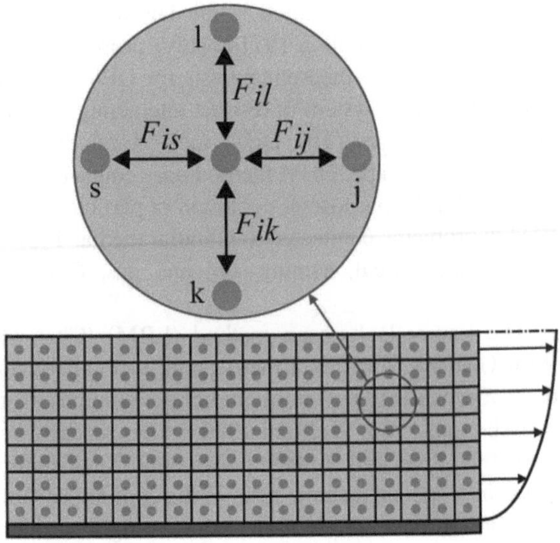

Fig. 4 Interaction forces in the DPD method

where m_i is the mass of particle "i"; $\dot{\mathbf{v}}_i$ is the particle acceleration as the time derivative of velocity; $\mathbf{F}_{ij}^C$, $\mathbf{F}_{ij}^D$, and $\mathbf{F}_{ij}^R$ are the conservative (repulsive), dissipative and random (Brownian) interaction forces that particle "j" exerts on particle "i", respectively, provided that particle "j" is within the radius of influence r_c of particle "i"; and $\mathbf{F}_i^{ext}$ is the external force exerted on particle "i", which usually represents gradient of pressure or gravity force as a driving force for the fluid domain. The total interaction force $\mathbf{F}_{ij}$ (Fig. 4) between the two particles is

$$\mathbf{F}_{ij} = \mathbf{F}_{ij}^C + \mathbf{F}_{ij}^D + \mathbf{F}_{ij}^R \tag{2}$$

The component forces can be expressed as (Filipovic et al. 2008b)

$$\begin{aligned} \mathbf{F}_{ij}^C &= a_{ij}\left(1 - \frac{r_{ij}}{r_c}\right)\mathbf{r}_{ij}^0 \\ \mathbf{F}_{ij}^D &= -\gamma w_D\left(\mathbf{v}_{ij} \cdot \mathbf{e}_{ij}\right)\mathbf{r}_{ij}^0 \\ \mathbf{F}_{ij}^R &= \sigma w_R \xi_{ij} \mathbf{r}_{ij}^0 \end{aligned} \tag{3}$$

In Eq. 3, $\mathbf{r}_{ij}$ is vector between particles i and j, r_{ij} is the distance between particles i and j, $\mathbf{r}_{ij}^0 = \frac{\mathbf{r}_{ij}}{r_{ij}}$ is the unit vector pointing in direction from j to i, a_{ij} is the maximum repulsion force per unit mass, γ stands for the friction coefficient, and σ is the amplitude of the random force. Also, w_D and w_R are the weight functions for dissipative and random forces, dependent on the distance r from the particle i; and ξ_{ij} is a random number with zero mean and unit variance. The interaction force is equal to zero outside the domain of influence, r_c (cut radius), hence $F_{ij} = 0$ for $r_{ij} > r_{c.}$

To reach a Gibbs–Boltzmann equilibrium state in DPD fluid system, the following relation between the amplitudes of the weight functions of dissipative and random forces, w_D and w_R, must hold

$$w_D = w_R^2. \tag{4}$$

The amplitude of the random force σ is related to the absolute temperature T

$$\sigma = (2k_B T\gamma)^{1/2} \tag{5}$$

where k_B is the Boltzmann constant. The weight functions is given in the following form by Boryczko et al. (2003)

$$w_D = \left(1 - \frac{r_{ij}}{r_c}\right)^2, \quad w_R = 1 - \frac{r_{ij}}{r_c}. \tag{6}$$

The particles used in this study represent both inhibition agents and surrounding coating material. It was achieved by taking into account different repulsion force coefficient a_{ij}. The additional interaction forces between particles of inhibition agents, are added. These attractive forces are given as

$$F_a = k_{sf}\left(1 - \frac{L_{sf}}{L_{sf}^{\max}}\right) \tag{7}$$

where L_{sf} is the distance of the inhibition particle from the substrate, and $L_{sf}^{\max}$ is the maximum length of inhibition particle attractive domain and k_{sf} is the effective spring constant.

2.2 FEM Modeling

The mass transport process for the inhibition is governed by convection-diffusion equation

$$\frac{\partial c}{\partial t} + v_x\frac{\partial c}{\partial x} + v_y\frac{\partial c}{\partial y} + v_z\frac{\partial c}{\partial z} = D\left(\frac{\partial^2 c}{\partial x^2} + \frac{\partial^2 c}{\partial y^2} + \frac{\partial^2 c}{\partial z^2}\right) + q^{wall} \tag{8}$$

where c denotes the concentration of inhibitors; v_x, v_y and v_z are the velocity components in the coordinate system x, y, z; D is the diffusion coefficient, assumed to be constant, of the transported material, and q^{wall} is the flux of the binding process for inhibitors which adhere to the substrate surface. A similar concept is used for calculation of the volume of inhibitors which are possible to diffuse on the scratched surface. The diffusivity coefficient and the wall binding flux are the fitting parameters. In the continuum approach particle-particle interaction cannot be modeled but the benefit is that a large substrate area is possible to be modeled.

2.2.1 Finite Element Navier-Stokes Equations

The FE equations of balance in a form commonly used (e.g. Bathe 1996; Huebner 1975; Kojic et al. 1998) for the global fluid domain are presented in this section. The Navier-Stokes equations and the continuity equation for incompressible fluid are

$$\rho\left(\frac{\partial v_i}{\partial t}+\frac{\partial v_i}{\partial x_j}v_j\right)=\frac{\partial \sigma_{ij}}{\partial x_j}+f_i^V \tag{9}$$

$$\frac{\partial v_j}{\partial x_j}=0 \tag{10}$$

where ρ is the fluid density, σ_{ij} are the stresses at a fluid point, v_i are the velocity components, and f_i^V is the (external) body force; summation is assumed for the repeated index "*j*", j = 1, 2, 3. The stresses and the constitutive equations for the Newtonian fluid are

$$\sigma_{ij}=-p\delta_{ij}+\tau_{ij} \tag{11}$$

$$\tau_{ij}=\mu\left(\frac{\partial v_i}{\partial x_j}+\frac{\partial v_j}{\partial x_i}\right) \tag{12}$$

where p is the pressure, τ_{ij} are the viscous stresses, and μ is the fluid viscosity.

A standard Galerkin procedure is used (Huebner 1975), where the Eqs. (9) and (10), with use of (11) and (12), are transformed into a FE incremental-iterative form (for iteration "i")

$$\begin{bmatrix}\frac{1}{\Delta t}\mathbf{M}+{}^{t+\Delta t}\widehat{\mathbf{K}}^{(i-1)} & \mathbf{K}_{vp}\\ \mathbf{K}_{vp}^T & 0\end{bmatrix}\begin{Bmatrix}\Delta\mathbf{V}^{(i)}\\ \Delta\mathbf{P}^{(i)}\end{Bmatrix}=$$
$$\begin{Bmatrix}{}^{t+\Delta t}\mathbf{F}^{ext(i-1)}\\ 0\end{Bmatrix}-\begin{bmatrix}\frac{1}{\Delta t}\mathbf{M}+{}^{t+\Delta t}\widehat{\mathbf{K}}^{(i-1)} & \mathbf{K}_{vp}\\ \mathbf{K}_{vp}^T & 0\end{bmatrix}\begin{Bmatrix}{}^{t+\Delta t}\mathbf{V}^{(i-1)}\\ {}^{t+\Delta t}\mathbf{P}^{(i-1)}\end{Bmatrix}+\begin{Bmatrix}\frac{1}{\Delta t}\mathbf{M}\,{}^{t}\mathbf{V}\\ 0\end{Bmatrix} \tag{13}$$

where $\Delta\mathbf{V}^{(i)}$ and $\Delta\mathbf{P}^{(i)}$ are the vectors of nodal velocity and pressure increments (nodal pressure vector is **P**), and ${}^{t+\Delta t}\widehat{\mathbf{K}}^{(i-1)}$ is the matrix

$$ {}^{t+\Delta t}\widehat{\mathbf{K}}^{(i-1)} = {}^{t+\Delta t}\mathbf{K}^{(i-1)} + \mathbf{K}_\mu \tag{14} $$

The matrices ${}^{t+\Delta t}\mathbf{K}^{(i-1)}$ and $\mathbf{K}_\mu$ are the convective and viscous matrices. The terms in the above matrices are

$$ [M_{KJ}]_{mm} = \int_V \rho N_K N_J dV \quad \text{no sum on } m \tag{15} $$

$$ \left[{}^{t+\Delta t}K_{KJ}^{(i-1)}\right]_{mn} = \left[\rho \int_V N_K N_{J,j} \, {}^{t+\Delta t}v_j^{(i-1)} dV\right]_{mm} + \rho \int_V N_K \, {}^{t+\Delta t}v_{m,n}^{(i-1)} N_J dV \tag{16} $$

no sum on m

$$ \left[(K_\mu)_{KJ}\right]_{mm} = \int_V \mu N_{K,j} N_{J,j} dV \quad \text{no sum on } m \tag{17} $$

$$ \left[(\mathbf{K}_{vp})_{KJ}\right]_{mm} = -\int_V \frac{\partial N_K}{\partial x_m} \widehat{N}_J dV \quad \text{no sum on } m \tag{18} $$

where $\widehat{N}_J$ are the interpolation functions for pressure; and m and n denote components corresponding to x, y, z coordinate directions. The components of the external force vector ${}^{t+\Delta t}\mathbf{F}^{ext\,(i)}$ are

$$ {}^{t+\Delta t}F_{Kj}^{ext(i-1)} = \int_V N_K \, {}^{t+\Delta t}f_j^V dV + \int_S N_K \, {}^{t+\Delta t}\sigma_{jl}^{(i-1)} n_l dS \tag{19} $$

where n_l ($l = 1, 2, 3$) are the components of the normal $\mathbf{n}$ to the element surface.

2.2.2 Fitting Parameters Procedure

One of the primary aims is to determine diffusion coefficient and diameter of nanocontainers so that computer simulations match experiments. A simplex optimization method developed by Nelder and Mead (1965) is used. This is a nonlinear procedure which involves only function evaluations (no derivatives). The best fitting minimizes the sum of squared residuals, a residual being the difference between an experimental value of the measured creepage distance and creepage distance provided by a simulation. Creepage distance is distance on the creep left and right side in the direction normal to slit on the plate.

3 Results

3.1 DPD Results

The software has been developed due to modeling and post-processing necessities. Figure 5 shows the web interface of the "Particle Methods" application.

The parameters of the model are:

- **DetlaT** is the time step used for DPD simulation.
- **Ext. force** is the external force which acts on all particles. At this point, the force in X direction is assumed only.
- **Division U** is the total number of particles in X direction.
- **Division V** is the total number of particles in Y direction.
- **Gamma** is the viscosity friction coefficient.
- **Step average** is the total number of steps for saving results.
- **Total steps** is the total number of time steps.
- **Rep. force. coefficient** is Repulsive Force Coefficient used for repulsive force for all particles around the microcapsule.

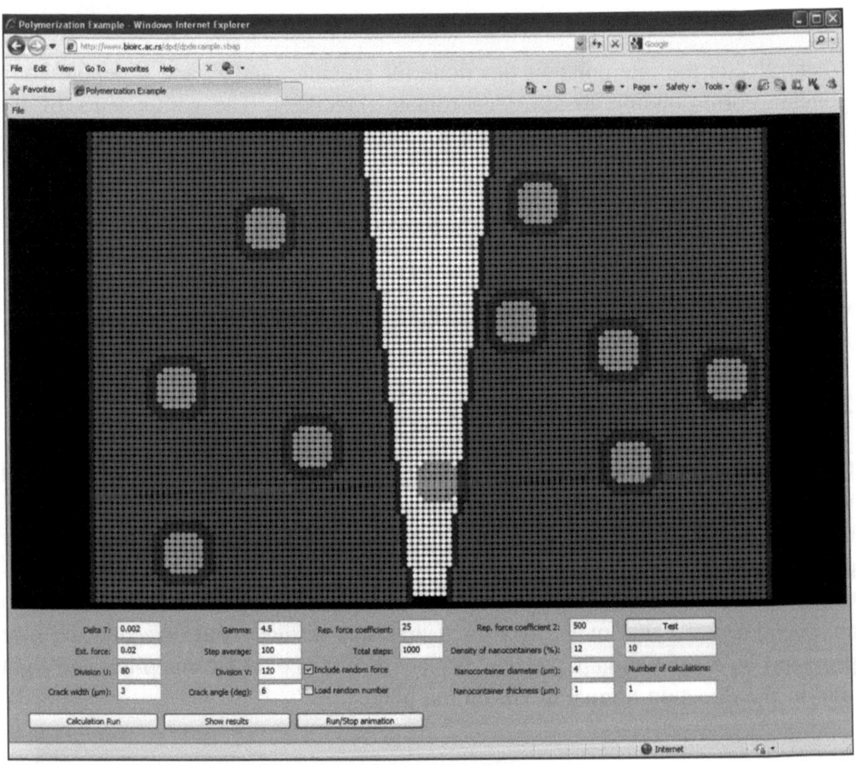

Fig. 5 The web application of the DPD simulation software

- **Rep. force. coeff. 2** is Repulsive Force Coefficient used for repulsive force for particles inside the microcapsule.
- **Ball radius** is the radius of the microcapsule.
- **Torus width** is the depth of microcapsule membrane.
- **Include random force** checking button is used for including/excluding random force in DPD calculation.

The simulation process in macro-scale refers to the multiple run of the developed software, which solely represents meso-metric coating slices, considering reasonable range of dimensions and parameters and analyzing the model by varying different parameters (e.g. randomly distributed nanocontainers in certain levels of the coating) in each execution. The sequence in Figs. 6, 7, 8 and 9 demonstrates the concept of multi-scale modeling.

Regarding the quality assurance of the coating system, one may assess whether or not the crack area is partly filled with enough agents to ensure the provision of a monolayer of inhibitor molecules on the metallic surface. With this aim, we consider if the crack area is partly filled with enough inhibitor particles representing adequate amount of molecules to cover the whole surface, also with respect to the third dimension, characterizing the 2.5D model. This analysis is dependent on the molecular size of the healing agents compared with the particle size, the solvability of the corrosion inhibitor in the corrosive environment, and the average amount of inhibitor that will remain in the area after elimination of the corrosive solvent. Assume 8-hydroxyquinoline (8-HQ) as a corrosion inhibitor material and a water based solvent as a corrosive material that has filled the crack area. According to the molecular size of 8-HQ as an inhibitor, assuming that all inhibitors that are solved in the corrosive solution will be uniformly distributed on the surface after removal of the solvent, it is concluded that if 0.01 percent of the corrosive solution is filled by corrosion inhibitors, the metallic surface will be covered by a mono-layer of

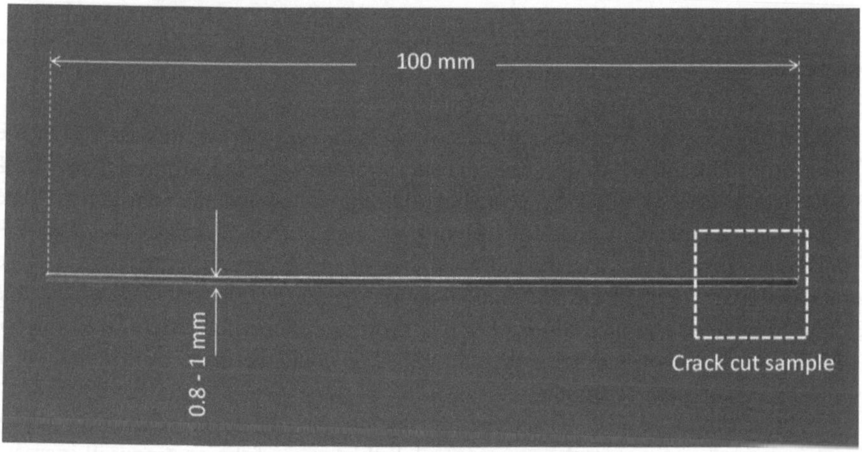

Fig. 6 Corrosion protection investigated on the artificial crack (*Top view*)

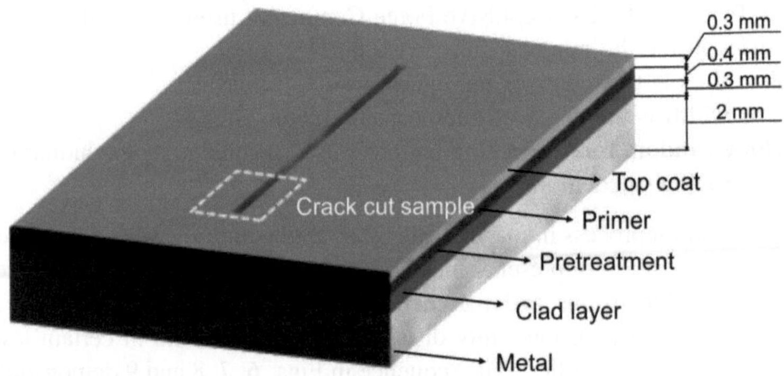

Fig. 7 The coating layers: top coat, primer, pretreatment, clad layer, metal. The crack was formed artificially

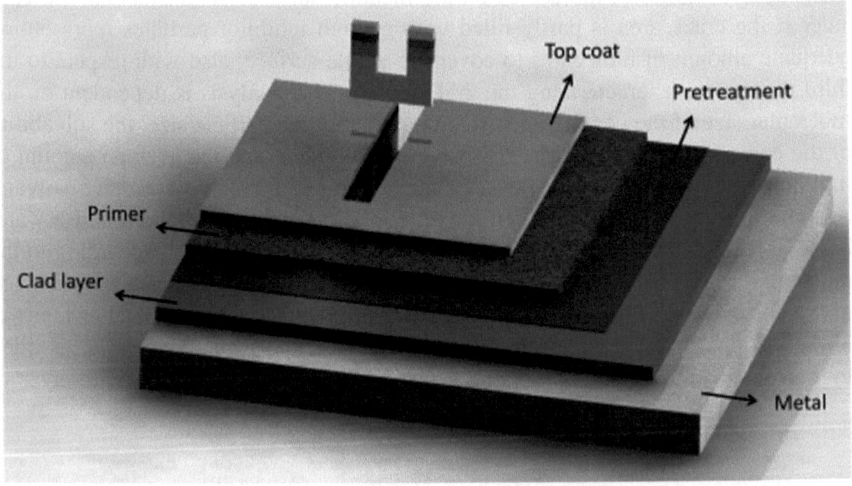

Fig. 8 The coating layers of the artificial crack (2)

8-HQ molecules. So, considering a factor of two, one can assume 0.02% as the lower limit. This means that in each execution of the software, if more than 0.02% of the crack area is filled by corrosion inhibitors, the metallic surface will be properly protected. Considering 100 times execution of the software, the risk analysis results in this case, for different sets of input, are presented. For a large crack with the size of 10 cm this number should increase 5000 times. However, this is not possible due to limitations of single-processor computers. Here, the concept of solution is shown and simulation in real size could be done by the help of a multi-processor supercomputer.

As it is shown in Fig. 10, increasing the primer (the layer of coating which contains nanocontainers in the current model) thickness of the coating will increase the reliability and the chance of getting sufficient number of nanocontainers through

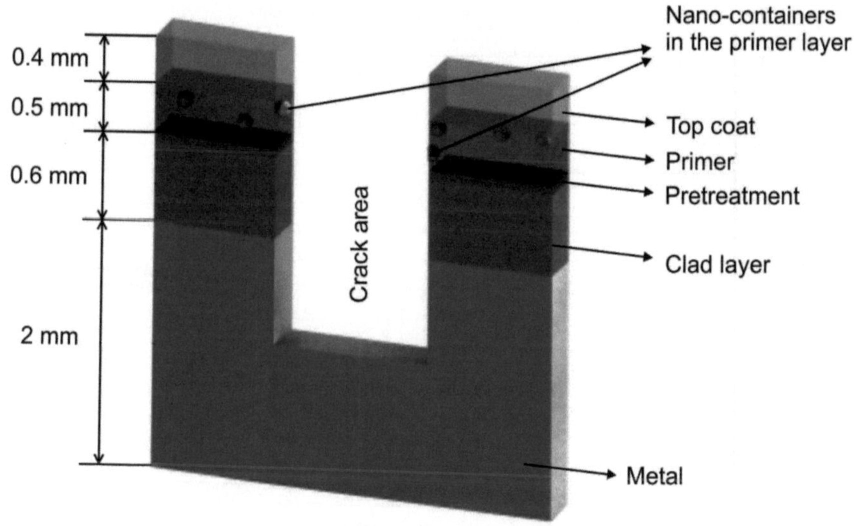

Fig. 9 The meso-scale sample case of the coating with random distribution of nanocontainers in the primer layer (2.5D sample)

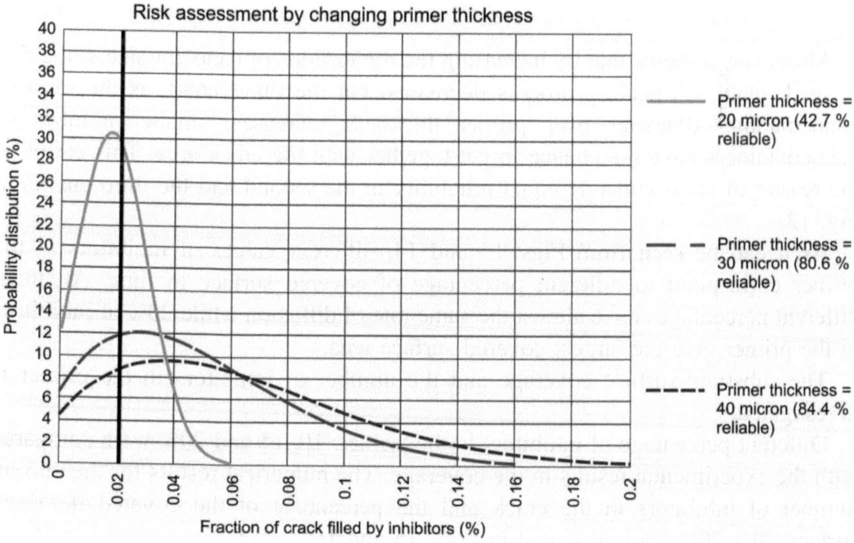

Fig. 10 Comparison of the reliability of the coating with different primer thicknesses

crack area to fill it by corrosion inhibitors properly. A linear increase in the primer thickness does not necessarily raise the reliability linearly. The same scenario occurs when increasing the density of nano containers in the coating. The results are shown in Fig. 11.

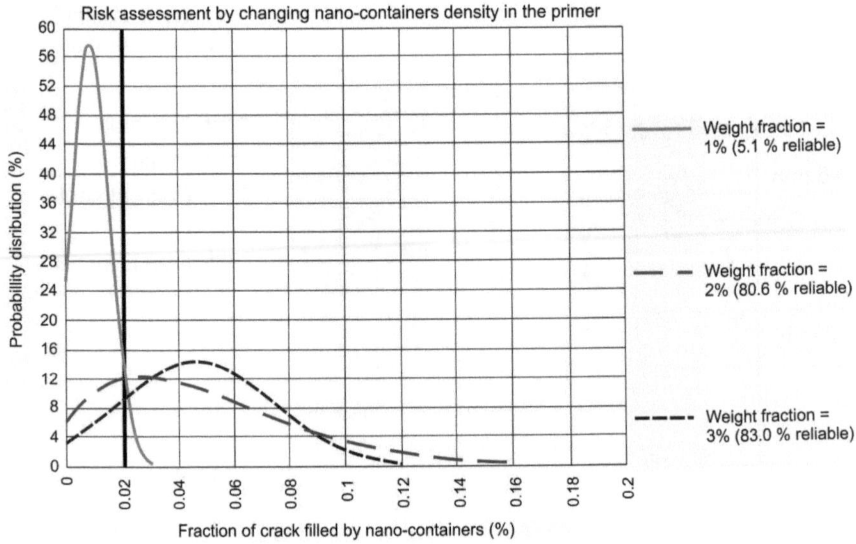

Fig. 11 Comparison of the reliability of the coating with different nanocontainer densities in the primer

Moreover, it seems that by increasing the dimensions of nanocontainers, the risk of inadequate self-healing process decreases. On the other hand, as the ratio of nanocontainer diameter over primer thickness increases, smaller number of nanocontainers have the chance to get together with the crack area. This could be the reason of approximately equal reliability in the second and the third curves in Fig. 12.

As it can be seen from Figs. 13 and 14, different values of inhibitors in the primer correspond to different percentage of covered surface in time. All three different percentages have almost the same rate of diffusion while 20% of inhibitors in the primer give the largest covered surface area.

The substrate surface coverage and the number of inhibitors in the primer is investigated.

Different percentage of inhibitors in the primer: 10, 15 and 20% were compared with the experimental results in the coverage. The numerical results for the current number of inhibitors in the crack and the percentage of the covered damaged surface after 200 s are presented in Figs. 15 and 16.

The second DPD model is created for the case of model with no water existing (Fig. 15). This model contains nanocontainers in the primer layer, which is scaled due to results' presentation. The inhibitors are released from nanocontainers which are randomly positioned inside the primer domain. The thickness of the primer layer is 500 nm and the diameter of the nanocontainer is 300 nm.

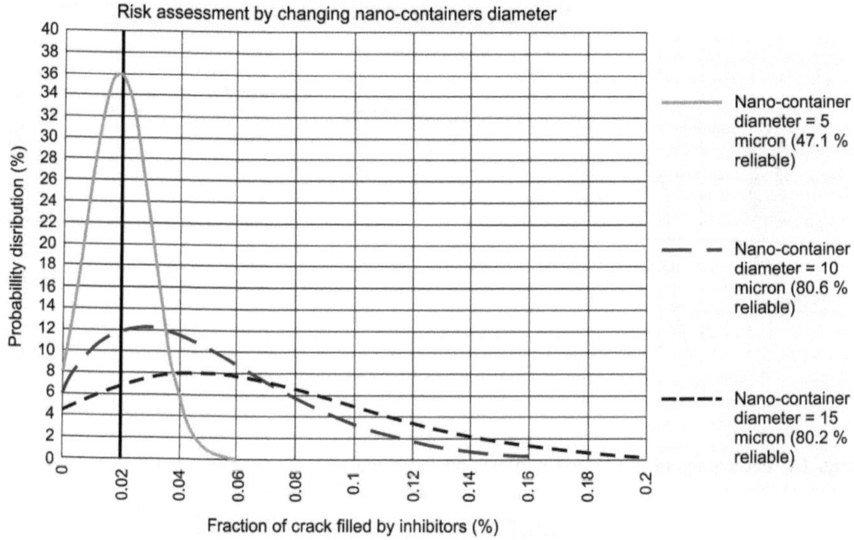

Fig. 12 Comparison of the reliability of the coating with different nanocontainer diameters

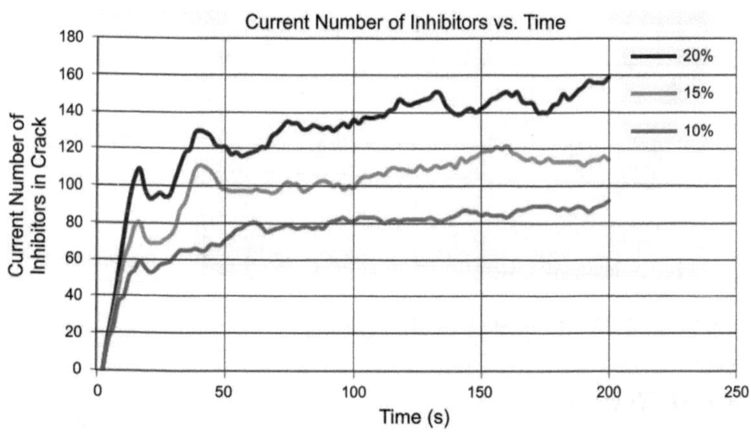

Fig. 13 Increase in the number of inhibitors on the metal surface within the crack during time obtained by a simple 2D analysis, for three percentage values of inhibitors in the primer: 10, 15 and 20%. Total time steps for DPD calculation was 100,000

The model is firstly matched with the volume needed for covering the scratch surface and time for diffusion of the inhibitors. The production of nanocontainers and inhibitors is too expensive and it is necessary to precisely establish the minimum percentage of inhibitors in the nanocontainers for the full protection of the treated material.

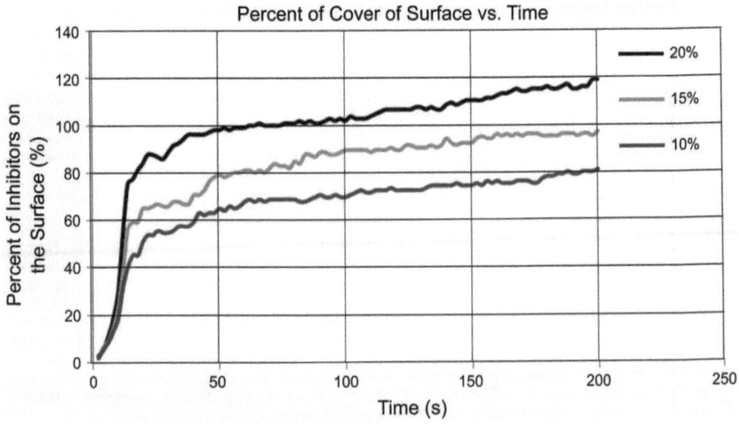

Fig. 14 Percentage of inhibitors' particles on the metal surface within the crack during time for three percentage values of inhibitors in the primer: 10, 15 and 20%

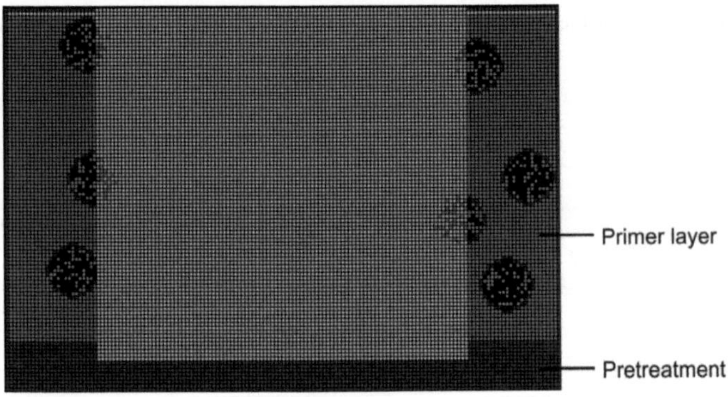

Fig. 15 Model with nanocontainers in the primer layer

3.2 FEM Results

The continuum model consists of mesh sized 80 × 80,000 = 6,400,000 3D finite elements where inhibitors are randomly prescribed as the influx boundary conditions. The scratch dimension is 0.1 × 90 mm, the primer layer is 3000 nm and the nanocontainer diameter is 350 nm. The percentage of inhibitor which is inside the nanocontainers is 15% and the percentage of nanocontainers in the primer or pre-treatment layers is approximately 10%. The convection velocity is assumed to be zero because diffusion process is more dominant. The binding flux was prescribed to be the unit with no water inclusion on the surface. Computed results during the time period of 24 h of inhibition process are shown in Fig. 16. It can be seen that almost 100% coverage is achieved in 12 h.

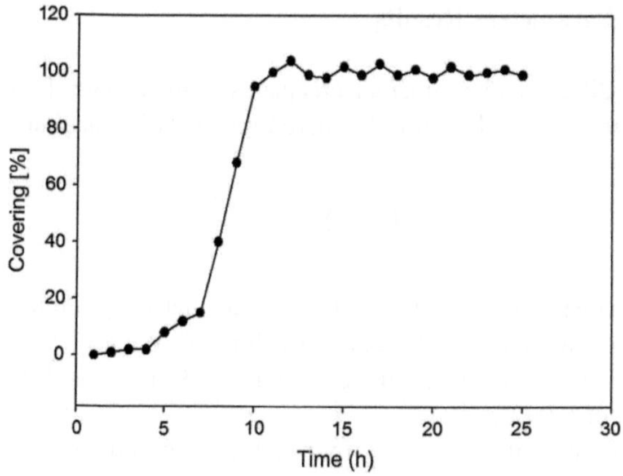

Fig. 16 Cover percentage of the inhibitors on the substrate surface during 24 h time interval (FEM model)

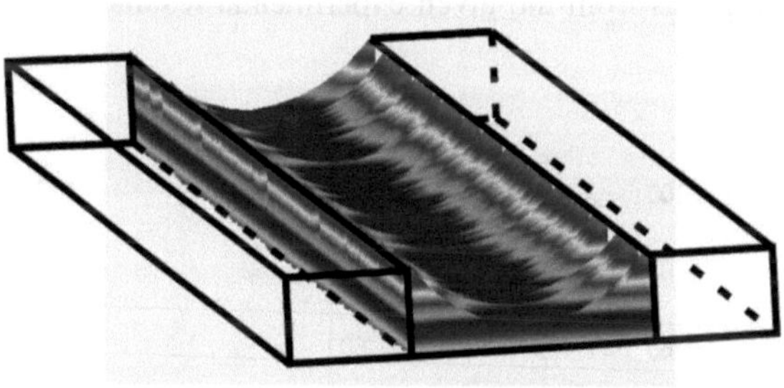

Fig. 17 Distribution of inhibitors on the scratch surface for t = 4 h. The width of the scratch is 0.1 mm and the length is 90 mm

Fig. 18 Coverage of the inhibitors on the substrate surface. The width of the scratch is 0.1 mm and the length is 100 mm

Distribution of inhibitors on the scratched surface for the time period of 4 h is presented in Fig. 17. The inhibitors' fluxes are randomly distributed along the plate, which causes higher coverage along the plate sides.

Coverage of the inhibitors on the substrate surface has been shown in Fig. 18.

3.2.1 Fitting Parameters Results

We have four different experimental creepages defined with 1200 points which represent1200 residuals. The sum of squared residuals is calculated as:

$$SE = \sum_{i=1}^{1200} (p_i - t_i)^2 \tag{20}$$

where p_i is the creepage distance for i-th point calculated by simulation and t_i.is the target (experimental) creepage distance for i-th point. We achieved minimum error (SE = 3082) with diffusion coefficient value 0.32 m^2/s and the diameter of nanocontainers of 0.092 nm.

At the upper panel of the Fig. 19, four pictures with compared creepage results obtained from simulations and the experimental plates have been presented. Red color represents the simulation results, and blue color denotes the experimental data. Slit length—value from 0 to 100 mm is given on the y-axis. Slit width and creepage values on the left and right side are given on the x-axis. For calculations, we used different plates from the given experimental results.

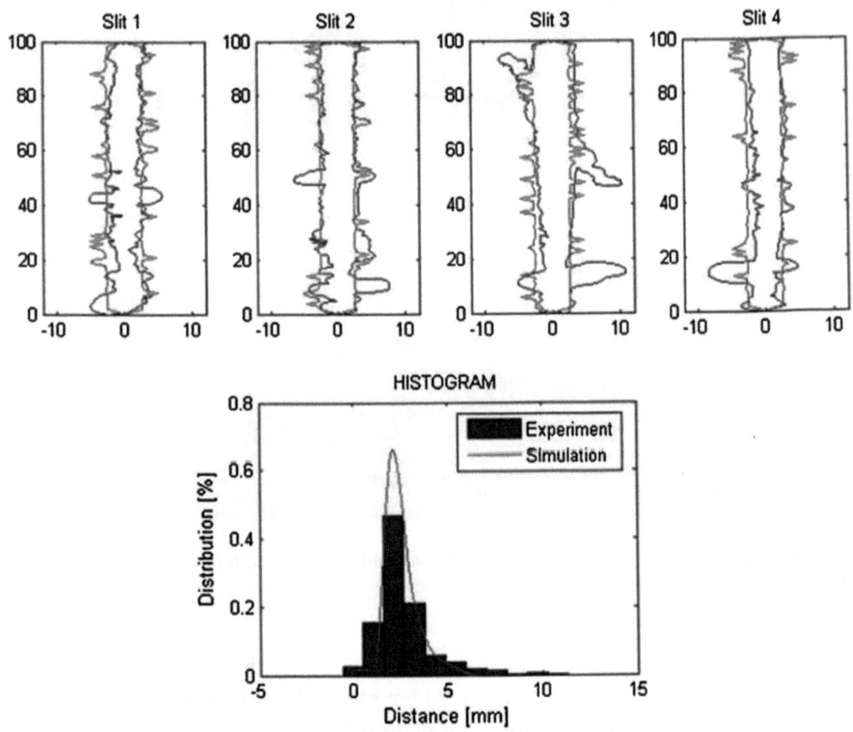

Fig. 19 Comparison of the experimental and computer simulation results

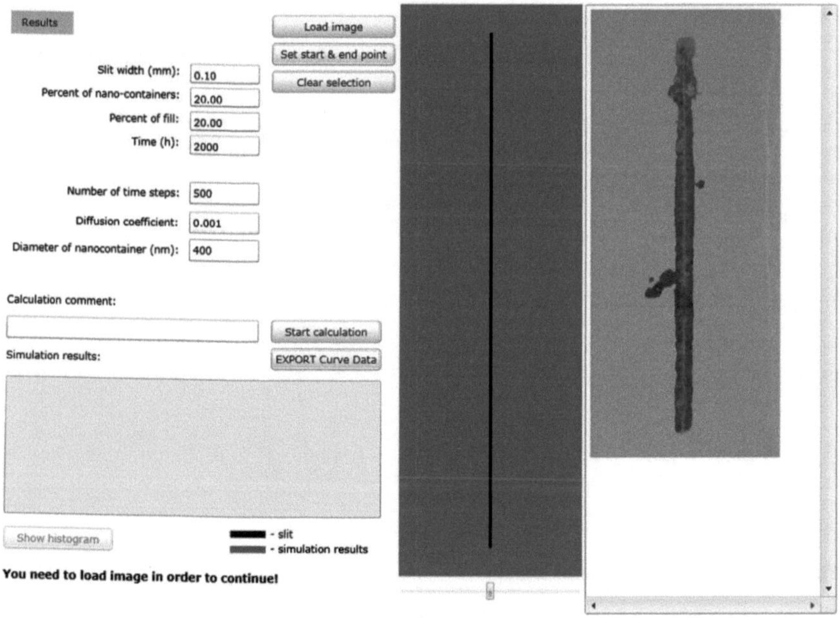

Fig. 20 Experimental data and simulation results for creepage

We compared the simulation and experimental results using histogram. Red line represents the simulation results while blue bars are used for the experimental ones. We represented frequency of repeating certain value—millimeters of obtained creepage (bottom of the Fig. 20).

We used the plate from the experimental data and ran calculation on site with parameters which can be seen in Fig. 21. After the calculation is finished, the results are loaded. The experiment is shown on the right side of Fig. 20. The simulation results are on the right panel. Black color is used for the slit, and red color for the simulation results—creepage.

Experimental creepage effects are caused by corrosion of the plate because scratching. The plates after scratching are exposed to corrosion after 30 days on the air. The distances from the scratch left and right are analysed with image processing method.

The output results have been shown in the histogram. By clicking the Show histogram button, Fig. 21 appears.

Average coverage after 500 h is presented in Fig. 22.

Average coverage after 2000 h is presented in Fig. 23.

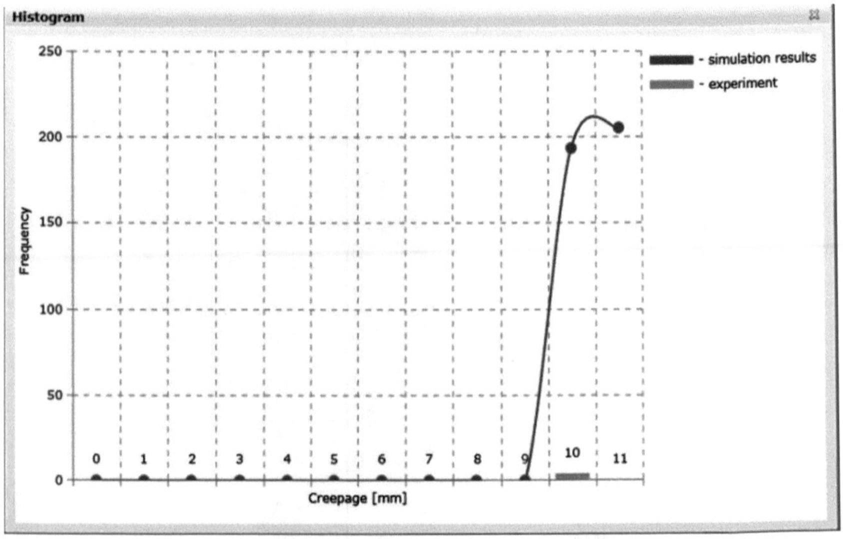

Fig. 21 Output results in the histogram

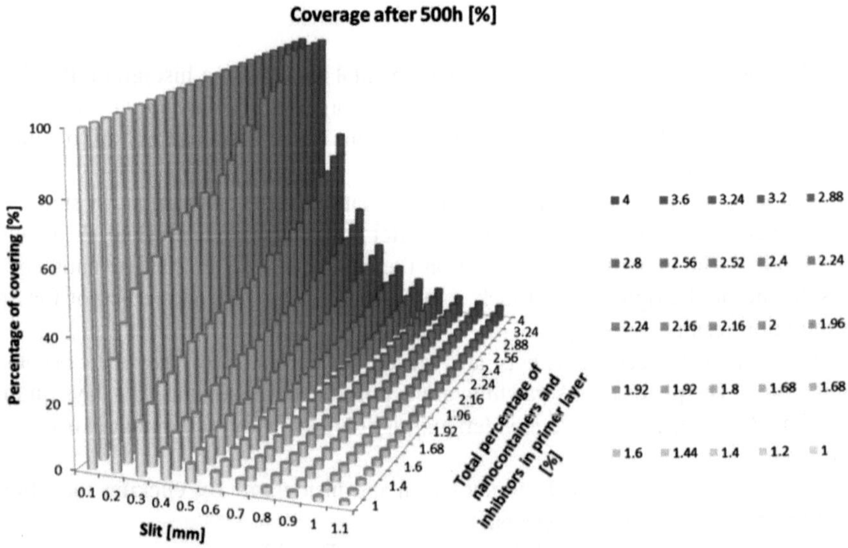

Fig. 22 Average coverage after 500 h

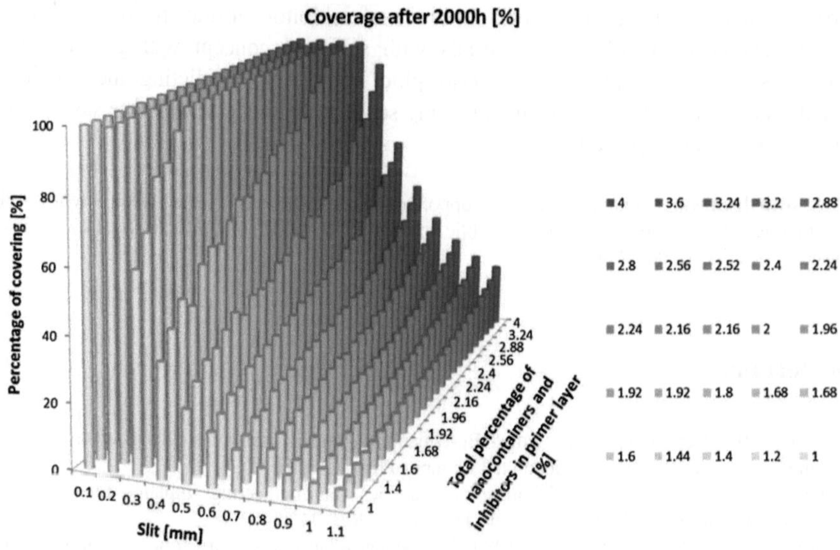

Fig. 23 Average coverage after 2000 h

4 Discussion

In this chapter bio-inspired self-healing nanocontainers are presented with help of computer simulation. Two different approaches were used, DPD and FE modeling methods, in order to simulate self-healing systems with nanocontainers and inhibitors on real industrial problems in the automotive, aerospace and biomedical industry. DPD simulations were tested on several examples. A high rate of the covered surface was determined in the beginning of the simulation process while the layer thickness corresponded to the inhibitors concentration in the primer layer.

The full three-dimensional Navier-Stokes equations were used for the finite element study. With these two methodologies a number of real cases from industry can be simulated. These complex computer models integrate fluid dynamics, structural mechanics and chemical reactivity.

5 Conclusions

Nanocoating interfaces represent a new technology for automotive, aerospace and biomedical industry. Two different modeling approaches, discrete and continuum are described in this chapter DPD and FEM. DPD method is discrete method with three forces: repulsive, dissipative and random forces, as well as additional forces which bound healing agents to a metal substrate. FEM is continuum modeling method with different diffusivity and fluxes. We described realistic case examples from industry with different concentrations of inhibitors inside the primer layer. Next generation of adaptive materials with self-heal concept will give a great progress in field of preventive catastrophic failure. By predicting the possible chances of damage by mesoscopic bridging scale modeling, considerable savings of materials can be expected.

Acknowledgements This research was supported by the Ministry of Education, Science and Technological Development of the Republic of Serbia, III41007, ON174028 and FP7 MUST project NMP3-LA-2008-214261.

References

Abe T (1997) Derivation of the lattice Boltzmann method by means of the discrete ordinate method for the Boltzmann equation. J Comput Phys 131:241

Agarwal R, Garcia JA (2015) Biomaterial strategies for engineering implants for enhanced osseointegration and bone repair. Adv Drug Deliv Rev 94:53–62

Agarwala VS, Delucia JJ, Balin LJ (1989) Microencapsulation crack arrestment compound and corrosion inhibitor for organic coatings. In: Scantlebury DJ, Kendig MW (eds) Advances in corrosion protection by organic coatings. ECS, Pennington, NJ

Alder BJ, Wainwright TE (1959) Studies in molecular dynamics. I Gen Method J Chem Phys 31:459–466

Balazs AC (2007) Modeling self-healing materials. Mater Today 19(9):18–23

Batchelor A, Chandrasekaran M (2004) Service characteristics of biomedical materials and implants, vol 3. Imperial College Press. ISBN: 1-86094-475-2

Bathe KJ (1996) Finite element procedures. Prentice-Hall, Englewood Cliffs, NJ

Bibbs BF, Kermasha S, Alli I, Mulligan CN (1999) Int J Food Sci Nutr 50:213

Boryczko K, Dzwinel W, Yuen D (2003) Dynamical clustering of red blood cells in capillary vessels. J Mol Model 9:16–33

Cagin T (1993) Mechanical response of high performance polymers. Mater Theory Model 291:321–324

Cavaleiro A, Hosson JTHM (2006) Nanostructured coatings. Springer. ISBN-10: 0-387-25642-3

Chen H, Chen S, Matthaeus WH (1992) Recovery of the Navier-Stokes equations using a lattice-gas Boltzmann method. Phys Rev A 45:5339

Dale H, Hallan G, Espehaug B, Havelin LI, Engesaeter LB (2009) Increasing risk of revision due to deep infection after hip arthroplasty. Acta Orthop 80:639–645

De Breij A, Riool M, Kwakman SHP et al (2016) Prevention of Staphylococcus aureus biomaterial-associated infections using a polymer-lipid coating containing the antimicrobial peptide OP-145. J Control Release 222:1–8

Enos DG, Guilbert CR, Norman CF, Kehr JA, Boyer CE (1998) Improving the damage tolerance and extending the service-life of fusion-bonded epoxy coatings. ICCRCS, Orlando, FL

Filipovic N, Kojic M, Tsuda A (2008a) Modeling thrombosis using dissipative particle dynamics method. Philos Trans R 366:3265–3279

Filipovic N, Ravnic DJ, Kojic M, Mentzer SJ, Haber S, Tsuda A (2008b) Interactions of blood cell constituents: experimental investigation and computational modeling by discrete particle dynamics algorithm. Microvasc Res 75:279–284

Fournier E, Devaney R, Palmer M et al (2014) Superelastic orthopedic implant coatings. J Mater Eng Perform 23:2464–2470

Frisch U, Hasslacher B, Pomeau Y (1986) Lattice-gas automata for the Navier-Stokes equations. Phys Rev Lett 56:1505

Gallo J, Holinka M, Moucha SC (2014) Antibacterial surface treatment for orthopaedic implants. Int J Mol Sci 15:13849–13880. doi:10.3390/ijms150813849
Goodman BS, Yao Z, Keeney M, Yang F (2013) The future of biologic coatings for orthopaedic implants. Biomaterials 34:3174–3183
Groot RD, Warren PB (1997) Dissipative particle dynamics: Bridging the gap between atomistic and mesoscopic simulation. J. Chem. Phys. 107(11):4423–4435
Guan H, van Staden Johnson WN, Loo Y (2011) Dynamic modelling and simulation of dental implant insertion process—a finite element study. Finite Elem Anal Des 47:886–897
Hansson S, Loberg J, Mattisson I, Ahlberg E (2010) Characterisation of titanium dental implants: II. Local biomechanical model. Open Biomater J 2:36–52
Hansson S, Loberg J, Mattisson I, Ahlberg E (2011) Global biomechanical model for dental implants. J Biomech 44:1059–1065
Hendrickson B, Leland R (1995) A multilevel algorithm for partitioning graphs. In: Supercomputing '95: Proceedings of the 1995 ACM/IEEE Conference, High Performance Networking and Computing. ACM, New York
He X, Luo LS (1997) Theory of lattice-Boltzmann method: from the Boltzmann equation to the lattice Boltzmann equation. Phys Rev E 56:6811
Huebner KH (1975) The finite element method for engineers. Wiley, New York
Jud K, Kausch HH, Williams JG (1981) Fracture-mechanics studies of crack healing and welding of polymers. J Mater Sci 16:204–210
Kart SO, Tomak M, Uludogan M, Cagin T (2004) Liquid properties of Pd-Ni alloys. J Noncryst Sol 337:101–108
Kheirkhah M, Fathi M, Salimijazi RH, Razavi M (2015) Surface modification of stainless steel implants using nanostructured forsterite (Mg_2SiO_4) coating for biomaterial applications. Surf Coat Technol 276:580–586
Kojic M, Filipovic N, Zivkovic M, Slavkovic R, Grujovic N (1998) PAK-F finite element program for laminar flow of incompressible fluid and heat transfer. Faculty of Mechanical Engineering, University of Kragujevac, Serbia
Kuharsky AL, Fogelson AL (2001) Surface-mediated control of blood coagulation: the role of binding site densities and platelet deposition. Biophys J 80:1050–1074
Landfester K (2001) The generation of nanoparticles in miniemulsions. Adv Mater 13(10):765–768
Lee LH (1991) Adhesive bonding 239-291. Plenum, New York
Lentino JR (2003) Prosthetic joint infections: bane of orthopedists, challenge for infectious disease specialists. Clin Infect Dis 36:1157–1161
Leon B, Jansen JA (2009) Thin calcium phosphate coatings for medical implants. Springer. ISBN: 978-0-387-77718-4
Loscalzo J, Schafer AI (2003) Thrombosis and hemorrhage, 3rd edn. Lippincott Williams & Wilkins, Philadelphia
Lu Y, Fan H, Stump A, Ward TL, Reiker T, Brinker CJ (1999) Aerosol-assisted self-assembly of mesostructured spherical nanoparticles. Nature 398:223–226
Nag S, Banerjee R (2012) Implant: materials for medical devices, vol 23. ASM International
Nelder J, Mead R (1965) A simplex method for function minimization. Comput J 7(4):308–313
Philippi PC, Hegele LA Jr, dos Santos LOE, Surmas R (2006) From the continuous to the lattice Boltzmann equation: the discretization problem and thermal models. Phys Rev E73:056702
Plimpton SJ (1995) Fast parallel algorithms for short-range molecular dynamics. J Comp Phys 117:1–19
Qian YH, d'Humi`eres D, Lallemand P (1992) Lattice BGK models for Navier–Stokes equations. Europhys Lett 17:479
Raphel J, Holodniy M, Goodman BS, Heilshorn CS (2016) Multifunctional coatings to simultaneously promote osseointegration and prevent infection of orthopaedic implants. Biomaterials 84:301–314
Shan X, He X (1998) Discretization of the velocity space in the solution of the Boltzmann equation. Phys Rev Lett 80:65

Shan X, Yuan XF, Chen H (2006) Kinetic theory representation of hydrodynamics: a way beyond the Navier-Stokes equation. J Fluid Mech 550:413

Srinivasa AR, Phares DJ (2004) Molecular dynamics with molecular temperature. J Phys Chem A 108:6100–6108

Stavrinidis B and Holloway DG (1983) Crack healing in Glass. Phys Chem Glasses 24(1):19–25

Stillinger FH, Rahman A (1974) Improved simulation of liquid water by molecular dynamics. J Chem Phys 60:1545–1557

Tyagi S, Lee JY, Buxton GA, Balazs CA (2004) Using nanocomposite coatings to heal surface defects. Macromolecules 37:9160–9168

Valstar ER, Nelissen RGHH, Reiber JHC, Rozing PM (2002) The use of Roentgen stereophotogrammetry to study micromotion of orthopaedic implants. ISPRS J Photogramm Remote Sens 56:376

van Ooij WJ, Zhu D, Palanivel V, Lamar JA, Stacy M (2002) Silicon Chem

Vanegas-Acosta CJ, Landinez PSN, Garzón-Alvarado AD, Casale RCM (2011) A finite element method approach for the mechanobiological modeling of the osseointegration of a dental implant. Comput Methods Programs Med 101:297–314

Verberg R, Dale AT, Kumar P, Alexeev A, Balazs AC (2007) Healing substrates with mobile, particle-filled microcapsules: designing a 'repair and go' system. J R Soc Interface 4(13):349–357

White SR, Sottos NR, Geubelle PH, Moore JS, Kessler MR, Sriram SR, Brown EN, Viswanathan S (2001) Autonomic healing of polymer composites. Lett Nat 409:794–797

Williams JR, Hocking G, Mustoe GGW (1985) The theoretical basis of the discrete element method. In: Proceedings of the NUMETA'85 conference, Swansea, UK, pp 897–906

Wolf-Gladrow DA (2000) Lattice gas cellular automata and lattice-boltzmann models: an introduction. Springer, Berlin

Wolfram S (1986) Cellular automaton fluids 1: basic theory. J Stat Phys 45:471

Zheludkevich ML, Shchukin DG, Yasakau KA, Mohwald K, Ferreira MGS (2007) Anticorrosion coatings with self-healing effect based on nanocontainers impregnated with corrosion inhibitor. Chem Mater 19:402–411

Zmistowski B, Karam JA, Durinka JB, Casper DS, Parvizi J (2013) Periprosthetic joint infection increases the risk of one-year mortality. J Bone Jt Surg Am 95:2177–2184

Nanomaterials for Skin Care

Suzana Miljković, Marija Tomić, Igor Hut and Svetlana Pelemis

Abstract There are a lot of cosmetic products on the market based on nanotechnology. Those products contain nanomaterials since they have many advantages such as improved delivery of active ingredients, stability and photostability of potentially unstable cosmetic ingredients, increased efficacy and tolerance of the skin for various UV filters. Nanomaterials contribute to easiness of application and aesthetic appearance of final products. Although they offer many possibilities, their use demands caution. Nanoparticles have a large surface to volume ratio leading to their reactivity and alteration in biological activity compared to the parent bulk materials. The shape and size of the particles are the cause if their toxic effects, rather than their chemical properties. There are various nanosystems currently in use, in pharmaceutical and cosmetic industry, and many new ones, waiting to be applied. In this article we are going to introduce those who find an application in personal care products such as: liposomes, niosomes, transfersomes, nanoemulsions, solid lipid nanoparticles, polymeric systems, nanocrystals, fullerenes and finally metal oxide nanoparticles.

Keywords Nanomaterials · Nanodispersed systems · Nanoparticles · Skin care · Toxicity

S. Miljković
Faculty of Pharmacy, Novi Sad, University Business Academy in Novi Sad, Trg Mladenaca 5, 21000 Novi Sad, Serbia

M. Tomić (✉) · I. Hut
Faculty of Mechanical Engineering, Department of Biomedical Engineering, University of Belgrade, Kraljice Marije 16, 11000 Belgrade, Serbia
e-mail: marija.m.tomic@gmail.com

S. Pelemis
Faculty of Technology, University of East Sarajevo, Karakaj 1, 75400 Zvornik, Bosnia and Herzegovina

D. Brabazon et al. (eds.), *Commercialization of Nanotechnologies–A Case Study Approach*, DOI 10.1007/978-3-319-56979-6_9

1 Introduction

Nanotechnology is a quite young scientific and technological discipline, although partial research in nanometre range is known for a long time. The studies of biological systems and nanoengineering of many materials such as colloidal dispersions, metal nanoparticles (NPs) or various catalysers have been done for centuries. Three thousand years ago the ancient Chinese were writing their scrolls in ink containing traces of nanomaterial (molecule C_{60}). The ancient Chinese have also used nanosized gold powder to produce the famous ruby red colour of their porcelain. The use of colloidal gold has a long history however the first broad study on its preparation and properties were published in mid-19th century. Therefore by using new techniques we are finally able to comprehend the interactions at atomic level.

NPs with particles from 1 to 100 nm in size have a large surface to volume ratio. This is the reason of altered biological activity compared to the original bulk materials (Yildirimer et al. 2011). Their size represents great advantage for cosmetic products since they improve the stability of potentially unstable cosmetic ingredients, like vitamins, antioxidants, unsaturated fatty acids, fragrances and others; increase stability and photostability, efficacy and tolerance of the skin for various UV filters; serve as delivery systems which increase the release of cosmetic ingredients in lower layers of the skin and make final products more aesthetically acceptable for the customers, easier for the application and with more pleasant feeling. Many different NPs were investigated and incorporated in cosmetic products and some of them are already on the market: metal oxide NPs, fullerenes, nanocrystals, polymeric nanocapsules, solid lipid NPs (SLN) and nanostructured lipid carriers (NLC), as well as many others.

Unfortunately, their size could be the main reason of side effects and toxicity, as well. The surface chemistry of this kind of materials can affect toxicity and it is possible to design their surface chemistry to enhance safety. So far, there is no evidence of NPs in beauty products being a health problem, but on the flip side, there is little conclusive proof that they are safe. Many experts believe nanomaterials should be treated and tested as new chemicals.

First cosmetic line with NPs (liposomes) on the market was Capture brand by Christian Dior, launched in 1986 (Müller-Goymann 2004). Today, after almost 30 years, many cosmetic lines and brands use some kind of nanomaterials in their products in order to improve their properties, efficacy and appearance. One of the largest cosmetic manufacturers in the world today, L'Oreal, includes nanotechnology improvement in many products and ranks 6th in US in the number of nanotechnology related patents (Sharma et al. 2012; Raj et al. 2012).

There are a lot of different nanosystems currently in use in drugs and cosmetic products. We are going to introduce those who find an application in personal care products. Generally, they could be divided to nanodispersed systems and NPs:

1. Nanodispersed Systems

   - Vesicular Delivery Systems—Liposomes, Niosomes, Transfersomes
   - Nanoemulsions
   - Solid lipid NPs.

2. Nanoparticles

   - Polymeric systems (nanocapsule, dendrimer)
   - Nanocrystals
   - Fullerenes (Buckyballs)
   - Metal oxide NPs.

2 Nanodispersed Systems

2.1 Vesicular Delivery Systems

2.1.1 Liposomes

Liposomes are biocompatible colloidal systems which provide many benefits to the skin (Dua et al. 2012). The wall of these vesicles, filled with water, is made of hydrophobic phospholipids organised in lipid bilayers. Phospholipids forming the liposomes' bilayer structures could be: phosphatidylcholine, phosphatidylethanolamine, phosphatidylinositol as well as stearylamine, cholesterol or lecithin. Liposomes can have only one (unilamellar, UL), or more (multilamellar, ML) bilayer structures separated by water layers which defines their size: usually, they are between 15 nm and a few micrometres (Fig. 1).

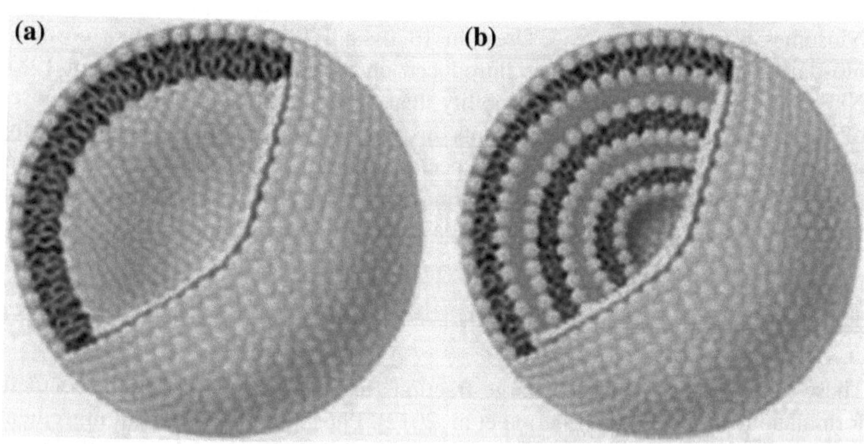

Fig. 1 **a** Unilamellar liposome; **b** Multilamellar liposome (Encapsula Nanosciences)

Phospholipids, including phosphatidylcholine, serve as a selective and protective barrier in the cellular membrane. These long natural molecules, characterized by hydrophilic head and two lipophilic tails, are essential constituents of the cell membrane. This amphiphilic character is responsible for organising the bilayer structure and forming the liposomes in water media (Morganti 2010).

Phospholipids are related and have a high affinity to the skin and therefore are diffused through the cutaneous layers easily. They offer many benefits to the skin: as a source of essential fatty acids, and as a creator of protective film on the skin surface. Since phospholipids prevent the water loss they increase the hydration of the stratum corneum (SC) simultaneously. Because of all these functions, phospholipids, as a part of liposomes, are not only excipient for different active ingredients, but an active component with a lot of benefits for the skin. This is one of the main advantages of liposomes compared to many other carriers in cosmetic products.

Liposomes are one of the most thoroughly studied types of colloidal nanocarriers (Pippa et al. 2013). They are used as delivery systems for many drugs, and many liposomal drug formulations are already approved for clinical use, most of them are anticancer drugs (Liu et al. 2010). Liposomes have the possibility to encapsulate both water-soluble and oil-soluble compounds. They facilitate penetration and give continuous and long-lasting effect of many cosmetic active ingredients in the skin, like vitamins, lipids, various extracts, hyaluronic acid, water, etc. (Patravale and Mandawgade 2008).

There are some drawbacks of liposomes, as stability, poor batch-to-batch reproducibility in a manufacturing process, insufficient size of active ingredients compartment, fast release of hydrophobic drugs, difficulties in sterilisation (Rai et al. 2008; Beija et al. 2012).

2.1.2 Niosomes

Niosomes were patented by L'Oreal in 1970s and 1980s, and the first cosmetic niosome-containing product was introduced on the market by Lancôme in 1987. Niosomes have higher chemical stability than liposomes, since they are composed of two-layers of non-ionic surfactants instead of phospholipids. Manufacturing process is much easier, requiring no special conditions, and costs are lower comparing to production of liposomes (Sharma et al. 2012). These carriers structure can improve stability, photostability and bioavailability of many ingredients.

2.1.3 Transfersomes

These lipid vesicles consist of large fractions of fatty acids appeared in cosmetic formulations in the 1990s (Sharma et al. 2012). Phospholipids are main ingredient, but transfersome contains 10–25% surfactant and 3–10% ethanol, as well (Dubey et al. 2006). They could contain cholesterol, sodium cholate as an additional surfactant and other ingredients. They are elastic liposomes, ultradeformable structures which can pass through pores less than one-tenth of their diameter. It means that 200–300 nm sized transfersome could penetrate the intact skin (Sharma et al. 2012).

2.2 *Nanoemulsions*

Due to numerous potential applications, in pharmaceutical, cosmetic, food and chemical industries, interest for nanoemulsions shows steady growth over the last years (Solans and Solé 2012). Although there is still some confusion about the differences and similarities, these colloidal systems could be divided into nanoemulsions, ultrafine emulsions with average droplet size ranging from 50–1000 nm, and microemulsions with particle diameter less than 100 nm (Sharma et al. 2012; Solans and Solé 2012). Anyway, there is a strict difference between nano and conventional emulsions, which usually have micrometre size droplets.

Nanoemulsions are macroscopic preparations which contain droplets of water or oil reduced to nanosize (20–500 nm) to preserve the transparency and the lightness of the formula (Fig. 2). Most emulsions in cosmetic products have water as a continuous phase. Contrary to conventional emulsions, nanoemulsions are stable against sedimentation or creaming, since their droplet size is very small. Another advantage is their transparent or translucent properties. Comparing to microemulsions, which are thermodynamically stable, nanoemulsions are prone to flocculation, coalescence or other forms of destruction. Careful selection of ingredients and method of preparation are of utmost importance in order to prevent instabilities (Solans and Solé 2012).

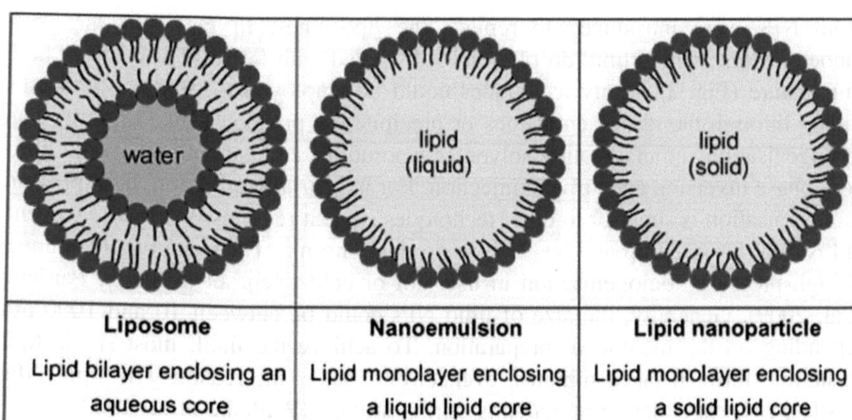

Fig. 2 Structure of the liposome, nanoemulsion and lipid nanoparticle (Cosmetic Web)

It should be emphasized that the term microemulsion is inappropriate, since this system is not an emulsion and its droplet sizes are much smaller than a micrometre range. Nevertheless, it is adopted in the scientific community for more than 50 years.

To produce a nanoemulsion, it is necessary to use either high-energy method (higher the energy input, smaller the droplet size), or much more efficient low-energy emulsification method, which uses the internal chemical energy of the system. Whatever method is used, similar droplet size could be achieved (Yang et al. 2012).

Nanoemulsions are already used in commercial personal care products—different skin and hair care products like sunscreens, shampoos, deodorants. They are stable, transparent and have desired properties—rapid penetration and hydrating activity. For dry hair care, cationic nanoemulsions are very important, since they improve tactile and combing properties and quality of the hair (Sonneville-Auburn et al. 2004). Nanoemulsions are included in development of aqueous-based nail lacquers: new conditioning nail enamel is based on W/O emulsion with nitrocellulose (Patravale and Mandawgade 2008).

There are many products and patents referring as microemulsions aimed for skin and hair care: for cleaning and regeneration of the hair, antiperspirants and deodorants, moisturisers, face and body creams and lotions, sunscreens, pre- and after- shave lotions, makeups etc. Interesting recent advance in hair colouring and lightening is US Patent 6315989 assigned to Revlon Consumer Products Corporation, for W/O microemulsion peroxide formulation. This composition reduces the bleaching and oxidative dying of hair to 5–10 min, and made it fast and easy procedure possible to do at home (Sharma et al. 2012).

2.3 Lipid Nanoparticles

Lipid NPs were introduced to replace the liposomes. In fact, they are W/O nanoemulsions, where lipid droplets are substituted with fat solid at room or body temperature (Fig. 2). Many techniques could be employed to produce lipid NPs, mainly through the use of emulsions or precipitation processes, like high pressure homogenisation, emulsification-solvent evaporation, emulsification-solvent diffusion, phase inversion and solvent injection. For industrial production, high pressure homogenisation is superior to other techniques since it requires no organic solvents and is easy for scale up and has a short production time. There are two techniques of high pressure homogenisation in use, hot or cold (Beija et al. 2012; Pardeike et al. 2009). Generally, the size of lipid NPs could be between 10 and 1000 nm, depending on the method of preparation. To achieve the final, most stable lipid crystalline form, it takes minutes, even days. That is the reason why lipid NPs, mostly SLN, expel active ingredients during storage (Pardeike et al. 2009).

These are biocompatible and biodegradable systems with possible size and surface modifications. They could even be stimuli-responsive systems.

Incorporation of either SLN or NLC into the final product is the same. There are three methods: the direct production of a final product containing only NPs in one-step process, the admixing NPs to previously prepared product or by adding viscosity enhancers to the aqueous phase of NPs to get a gel form. Active ingredients sensitive to degradation, like retinol, tocopherol or coenzyme Q10, will be protected and stabilised after the incorporation in lipid NPs (Jenning and Gohla 2001). It is possible to control the way and duration of the release of active ingredients from lipid NPs (Wissing and Müller 2003).

When topical product is applied to the skin, lipid NPs create a uniform occlusive film on the surface. To form a monolayer film on the skin, 4 mg of the product with 4% of lipid NPs of about 200 nm size is necessary (Wissing et al. 2001). Lipid NPs are small and perform high density packing on the surface of the skin, with small spaces in between. Lipid particles of 200 nm size have an occlusion factor of 50%, the opposite to particles with diameter greater than 1 μm, where an occlusion factor is only 10% (Fig. 3). This occlusive film prevents dehydration of the skin, retards the penetration of UV filters in sunscreens, and offers whitening effects, lubrication/emolliency and controlled release properties (Mehnert and Mader 2001; Souto and Müller 2008).

Different cosmetic active ingredients and drugs could be incorporated into or on the lipid NPs, but they have to be lipophilic (Fig. 4). There are reports in scientific literature that drugs like chloramphenicol, dexamethasone, hydrocortisone, indomethacin, metronidazole, progesterone etc. as well as the cosmetic active ingredients like benzyl nicotinate, ketoconazole, ascorbyl palmitate, beta-carotene and coenzyme Q10 were included in lipid NPs (Souto et al. 2007; Doktorovova and Souto 2009).

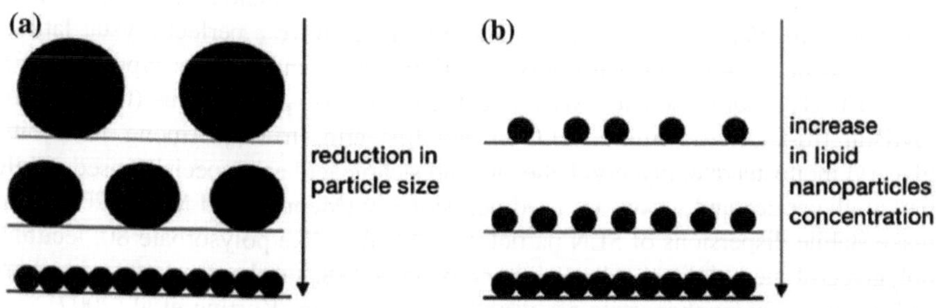

Fig. 3 Occlusion factor of lipid NPs depends on various factors: for identical content of lipids, a decrease of particle dimensions provides an increase of the number of the particles and the film becomes thicker (*left*) causing an increase of the occlusion factor. If the number of the particles is defined an increase of a concentration of the lipids causes an increase of the number of the particles and film thickness (*right*), providing better occlusion effect as well (from Papić-Obradović et al. 2009, with permission of the author)

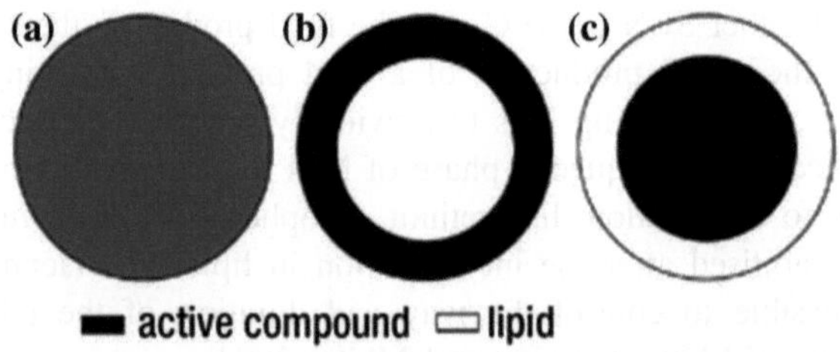

Fig. 4 An interaction of a drug and SLN: **a** homogenously solvated drug or a drug dispersed inside the whole lipid matrix; **b** an active substance in a capsule (homogenously or by cluster formation); **c** a drug in the middle of a particle (homogenously or by cluster formation) (from Papić-Obradović et al. 2009, with permission of the author)

Fig. 5 The structure organisation of SLNs (*left*) and NLCs (*right*) (from Papić-Obradović et al. 2009, with permission of the author)

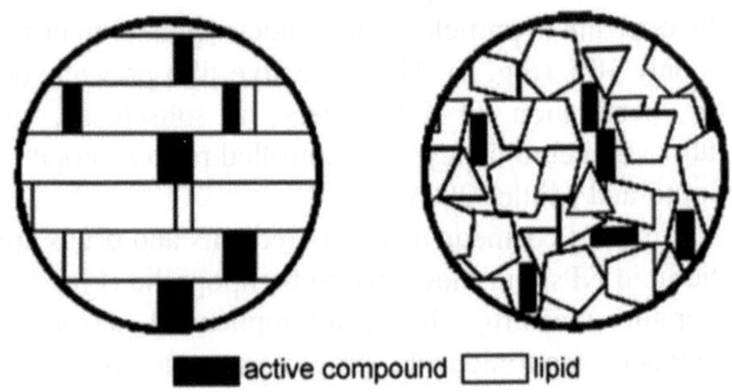

2.3.1 Solid Lipid Nanoparticles (SLN)

SLN contain lipid matrix solid at body temperature (0.1–30% w/w), water and sometimes surfactant (0.5–5%). The diameter of SLN is in nano range—between 40 and 1000 nm (Pardeike et al. 2009). These particles have a perfect crystal lattice, and are made of physiological lipids like triglycerides and hard fat types (Fig. 5).

The triglycerides used for SLNs are tricaprin (tri-C_{10}), trilaurin (tri-C_{12}), trimyristin (tri-C_{14}), tripalmitin (tri-C_{16}) and tristearin (tri-C_{18}).Among hard fats, glyceryl monostearate, glyceryl behenate and stearic acid are especially used, while partial glycerides and waxes are used occasionally (Mehnert and Mader 2001). To make stabile dispersions of SLN particles, emulsifiers like polysorbate 80, lecithin, polyglycerol methylglucose distearate, poloxamer 188, saccharose fatty acid esters and sodium cocoamphoacetate are usually used (Schafer-Korting et al. 2007).

There are only few commercial products with SLN on the market, due to many drawbacks of these active ingredient carriers. Disadvantages of SLNs are poor stability and batch-to-batch reproducibility, inadequate control of drug release, insufficient active ingredient loading, and possible presence of residual chemicals from synthesis processes (Beija et al. 2012). Before their mass incorporation into cosmetic products, it is necessary to perform more stability tests, to investigate the courses of SLNs interaction with the lipids of SC and how they interfere with the delivery of active ingredients. Famous Chanel's Allure perfume has incorporated

SLNs and the result was a delayed release of perfume over a satisfying period of time (Wissing et al. 2000).

2.3.2 Nanostructured Lipid Carriers (NLC)

NLC represent a mixture of solid and liquid lipids, with typical ratio between 70/30 to 99.9/0.1 and fat content at least 95%. On the opposite of the SLN suspensions, NLC hold less water. The matrices of NLC have distorted structure with much more space for cosmetic actives or drug load, (Fig. 5) (Sharma et al. 2012; Müller et al. 2007). Also, NLC have an improved stability and rarely expulse actives during storage like SLN.

The first commercial products with NLC were launched on the market in October 2005. Those were cream and serum with coenzyme Q10. Today, there are more than 30 commercial cosmetic products on the market with numerous active ingredients incorporated in NLC: avocado oil, black currant seed oil, kukuinut oil, *Macadamia ternifolia* seed oil, *Manoi tiaretahiti*, and ω-3 and ω-6 unsaturated fatty acids (Pardeike et al. 2009).

3 Nanoparticles

3.1 Polymeric Nanocapsules

Polymeric nanocapsules are spherical hollow structures made of polymers like: poly-(ε-caprolactone) (PCL), poly-(glycolic acid) (PGA), poly-(alkylcyanoacrylates), poly-(butylcyanoacrylates), poly-(ethylcyanoacrylates), poly-L-lactide (PLA), poly-(lactide-co-glycolide) (PLGA), poly-(alkylene adipate), polyvinyl acetate (PVA), cellulose acetate phthalate, cellulose acetate butyrate and poly-(ε-caprolactone)-block-poly-(ethylene glycol). PLGA have the satisfactory biocompatibility and represents the one of the Food and Drugs Administration (FDA) approved biodegradable polymers for biomedical applications. It is possible to manipulate with size, shape and mechanical properties, and accomplish the controlled release of active ingredients (Beija et al. 2012). Compared with liposome structures, these nanocapsules are more stable, as they are held together by strong covalent bonds. There is a possibility to use nanocapsules to enclose drug of poor solubility and permeability, to make products with controlled drug release and long circulation half-life. Unfortunately, they do not offer biocompatibility as high as liposomes (Liu et al. 2010). The disadvantage could be the potential residual chemicals from the production process (Beija et al. 2012).

In the polymer structure active ingredients are encapsulated and protected by a polymer shell (Fig. 6). Sensitive active substances prone to oxidation or degradation when exposed to oxygen, light or heat, retain their original properties, and incompatible substances are isolated and protected when entrapped in polymeric nanocapsules.

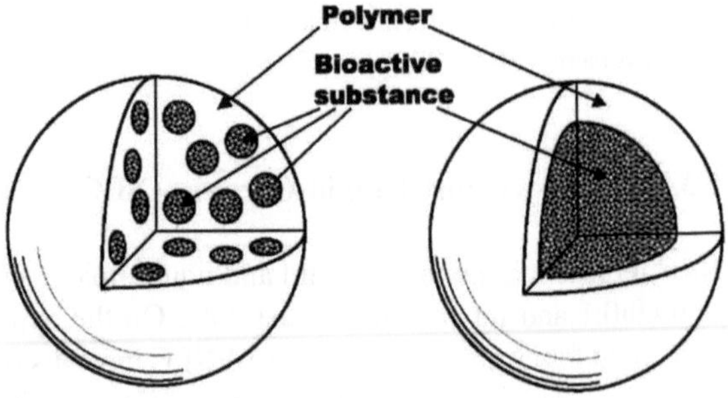

Fig. 6 Different possibilities of incorporating an active substance in a polymer nanoparticle

The core of the polymeric nanocapsules is filled with oil, and usually, many lipophilic active ingredients are dissolved in it, like vitamin E (α-tocopherol), various UV filters (ethylhexyl para-methoxycinnamate), fish or hazelnut oil, bisabolol, farnesol, trigicerides rich in linoleic and/or linolenic acids and many other ingredients and their derivatives.

They are not able to penetrate deep into the skin, although their particle sizes are less than 100 nm. Since these carriers are very rigid and form a film on skin surface, they cannot pass through intercellular space in SC, but stay on the surface. In some studies they showed certain affinity for hair follicles.

Polymeric nanocapsules can be modified on many ways to achieve desirable properties and fulfil different aims, and they are able to respond to the environmental changes and according to that, release the payload. With these nanocarriers, it is possible to control burst or release of fragrances, vitamins or nutraceuticals upon the application of the product. For example, in perfumes, where the aim is to achieve long-lasting fragrance by preventing dilution or washing off, the fragrance is encapsulated into the polymeric nanocapsules which are functionalised with a high cationic charge density on their surface. Normal, healthy skin has negative charge on the surface, and establishes strong adhesion with cationic nanocapsules and gives the perfume prolonged activity. Many sunscreens could be entrapped into the polymeric nanocapsules (octyl methoxycinnamate-OMC, benzophenone-3, octyl salicylate) (Olveira-Martinez et al. 2005). The advantage of this kind of vehicles is the possibility to form a uniform, protective layer on the skin surface while preventing penetration of the load deeper into the skin (Luppi et al. 2004; Hwang and Kim 2008). Polyvinyl alcohol (PVA)-fatty acid nanocapsules decrease percutaneous absorption of benzophenone-3.

3.1.1 Hydrogels

This is a network of hydrophilic polymers that swell in water or biological fluids without dissolving (Morales et al. 2009). For example, hydrogel in the facial mask reacts to the increase of the temperature, shrink and release the active ingredient.

3.1.2 Dendrimers

One of the first dendrimers was synthesized in 1982 (Crommelin and Florence 2013). Dendrimers are unimolecular, spherical micellar nanostructures, around 20 nm in size, organized by symmetrical branches surrounding centre core and holding a high density of functional end groups, carboxyl, hydroxyl and amine, on the periphery (Fig. 7) (Sharma et al. 2012; Raj et al. 2012). The branches could be organised on the premium-first level branches, and make second, third etc. generation of the dendrimers (Beija et al. 2012; Papić-Obradović et al. 2009). They are stable and biocompatible, and offer a number of morphologies depending upon application. The surface of the dendrimers could be multifunctionalised.

The important advantage of dendrimers as a carrier for drug and cosmetic active ingredients is the possibility to entrap different ingredients at the same time, including both hydrophilic and hydrophobic drugs. They can be used either for encapsulation of bioactive ingredient inside the inner cavities, or for attachment of drug at the periphery functional end groups via covalent bonds (Sharma et al. 2012; Kontogiannopoulos et al. 2012). Incorporation of the drug in this kind of drug delivery system provides many advantages, such as increased water solubility of the drug actives, extending the drug circulation time, controlled release of drug, protection of sensitive molecules and prolonged stability (Gillies and Frechet 2005).

Disadvantages of dendrimers are tedious and very complex synthesis, resulting in an expensive product and limited possibilities in manufacturing. Instead of dendrimers, there are ideas to use hyperbranched polymers with similar properties, but cheaper and easier synthesis process (Kontogiannopoulos et al. 2012). Dendrimers are used in skin, hair and nail care cosmetic products.

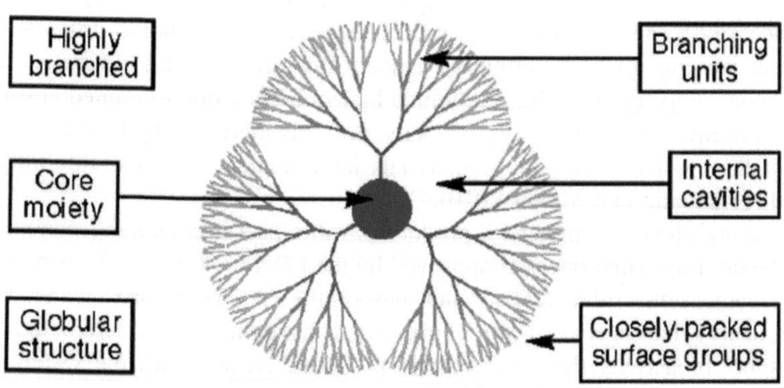

Fig. 7 The main elements of dendritic structure (from Papić-Obradović et al. 2009, with permission of the author)

3.2 Nanocrystals

Nanocrystals are one of the most interesting and promising drug delivery technologies in the last years. Opposite to other nanosystems, nanocrystals consist mostly of a pure drug (Möschwitzer 2012). Usually, they use water as a medium and stabilizers for the colloidal state and they represent a form of nanosuspensions.

Nanocrystals are formed by aggregation of many atoms. These structures are clusters with a typical size of 10–600 nm. To prevent the formation of larger aggregates, clusters must be stabilized. Most often, nanocrystals are produced with one of these procedures: ball milling or high pressure homogenisation, in water or non-aqueous media. It could be used spray-drying or lyophilisation in combination with high pressure homogenisation in which case nanocrystals have better properties: smaller size, better physical stability, and shorter production process (Möschwitzer 2012; Keck et al. 2008). Many research groups and pharmaceutical companies try to develop their one way for nanocrystals production, but they are not widely used yet, because they require corresponding production equipment and processing expertise (Möschwitzer 2012).

Nanocrystals are very simple to add into cosmetic products, creams, lotions, liposomal dispersions. First, they are dispersed in water making the nanosuspension, which is admixed within the preparation.

Nanocrystals promote passive transport and penetration of poorly soluble cosmetic active ingredients from formulation to the skin. They increase saturation solubility of ingredients in a water phase and make higher concentration gradient between product applied on the skin and the skin itself. Molecules of an active cosmetic ingredient penetrating from the water phase into the skin are being replaced with other molecules dissolving from nanocrystals during preparation (Shegokar and Müller 2010). Nanocrystals can reduce the applied dose of active ingredient, since they perform prolonged action (Sharma et al. 2012).

Since 1990s, nanocrystals are a model to deliver hydrophobic drugs and cosmetic active ingredients. Currently, nanocrystals are used for dermal, oral and intravenous delivery. They have a limited bioavailability due to limited absorption. It is a consequence of dissolution velocity, and it is increased by increased surface area of the crystals. Amorphous materials have higher saturation solubility than crystals (Shegokar and Müller 2010).

There are already commercial products on the market containing nanocrystals, like several drugs (first one was approved by the FDA in 2000), and cosmetic, skin care creams with rutin (2007) and hesperidin (2008), poorly water soluble antioxidants (Sharma et al. 2012). Those creams have much more active ingredients and higher bioactivity than traditionally produced creams with the water-soluble derivative of rutin—rutin glucoside.

3.3 Fullerenes

Fullerene was discovered by Kroto/Smalley research team in 1985 as the third allotrope modification of carbon, besides the diamond and graphite (Kroto et al. 1985). It consists of sixty perfectly symmetrically arranged carbon atoms (C_{60}) in a structure like buckyballs. The first production of C_{60} was done at the University of Arizona in 1990. by the Kratschmer/Hoffman research team and the C_{60}'s first image, with atomic resolution, was taken at the University of Belgrade in 1992. by the Koruga/Hameroff research team. Fullerenes have been discovered in nature, and have recently been proven to exist in interstellar space (Matija et al. 2013).

In last couple decades C_{60} molecule has been intensely investigated for the possibility of using it in many applications, particularly in medicine and cosmetics. However, safety issues and its toxicity have been debatable. Fullerenes could be toxic and genotoxic (Oberdorster et al. 2004). Research results proved it was toxic to the vascular endothelial cells (Hideyuki and Naoharu 2006).

Recently, fullerene (C_{60}) and its derivatives served as active compounds in the preparation of anti-aging skin formulations. They are incorporated for their wide range of biological activities, including potent ROS scavenging and potential antioxidant functions (Lens 2009). High amounts of free radicals (superoxide anion, hydroxyl radicals and lipoperoxide), are produced following exposure to UV radiation. These highly reactive free radicals have a devastating effect on nearby living tissues. Burangulov et al. developed a cosmetic product with fullerene clusters as an active ingredient to prevent or retard free radical destroying activities in the skin (Burangulov et al. 2006).

There are many patents connected with the fullerene use in personal care products. In last decades, many influential cosmetic companies in the world claim their products contain NPs like fullerenes, nanotubes, quantum dots etc. (Morganti 2010). Fullerene is used, mostly as a free radical-scavenger, in different personal care products, like moisturiser face creams, sunscreens, as whitening, anti-aging, and cellulite control preparations (Rania et al. 2007).

The modifications of fullerene molecules provide the possibility to produce safe and efficient active ingredient, of diameter size between 4.7–9.5 nm, for medical and cosmetic application (Papić-Obradović et al. 2009). The structure could be improved if the C_{60} is surrounded by the shell of water crystalline, which makes a $C_{60}(OH)_x$ complex (Fig. 8). This layer of water isolates fullerene structure and prevents any toxic effects of C_{60}, but at the same time it keeps all beneficial effects for the biomolecules (Matija et al. 2013). This hydrogenated fullerene becomes water soluble and simple to formulate. Functionalized fullerenes are currently being investigated as drug delivery systems, in targeted imaging in diagnostics, and as free radicals scavenging and antioxidant active ingredient. They are one of the innovations in skin treatment and care. Prerequisite for their incorporation in skin care products were investigations of their toxicity, fortunately fullerene derivative harmonized $C_{60}(OH)_{24}$ showed no toxicity at all in test on both human skin (HDF) and liver carcinoma (HepG2) cells (Sayes et al. 2004).

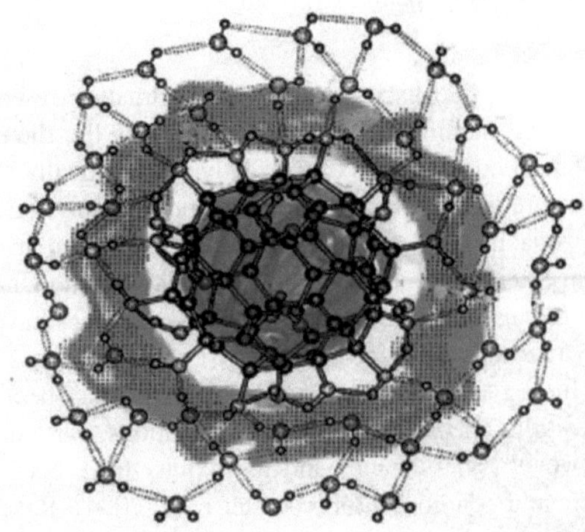

Fig. 8 The structure of a fullerene derivative harmonized $C_{60}(OH)_{24}$ (from Koruga 2008, with permission of the author)

3.4 Metal Oxide Nanoparticles

NPs of zinc oxide (ZnO) and titanium dioxide (TiO_2) for sunscreens and lotions were the first, and are still the most important, applications of nanotechnology in skin care. Traditional mineral particle use in sunscreens formulations gave a noticeable white appearance, while mineral titanium dioxide-based and zinc oxide-based nanomaterials allow making of products with no white appearance on the skin surface at all due to their small size. In this case the physical law of reducing the particle size below one fourth the wavelength of light was used therefore zinc oxide and titanium dioxide become invisible. At the same time, these materials maintain their sun protectant properties (Fig. 9). If anything, the smaller particle size enhances protection by fitting into small irregularities in the skin (Herman 2010). For ZnO NPs, this threshold is 30 nm (Cross et al. 2007) and should be even smaller for TiO_2. Further reduction of the NP size entails the risk of rendering NPs skin penetrable, with potential toxic consequence.

They are efficient, photostable UV filters that act by absorbing UVB and UVA radiation and re-emitting of absorbed energy as less damaging UVA like visible fluorescence or heat (Antoniou et al. 2008). The basic mechanism of action of titanium dioxide and zinc oxide is the reflection of radiation. They act as a mechanical protection which is the reason of their safe applications in sunscreen products. Formulations that utilize TiO_2 or ZnO as the only active sunscreen agents provide photoprotective properties and reduce risk of irritation comparing to other sunscreens. To formulate an ideal sunscreen it is necessary to block UVA/UVB radiation, to use non-toxic ingredients and accomplish an aesthetic appealing. Uncoated TiO_2 absorbs photons of light and emits an excited electron, which can be transferred to free radicals and absorbed into dermal layers, resulting in oxidative damage (Dunford et al. 1997). To stop the formation of reactive oxygen species

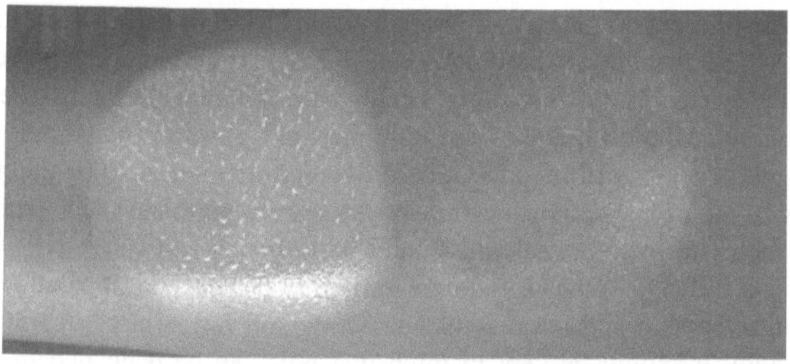

Fig. 9 The difference between non-reduced (*left*) and reduced particle size of zinc oxide and titanium dioxide in a sun protectant cream

(ROS) and prevent the agglomeration of particles, TiO_2 and ZnO are generally coated with aluminium oxide, silicon dioxide or silicon oils (Katz 2007).

Sunscreen products that contain NPs of ZnO and TiO_2 provide both dispersibility and attractive skin feel. To enhance their efficiency, the standard ZnO and TiO_2 UV protection systems can be modified. For example, carnauba wax-loaded NPs have been shown to improve the sun protection factor of TiO_2 in aqueous media without the use of complex formulations (Villalobos-Hernandez and Müller-Goymann 2006). It is assumed that the mechanism of electron transfer taking place from the π delocalized organic compounds, i.e., cinnamates, to the empty conductive bands of the TiO_2 aggregates is responsible for this effect. This transfer may be more effective when the TiO_2 crystals are encapsulated or bonded by the carnauba wax and not when these crystals are attached to the surface of the wax particles.

4 Toxicity of Nanoparticles

Although NPs have many possibilities, their use demands caution.

Toxic effect of NPs originates from their ability to adversely affect normal physiology and to directly destroy the normal structure of organs and tissues of humans and animals. It is widely accepted that toxicity depends on physico-chemical parameters such as particle size, shape, surface charge and chemistry, composition, and subsequent NPs stability. The exact mechanism is not clear yet however recent literature suggests oxidative stress and pro-inflammatory gene activation could be responsible. Beside the particle-related factors, the applied dose, route of administration as well as the extent of tissue distribution could be important parameters in nano-cytotoxicity (Yildirimer 2011).

It is already well known that the reactivity of NPs comes from their large surface area compared to the bulk material. It is not the chemical properties but rather the shape and size of the particles which are responsible for their toxic effects. The attention for the effects on the nature and health for those who work in the production of these materials is minimal. From 1.2 billion dollars spent for research in nanotechnologies only 1% is spent on healthy surrounding and safety research. The applications of nanotechnologies are unlimited but governments should seriously consider the development of safety platforms (Bawarski et al. 2008). For now the pathways of uptake and translocation of NPs are determined and explained but many of them are hypothetical and should be further investigated. NPs may enter the body via different routes of administration: oral ingestion, inhalation, penetration through the skin and intravascular injection and furthermore distribute to any organ system. It should be kept in mind that NPs can be deposited in lungs and could avoid elimination by phagocytosis.

Despite the theoretically brilliant logic, animal and human studies have showed us other possibilities and highlighted the problem of feasibility of interpretation of organ toxicity and the pre-determined dose; also, there are problems of extrapolating in vitro concentrations into in vivo situations which can be subdivided into two points; firstly, it should be determined how successfully any administered NP dose is reaching the target tissue and secondly, to find out if there are any biochemical changes induced by NPs in vivo since they could gone unnoticed in isolated cell-based studies. With the potentially disastrous consequences in mind, new ways of predicting as yet unpredictable, nondose-dependent actions of NPs in vivo must be sought (Yildirimer et al. 2011).

Apart from the applied dose, another, so far underexplored area of nanotoxicity relates to the route of particle administration. Regardless to the administrated dose, the route of application could influence toxicity of NPs in many ways. The fate of NPs in the body (biodistribution, accumulation, metabolism and excretion) will be different according to the route of administration as well as its toxicity.

Topically applicable substances such as sunscreen preparations and cosmetics already rely on the use of nanoformulations of titanium-dioxide and zinc-oxide by exploiting their ability to block ultraviolet radiation. In the future, their potential to penetrate through the skin could be exploited for transdermal drug delivery systems. Therefore, the transition properties and routes, as well as the potential dermal or systemic toxicity need to be evaluated. After intravenous and oral NP administrations systemic effect appears more rapidly compared to transdermal administration. When they come within the circulation, most substances are going through the first-pass metabolism within the liver. Their further destiny is weather to accumulate in the liver or distribute via the systemic circulation to end organs including the brain (Yildirimer et al. 2011).

The penetration of NPs into deeper layers of epidermis and dermis and even into systemic circulation is unwanted but still possible. It is assumed that small particles (such as TiO_2 particles which are 10–20 nm size as sun protection filter) may penetrate in smaller concentrations deeper into skin via hydrophilic channels of 50 nm in diameter. The problem is that those small concentrations could interact with immune system because TiO_2 particles are not biodegradable. Regarding biodegradability and size of the particles lipid NPs are safe because their diameter is around 200 nm. Even if they reach circulation the organism can endure them since they are natural substances and act like parenteral nutrition (Papić-Obradović et al. 2009).

4.1 Toxicity of ZnO and TiO_2 Nanoparticles

Many concerns have been raised regarding the possible systemic absorption of TiO_2 and ZnO NPs, which may arise as a consequence of NP distribution on the human skin. The ideal situations could be if the NPs should remain on the surface of the skin or in the superficial layers of skin (SC). In order to prevent systemic absorption of the nanoparticles their penetration into deeper parts of the skin, particularly down to the living cells in basal layer, should be avoided. Typical experiments testing the penetration of NPs into skin employ a tape stripping technique, in which SC is consecutively removed (Weigmann et al. 2001).

Direct imaging of the sunscreen NPs in skin in vivo could become public safety issue. Further, some research shows that, for mineral sunscreens, TiO_2 does not seem to penetrate the epidermis (Schulz et al. 2002) and ZnO has only limited systemic absorption, if any.

Nano TiO_2 and ZnO are permitted by the FDA and the Environmental Working Group (EWG) agrees with the FDA's assessment (Herman 2010) in part due to better features of nanomaterials compared to alternatives and due to findings that mechanical filters of mentioned nanomaterials measurably improved protection. EWG could not find evidence that NPs used as physical filters in sunscreens could penetrate the skin. It was clear to EWG that sunscreen containing this kind of sun protection materials was greatly improved.

Before reaching a general conclusion that titanium dioxide and zinc oxide NPs do not penetrate beyond the SC, it is necessary to underline that there are papers presenting a slightly different conclusion (Wiechers 2009).

One of the studies used zinc oxide of various non-specified particle sizes on normal and psoriatic human subjects. Levels of zinc were measured after application on the human skin and in vitro on pig skin. In in vivo experiments, there was no increase in plasma levels but in the in vitro experiments, penetration of zinc oxide into skin was less than 1% of the applied dose and majority of investigated material was recovered in the SC (SCCNFP 2003).

Another study on human skin in vitro involved zinc oxide particles of 15–30 nm. Less than 0.03% of the applied zinc was found in the receptor solution, with no particles detected in the epidermis or the dermis (Cross et al. 2007).

The study of titanium dioxide particles coated with dimethicone or silicone dioxide, 30–60 nm, and zinc oxide (uncoated, <160 nm) applied in vitro on pig skin found no penetration beyond the SC (Gamer et al. 2006). Based on these and many other results Nohynek et al. stated that "most available theoretical and experimental evidence suggests that insoluble NPs do not penetrate into or through normal as well as compromised human skin" (Nohynek et al. 2007).

In vitro studies demonstrated cell type-dependent TiO_2 toxicity interfering with the cellular functions such as cell proliferation, differentiation, mobility and apoptosis (Kiss et al. 2008; Pan et al. 2009). Such adverse effects, however, could not be replicated in vivo. In order to assess penetration capabilities, dermal infiltration studies have been carried out on human volunteers using different investigative techniques. Lademann et al. investigated the penetration after repeated administration of TiO_2-containing sunscreen on the volunteers' (Lademann et al. 1999). Methods such as tape stripping and histological appraisal of skin biopsies revealed that TiO_2 was able to penetrate into the open part of a hair follicle as opposed to the viable layers of the epidermis or dermis. It is important to note, the amounts of titanium dioxide in any investigated follicle was less than 1% of the total amount of sunscreen. Surface penetration via hair follicles or pores was also suggested by a study conducted by Bennat and Müller-Goymann where skin permeation was greater when sunscreen was applied to relatively hairy skins (Bennat and Müller-Goymann 2000). After 15 tape strippings of sunscreen Mavon et al. demonstrated near total recovery and no TiO_2 deposition in hair follicles or skin layers (Mavon et al. 2007). Probably the different amount of TiO_2 permeate in deeper layers of the skin and their toxicity correlate with the nanoparticles properties, surface coatings and functionalization of TiO_2 as well as with the number of follicular pores within the skin facilitating particle uptake.

5 Conclusion

Nanotechnology is a broad field with many potential applications in beauty. With so much emphasis on efficacy for skin and hair products, there is no doubt those different sorts of NPs will continue to be exploited by formulators for the elevation of brand performance.

The cosmetic industry has greatly improved nanotechnology, evidenced by the numerous nanotechnology patents held by cosmetic companies. Depending on their intended use, NPs can be designed to enhance or retard the penetration of active ingredients into the epidermis or deep dermis and beyond. The future trend for delivering actives may be improved NP systems that can deliver their active molecules via triggered-release mechanisms.

As the number of cosmetic products containing nanomaterials grows, consumers' transdermal exposure and potential adverse health effects becomes serious reason of concern. Therefore, it is necessary to improve methods of characterisation, quality, safety and efficacy tests for these products, and to provide regulation to ensure consumers protection. European Union made an important step forward in that direction with the newest Directive 1223/2009, demanding that "*For every cosmetic product that contains nanomaterials, a high level of protection of human health shall be ensured*" (Regulation of the European parliament 2009). It means that detailed specification of nano-ingredient and proofs regarding safety of the consumers are required for the nanomaterials containing cosmetic products.

In cosmetics, nanomaterials are mostly used in sun care products where the particle size ensure for high UVB and UVA attenuation but good transparency in the visible region is between 40 and 60 nm (Hewitt 2002). Based on current experimental evidence it is highly unlikely that NPs like titanium dioxide and zinc oxide used in sun care products will penetrate into epidermis.

Finally, to achieve penetration the product must be massaged into the skin in order to open the orifice surrounding the hair follicles. However, the nanomaterials used in cosmetic products such as metal oxides are too rigid and too big to penetrate, and are removed from the skin by sebum flow, hair growth and desquamation, and we humans are simply not hairy enough to allow them to penetrate (Wiechers 2009).

According to previously mentioned data, we can realize endless opportunities for nanotechnology in cosmetics, but this is also not true. At the moment the only reason to incorporate those particulate systems in formulations are purely cosmetic. Customers prefer transparent sunscreen formulations with no white appearance on the skin and for that, nano-sized particles are required. The skin moisturising, elasticity and lubrication benefits offered by the lipid nanomaterials are really secondary benefits, for which many other cosmetic ingredients are available that could offer more effect for less cost (Wiechers 2009).

Therefore, both the risks and the benefits of nanotechnology in cosmetics are probably exaggerated but it remains strange that there is a perceived risk for those topically applied NPs such as the metal oxides for which there is no evidence of skin penetration, whereas there is no perceived risk for those topically applied nanomaterials such as elastosomes and flexible liposomes for which there is evidence of skin penetration (Wiechers 2009).

References

Antoniou C, Kosmadaki MG, Stratigos AJ, Katsambas AD (2008) Sunscreens-What's important to know. J Eur Acad Dermatol Venereol 22:1110–1118

Bawarski EW, Chidlowsky E, Bharali JD, Mousa AS (2008) Emerging nanopharmaceuticals. Nanomedicine: NBM 4:273–282. doi:10.1016/j.nano.2008.06.002

Beija M, Salvayre R, Lauth-de Viguerie N, Marty JD (2012) Colloidal systems for drug delivery: from design to therapy. Trends Biotechnol 9(39):485–496

Bennat C, Müller-Goymann CC (2000) Skin penetration and stabilization of formulations containing microfine titanium dioxide as physical UV filter. Int J Cosmet Sci 22:271–283

Burangulov N, Moravsky AP, Kulikova YV, Dyachuk GI, Loutfy RO (2006) Cosmetic compositions containing fullerene clusters. European patent 1,653,920, 5 Jan 2006

Cosmetic web. http://www.cosmeticweb.co.za/pebble.asp?relid=16774&sec=aes. Accessed 11 Mar 2015

Crommelin DJA, Florence AT (2013) Towards more effective advanced drug delivery systems. Int J Pharm. doi:10.1016/j.ijpharm.2013.02.020

Cross SE, Innes B, Roberts MS, Tsuzuki T, Robertson TA, McCormick P (2007) Human skin penetration of sunscreen nanoparticles: in vitro assessment of a novel micronized zinc oxide formulation. Skin Pharmacol Physiol 20:148–154

Doktorovova S, Souto EB (2009) Nanostructured lipid carrier-based hydrogel formulations for drug delivery: a comprehensive review. Expert Opin Drug Del 6:165–176

Dua JS, Rana AC, Bhandari AK (2012) Liposome: methods of preparation and applications. Int J Pharm Studies Res 32:14–20

Dubey V, Mishra D, Asthana AJ (2006) Transdermal delivery of pinea lhormone: melatonin via elastic liposomes. Biomaterials 27:3491–3496

Dunford R, Salinaro A, Cai L, Serpone N, Horikoshi S, Hidaka H, Knowland J (1997) Chemical oxidation and DNA damage catalysed by inorganic sunscreen ingredients. FEBS Letters 418:87–90

Encapsula NanoSciences. http://www.encapsula.com/

Gamer A, Leibold E, van Ravenzwaay B (2006) The in vitro absorption of microfine ZnO and TiO_2 through porcine skin. Toxicol In Vitro 20(3):301–307

Gillies ER, Frechet JMJ (2005) Dendrimers and dendritic polymers in drug delivery. Drug Discov Today 10:35–43

Hideyuki Y, Naoharu I (2006) Cytotoxicity of water-soluble fullerene in vascular endothelial cells. Am J Physiol Cell Physiol 290:C1495–C1502

Herman S (2010) Nanobeauty. GCI Mag, 3 Aug 2010

Hewitt J (2002) Inorganic sunscreens. In: Schlossman ML (ed) The chemistry and manufacture of cosmetics, 3rd edn. Allured Publishing Book, Carol Stream, Illinois USA, 1 Chapter 32, pp 527–551

Hwang SL, Kim JC (2008) In vitro hair growth promotion effects of cosmetic preparations containing hinokitiol-loaded poly(epsilon-caprolacton) nanocapsules. J Microencapsul 25: 351–356

Jenning V, Gohla SH (2001) Encapsulation of retinoids in solid lipid nanoparticles (SLN). J Microencapsul 18:149–158

Katz LM (2007) Nanotechnology and applications in cosmetics: general overview. ACS Symp Ser 961:193–200

Keck CM, Kobierski S, Mauludin R, Müller RH (2008) Second generation of drug nanocrystals for delivery of poorly soluble drugs: smart crystals technology. Dosis 2:124–128

Kiss B, Biro T, Czifra G, Toth BI, Kertesz Z, Szikszai Z et al (2008) Investigation of micronized titanium dioxide penetration in human skin xenografts and its effect on cellular functions of human skin-derived cells. Exp Dermatol 17:659–667

Kontogiannopoulos KN, Assimopoulou AN, Hatziantoniou S, Karatasos K, Demetzos C, Papageorgiou VP (2012) Chimeric advanced drug delivery nano systems (chi-aDDnSs) for shikonin combining dendritic and liposomal technology. Int J Pharm 422:381–389

Koruga Đ (2008) Golden mean harmonized water and aqueous solution. US Patent Application US20080219917, Field 02/04/2008, PCT Application WO2008097922, 14 Aug 2008

Kroto HW, Heath JR, O'Brien SC, Curl RF, Smalley RE (1985) C_{60}: buckminsterfullerene. Nature 318:162–163

Lademann J, Weigmann H, Rickmeyer C, Barthelmes H, Schaefer H, Mueller G et al (1999) Skin Pharmacol Appl Skin Physiol 12:247–256

Lens M (2009) Use of fullerenes in cosmetics. Recent Pat Biotechnol 3:118–123

Liu Y, Pan J, Feng SS (2010) Nanoparticles of lipid monolayer shell and biodegradable polymer core for controlled release of paclitaxel: effects of surfactants on particle size, characteristics and in vitro performance. Int J Pharm 395:243–250

Luppi B, Cerchiara T, Bigucci F, Basile R, Zecchi V (2004) Polymeric nanoparticles composed of fatty acids and polyvinyl alcohol for topical application of sunscreens. J Pharm Pharmacol 56:407–411

Matija L, Tsenkova R, Munćan J, Miyazaki M, Banba K, Tomić M, Jeftić B (2013) Fullerene based nanomaterials for biomedical applications: engineering, functionalization and characterization. Adv Mater Res 633:224–238

Mavon A, Miquel C, Lejeune O, Payre B, Moretto P (2007) In vitro percutaneous absorption and in vivo stratum corneum distribution of an organic and mineral sunscreen. Skin Pharmacol. Physiol. 20:10–20

Mehnert W, Mader K (2001) Solid lipid nanoparticles: production, characterization and applications. Adv Drug Deliv Rev 47:165–196

Morales ME, Gallardo V, Clarés B, Garcia MB, Ruiz MA (2009) Study and description of hydrogels and organogels as vehicles for cosmetic active ingredients. J Cosmet Sci 60:627–636

Morganti P (2010) Use and potential of nanotechnology in cosmetic dermatology. Clin Cosmet Investig Dermatol 3:5–13

Möschwitzer JP (2012) Drug nanocrystals in the commercial pharmaceutical development process. Int J Pharm. doi:10.1016/j.ijpharm.2012.09.034

Müller-Goymann CC (2004) Physicochemical characterization of colloidal drug delivery systems such as reverse micelles, vesicles, liquid crystals and nanoparticles for topical administration. Eur J Pharm Biopharm 58:343–356

Müller RH, Petersen RD, Hommoss A, Pardeike J (2007) Nanostructured lipid carriers (NLC) in cosmetic dermal products. Adv Drug Deliv Rev 59:522–530

Nohynek GJ, Lademann J, Ribaud C, Roberts MS (2007) Grey goo on the skin? Nanotechnology, cosmetic and sunscreen safety. Crit Rev Toxicol 37:251–277

Oberdorster G, Oberdorster E, Oberdorster J (2004) Nanotoxicology: an emerging discipline evolving from studies of ultrafine particles. Environ Health Perspect 112:1058–1062

Olveira-Martinez BI, Cazares-Delgadillo J, Calderilla-Fajardo SB, Villalobos-Garcia R, Ganem-Quintanar A, Quintanar-Guerrero D (2005) Preparation of polymeric nanocapsules containing emulsification-diffusion technique: penetration across the stratum corneum. J Pharm Sci 94:1552–1559

Pan Z, Lee W, Slutsky L, Clark R, Pernodet N, Rafailovich MH (2009) Adverse effects of titanium dioxide nanoparticles on human dermal fibroblasts and how to protect cells. Small 5(4):511–520

Papić-Obradović M, Miljković S, Matija L, Munćan J, Koruga Đ (2009) Nanotehnologije u farmaciji. U: Osnove nanomedicine. Embriologija, farmakologija, nanotehnologija. DonVas/Nauka Beograd, pp 89–174. [In serbian]

Pardeike J, Hommoss A, Müller RH (2009) Lipid nanoparticles (SLN, NLC) in cosmetic and pharmaceutical dermal products. Int J Pharm 366:170–184

Patravale VB, Mandawgade SD (2008) Novel cosmetic delivery systems: an application update. Int J Cosmet Sci 30:19–33

Pippa N, Merkouraki M, Pispas S, Demetzos C (2013) DPPC: MPOx chimeric advanced drug delivery systems (chi-aDDnSs): physicochemical and structural characterization, stability and drug release studies. Int J Pharm 450:1–10

Rai S, Paliwal R, Gupta PN, Khatri K, Goyal AK, Vaidya B, Vyas SP (2008) Solid lipid nanoparticles (SLNs) as a rising tool in drug delivery science: one step up in nanotechnology. Curr Nanosci 4:30–44

Raj S, Jose S, Sumod US, Sabitha M (2012) Nanotechnology in cosmetics: opportunities and challenges. J Pharm Bioallied Sci 4(3):186–193

Rania B, Rainer MV, Muhammad N, Matthias R, Zoltan S, Christian WH et al (2007) Medicinal applications of fullerenes. Int J Nanomed 2:639–649

Regulation (EC) of the European parliament and of the council on cosmetic products (2009) Official Journal of the European Union L 342/59, No 1223/2009, 30 Nov 2009

Sayes CM, Fortner JD, Guo W, Lyon D, Boyd AM, Ausman KD, Tao YJ, Sitharaman B, Wilson LJ, Hughes JB, West JL, Colvin VL (2004) The differential cytotoxicity of water-soluble fullerenes. Nanoletters 4(10):1881–1887

SCCNFP (2003) Opinion of the scientific committee on cosmetic products and non-food products intended for consumers concerning zinc oxide. COLIPA No. S76. European Commission. http://ec.europa.eu/food/fs/sc/sccp/out222_en.pdf

Schafer-Korting M, Mehnert W, Korting HC (2007) Lipid nanoparticles for improved topical application of drugs for skin diseases. Adv Drug Deliv Rev 59:427–443

Schulz J et al (2002) Distribution of sunscreens on skin. Adv Drug Deliv Rev 54(Suppl 1):S157–S163

Sharma A, Kumar SM, Mahadevan N (2012) Nanotechnology: a promising approach for cosmetics. Int J Recent Adv Pharm Res 2(2):54–61

Shegokar R, Müller RH (2010) Nanocrystals: industrially feasible multifunctional formulation technology for poorly soluble actives. Int J Pharm 399(1–2):129–139

Solans C, Solé I (2012) Nano-emulsions: formation by low-energy methods. Curr Opin Colloid Interface Sci 17:246–254

Sonneville-Aubrun O, Simonnet JT, L'Alloret F (2004) Nanoemulsions: a new vehicle for skin care products. Adv Colloid Interface Sci 108–109:145–149

Souto EB, Almeida AJ, Müller RH (2007) Lipid nanoparticles (SLN, NLC) for cutaneous drug delivery: structure, protection and skin effects. J Biomed Nanotechnol 3:317–331

Souto EB, Müller RH (2008) Cosmetic features and application of lipid nanoparticles (SLN, NLC). Int J Cosmet Sci 30:157–165

Villalobos-Hernandez JR, Müller-Goymann CC (2006) Sun protection enhancement of titanium dioxide crystals by the use of carnauba wax nanoparticles: the synergistic interaction between organic and inorganic sunscreens at nanoscale. Int J Pharm 322:161–170

Weigmann HJ et al (2001) Correlation of the local distribution of topically applied substances inside the stratum corneum determined by tape-stripping to differences in bioavailability. Skin Pharmacol Appl Skin Physiol 14(Suppl 1):98–102

Wiechers JW (2009) Nanotechnology and skin delivery: infinitely small or infinite possibilities? JW Solutions

Wissing SA, Mader K, Müller RH (2000) Solid lipid nanoparticles (SLN)—as a novel carrier system offering prolonged release of the perfume Allure (Chanel). In: International symposium on controlled release of bioactive materials, vol 27, pp 311–312

Wissing SA, Müller RH (2003) Cosmetic application for solid lipid nanoparticles (SLN). Int J Pharm 254:65–68

Wissing S, Lippacher A, Müller RH (2001) Investigations on the occlusive properties of solid lipid nanoparticles (SLN). J Cosmet Sci 52:313–324

Yang Y, Marshall-Breton C, Leser ME, Sher AA, McClements DJ (2012) Fabrication of ultrafine edible emulsions: comparison of high-energy and low-energy homogenisation methods. Food Hydrocoll 29:398–406

Yildirimer L, Thanh NTK, Loizidou M, Seifalian AM (2011) Toxicological considerations of clinically applicable nanoparticles. Nano Today 6:585–607

Characteristics and Applications of Silver Nanoparticles

Fatima Zivic, Nenad Grujovic, Slobodan Mitrovic, Inam Ul Ahad and Dermot Brabazon

Abstract This review provides an insight into the level of knowledge about the properties of silver nanoparticles, their already existing applications and possible further developments, as well as their effects on different behaviour and properties of the products wherein they are used. This chapter reviews the application fields of nanosilver, starting from basic silver properties and influential parameters in definition of silver nanoparticles (Ag NPs). Toxicity of Ag NPs is observed from various aspects in relation to cell toxicity and relevant mechanisms of their behaviour within tissue environment. Range of sizes and surface chemistry of Ag NPs and possible effects, as well as known in vivo effects are reviewed, based on the already established research results. Antibacterial properties of Ag NPs and relevant mechanisms of action are presented. Application areas where commercialization of nanosilver has started are presented: medicine (wound dressings, drug delivery, biosensors and medical diagnostics, orthopedics), food and textile industries, and water disinfection systems. Environment related issues have been considered and important conclusions derived from established results are presented. Each application sector comprises descriptions of basic mechanisms related to Ag NPs, gained benefits, but also possible risks and recognised limitations in application of Ag NPs. Future directions as recognised in specific research groups dealing with some of the sectors are listed. Exquisite properties of Ag NPs, especially antibacterial and

F. Zivic (✉) · N. Grujovic · S. Mitrovic
Faculty of Engineering, University of Kragujevac, Kragujevac, Serbia
e-mail: zivic@kg.ac.rs

N. Grujovic
e-mail: gruja@kg.ac.rs

S. Mitrovic
e-mail: boban@kg.ac.rs

I.U. Ahad · D. Brabazon
Advanced Processing Technology Research Centre, Dublin City University, Dublin 9, Ireland
e-mail: inamul.ahad@dcu.ie

D. Brabazon
e-mail: dermot.brabazon@dcu.ie

D. Brabazon et al. (eds.), *Commercialization of Nanotechnologies–A Case Study Approach*, DOI 10.1007/978-3-319-56979-6_10

optical properties, along with availability and lower cost of fabrication, processing and storage, compared to other noble elements makes them very promising for numerous future applications. However, proven cell toxicity must be further studied and methods to overcome these adverse effects on tissue in general are subject of current research. One important application field is drug delivery for targeted cancer cell destruction that is expected to show particular results very soon. Even with all the limitations that need to be imposed on usage of Ag NPs, they are promising nanoagent for novel advancements in different areas.

Keywords Engineered nanoparticles · Nanosilver · Ag NPs · Toxicity · Wound dressings · Drug delivery · Biosensors · Orthopedics · Food industry · Textile industry · Water disinfection systems · Environment risks · Health risks

1 Introduction

Noble elements have been the subject of interest since ancient times. Gold, silver and platinum have found various uses, from physical applications in different products and up to the well known economic worth as monetary currencies throughout history as gold and silver coins in medieval world, or national gold reserves until recent times, and have been highly valued ever since. Silver vessels were used to preserve drinking water in ancient Greek and Rome. It is clear even from their "noble" name that this group of chemical elements is distinctively separated from all other existing elements. Silver was used as a medical treatment as far as 8th century, for purification of blood, heart palpitations and prevention of bad breath (Wadhera and Fung 2005). Nowadays, the use of gold and silver has entered the exciting field of nanotechnology.

Nanoparticles (NPs) of different origin have various applications and the very beginning of their use opened up concerns regarding their influences beyond desired effects. Commercialisation of different nanoproducts, containing different nanoparticles has started for some time now, but it is still at the very beginning precisely due to the fact that many properties are still under question. One repository of commercialised nanoproducts can be found at http://www.nanotechproject.org/cpi/, Nanotechnology Consumer Products Inventory (CPI), created within the Project on Emerging Nanotechnologies in 2005 (Vance et al. 2015). Nanoparticles used for medical purposes will influence the cell functioning when inside the tissue. When released into the environment (e.g. soil, water), different effects can be expected. Metal nanoparticles are especially the subject of concern because they are usually not degradable and cannot be easily incorporated into the natural surroundings. And it is very important to determine their additional effects in such cases, like pathways for degradation, chemical affinities, life time of the particles etc. Nanoparticles of noble elements have already proven their usefulness but the related prolonged effects of such use are still not investigated. More than half of all engineered nanoparticles used in commercial products today are silver nanoparticles (Ag NPs) with the largest share of 55% in comparison to other nanoparticles, the next being nanoparticles of TiO_2 (14%), Au (9%), SiO_2 and ZrO_2 (both 6%), Pt (4%), Pl (2%), CeO_2 (1%), Cu (1%), Ir (1%) and Pd (1%), as per Gatti and Montanari (2015). Another survey of commercial products also pointed silver as the most used nanomaterial in consumer products and enlisted 435 products containing nanosilver as per companies' information (Vance et al. 2015) or one earlier study 390 commercial products in 2013 (Martirosyan and Schneider 2014). Silver has been known for its antimicrobial effects for a long time, but there are certain limitations regarding the amount, and the size of silver nanoparticles. Above certain levels, Ag NPs would provoke unwanted effects in the surrounding cells. Very small sizes of particles have quite different effect than even a bit larger ones. It has been determined that there is a frame for their proper use, but even in that case it is rather unknown if any further effects will show. It is essential that any use of nanoparticles is investigated from aspects of provoking any harmful outcome, either through their interaction on cellular levels or some other molecular binding.

Existing and possible applications of Ag NPs are broad in scope. Silver nanoparticles have been increasingly used in variety of consumer products, such as cosmetics, deodorants, toothpaste, fabrics, food packaging, filters for water purification and new products are rapidly appearing. Several research directions are pursued related to: medicine (antimicrobial coatings, reinforcement of the matrix material in orthopedics or cardiovascular implants, neurosurgical catheters, drug delivery systems, wound dressings, diagnostic imagining, etc.), cosmetics, toothpaste, food (packaging, food additives, sensors), textile (medical cloths, socks), biosensors and bioelectronics, chemical sensors, semiconductor products, optics and lasers, disinfection of industrial wastewater, construction materials (paints, antibacterial ceramic tiles or keyboards, glues), cleaning (soaps, detergents), home appliances (refrigerators) and many other products emerging frequently (Soenen et al. 2015; Murphy et al. 2015; Haider and Kang 2015; Wei et al. 2015; Rizzello and Pompa 2014; Martirosyan and Schneider 2014; Yu et al. 2013; Kendall and Holgate 2012; Chaloupka et al. 2010; Abou El-Nour et al. 2010; Rai et al. 2009; Bhattacharya and Mukherjee 2008; Kumar et al. 2008). There are two general directions of Ag NPs application in medicine: as therapy and for diagnostics. The majority of commercially available therapeutic products are related to wounds and burns healing due to silver antibacterial effect. Well known "silver water" that can be purchased in almost all health food stores is widely used as "alternative therapy". Socks, towels and work suits can already be found to prevent odour and bacteria growth. In medicine, silver nanoparticles have been tested as addition to different cements or fillers, thus changing their final properties (Miola et al. 2017; Lewis 2016; Zivic et al. 2012, 2013). Ceramic tiles, linings on refrigerators and kitchen surfaces, air conditioners and purifiers, and vacuum cleaners have started to appear advertising antibacterial effects of added silver nanoparticles. Ag NPs are added to traditional paints to provide antibacterial paints for public rooms such as hospitals, schools, offices, halls, etc. (Kumar et al. 2008; Vivekanandhan et al. 2012; Rajeshkumar et al. 2012). Materials of food containers (food packaging industry) have been reinforced by Ag NPs to provide longer periods of preserved food due to prevention of bacteria growth. For diagnostic purposes, some new findings open up promising directions in early detection of cancer by non-invasive methods, such as Surface-Enhanced Raman Spectroscopy (SERS) and by using silver nanoparticles (Li et al. 2015). Silver nanoparticles have novel optical properties and high electrical conductivity and these properties have been used in conductive ink or paste, devices for molecular diagnostics or photonics.

However, even with the fact that the World Health Organization does not consider silver as toxic except when high doses are used, it is the fact that nanoparticles behave completely differently than any other known material and there are still many unanswered questions. In general, two aspects of safety are important in relation to nanoparticles influence on humans. The first one is related to the incorporation of nanoparticles in a material to change its properties and how that affects the system and the users. The second is a production of nanoparticles themselves and the occupational risks and safety of the workers in such industries. Production of nanoparticles has still not widely started, but safety aspects of such plants are of the utmost importance since they potentially could have much more larger harmful effects than any other 'dirty' industry. And high risks of such industries are urgent to determine because they are largely not understood due to the many existing questions and issues in relation to nanoparticles behaviour. Silver is known for years as a very human-friendly and beneficial material, but behaviour of silver nanoparticles is significantly different than the behaviour of bulk material.

Information about bulk metal properties or properties at macro level is not sufficient to explain all resulting influences on the nano level or behaviour of the micro and nanoparticles. Physicochemical properties, surface area, ion concentrations in the surrounding zones and their behaviour, atomic and electronic configurations together with structural, crystalline and magnetic properties are only some of the characteristics that must be considered during the investigation of nanoparticles behaviour, beside the level of its presence (dose or concentration). Quantum physics laws start to dominate under the 100 nm size and properties of such particles must include different mechanical, chemical, magnetic, electrical and optical properties and novel material behaviour has been observed at nanoscale. For nanoparticles, physicochemical properties in relation to the environment dominate the behaviour and further nano-toxicity effects. Physicochemical characteristics of the nanoparticles, such as the size and shape which determines surface area, chemical composition, aggregation, agglomeration, sedimentation, dissolution, surface charge and coating, or induced functionalization in some way, are some important properties. Nano-scale behaviour is complex because it is widely unknown on one side, and on the other side it further provokes also unknown effects in the living tissue, especially from aspect of long term effects, such as genotoxicity or effects on immune system, for example. Metallic materials and metal nanoparticles are rather exclusively subject of study in physics and engineering. Biology responses of any living tissue to foreign influences are studied in organic chemistry, medicine and biology. These quite different large areas have only begun to cooperate with each other more closely in recent decade and that is the important precondition for any profound investigations related to the behaviour of nanoparticles. Comprehensive characterisation of nanoparticles behaviour, whether inside some final product or their effects on the surroundings, can be done only by multidisciplinary efforts. However, most often, investigations are done focused on one area, not involving the other fields. For example, investigations of structural modifications of some material achieved by the addition of nanoparticles, are often observed only from aspects of how to uniformly distribute nanoparticles, how to prevent agglomeration, which modifications of the crystal lattice will be obtained, how it affects optical or magnetic properties, etc. Nanoparticles production is mainly observed from aspects of the synthesis process and their storage until final use. All such activities usually do not consider medical knowledge about their effects on cells and tissues, from the beginning. Only after some material is made, investigations start on the compatibility of such material with the living tissue, and might result in its total withdrawal, due to some determined harmful effect. It is necessary to connect knowledge from different areas of research in the beginning of the material/system design, in order to systematically cover all necessary aspects and such approach has started in recent times.

Application of nanoparticles is novel field and lacks standardization in general, meaning that even the definition of nanoparticles themselves is not standardized. This is one of the barriers in comparison of available data so far, since different experimental conditions usually results in quite different responses not comparable to each other. This is especially important from aspects of nanoparticles toxicity. In 2004, MINChar initiative (www.characterizationmatters.org) was established based on discussions between various stakeholders of the nano-toxicology issues: researchers, academia, industry, governmental bodies, aiming to define minimum material characterisation necessary for nano-toxicology studies. They recommended a key set of physical and chemical parameters and information that should be minimally included, in order to properly describe nanoparticles to enable nano-toxicology studies, as following:

- Size range of particle
- Shape of particles
- Conditions of agglomeration and aggregation
- Composition of particles, such as chemical composition and crystal structure
- Surface composition
- Purity with percentage of impurities and their definition
- Surface area
- Surface chemistry
- Surface charge
- Stability over time: dynamic stability, preparation, storage, handling, etc.
- Media relevant information, such as changes in material properties when within a certain environment/media or in contact with it
- If possible to define the dose-relevant behaviour in relation to end application of nanoparticles (e.g. relevant concentrations)

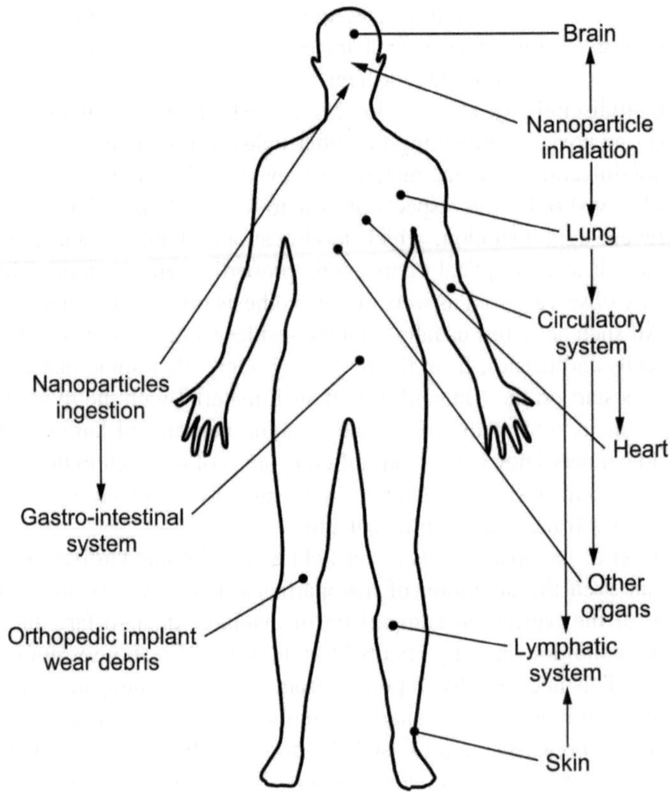

Fig. 1 Possible pathways for nanoparticles to enter the body (Buzea et al. 2007; Theodorou et al. 2014)

Very good review of different toxicology effects of nanoparticles in general and their routes to enter the body is given in Buzea et al. (2007). Possible pathways for nanoparticles to enter the body are shown in Fig. 1. Nanoparticle inhalation: Respiratory system is perhaps the first one exposed to different nanoparticles that might provoke asthma, bronchitis, neurological diseases, heart disease, cancer and many other. Nanoparticles that can be inhaled can originate from manufacturing process, aerosols, sprays, air filters or airborne Ag NPs in ambient. Nanoparticle ingestion: Gastro-intestinal system exposure to nanoparticles can provoke cancer, but can also be a doorway to exposure of other organs. Ingested nanoparticles can originate from food packaging and containers, water filters, dietary supplements, toothpaste etc. Circulatory system can be exposed to nanoparticles via drug delivery systems or different medical devices containing nanoparticles either within the bulk or in a form of thin coatings and films, but also through all other organs exposure. Due to circulation of blood through all organs, it can provoke many different diseases, such as auto-immune diseases, cancer, heart disease, blood pressure imbalance, neurological diseases—Alzheimer or Parkinson, kidney or liver diseases etc. Skin exposure to nanoparticles results in their migration to other organs, such as lymphatic system and similar to circulatory system it may provoke different harmful effects.

Kendall and Holgate (2012) summarized remaining challenges in application of nanoparticles in general, as following:

- Detection of nanoparticles presence in a material and environment,
- Limited info about human tissue response to nanoparticles embedded in consumer products, medical devices, environment and biological systems,
- Mechanics of the cardiovascular events resulting from exposure to nanoparticles,
- Key properties of the nanoparticles that determine toxic mechanisms and behaviour. So far, surface area and surface chemistry are two distinct characteristics,
- Animal studies in relation to nanoparticles toxic effects and especially genotoxicity and possible prediction of risks in humans,
- Effects of airborne nanoparticles on the vulnerable groups (children, elderly, persons with chronic diseases, etc.),
- Effects of long-time exposure are unknown, especially if related to cancer, neurodegenerative diseases and fibrosis.

2 Basic Properties of Silver

Silver belongs to a group of noble metals. Symbol of silver is Ag and belongs to transition metals. It is 47th element in a periodic table, with atomic weight of 107.87. Its melting temperature is 961.78 °C and boiling temperature is 2162 °C. Silver has very high malleability and ductility and in its pure form its electrical conductivity and thermal conductivity are the highest of all metals. Pure silver does not react with water or air, but it reacts when in contact with compounds containing sulphur or chloride. Naturally, silver is in metallic state, and very rarely as sole metal, but in combination with sulphur, chlorine, arsenic, or antimony and very often in ores containing copper, lead, gold, and zinc. The beginning of silver mining dates to 5000 years ago in Anatolia until 1200 B.C., when the silver production was located in Greece, near Crete. Around 100 A.D., Spain became the main supplier for the Romans and silver was the main currency in trading of spices from Asia. But the most significant event for the production of silver was the Columbus discovery of the New World in 1492. Many new mines were found and Bolivia, Peru and Mexico produced around 85% of silver in a world. Today, the mining of silver is distributed to several large mines and producers. Mexico is still the largest producer of silver, followed by Peru and China. In Europe, Poland accounts for the largest production of silver, comparable with Russia. Polish company KGHM Polska Miedz S.A., founded in 1961, produces the largest quantities of silver in the world.

Together with gold, as a noble metal, silver has been used from ancient times. Ancient Egyptians thought that skin of gods is made of gold and their bones of silver and for them silver was more rare 'white metal' than gold and it was used for very special jewellery. Ancient Greeks used it for coins, ornaments and vases. Native American Indians used edible silver foils to cover food and they also made exquisite jewellery. Edible silver foils have been used also in Europe and Asia, from ancient times to cover food for preservation and were accounted to have antibacterial properties, even though there are some recent claims that these foils are just inert. Silverware and silver water containers and food vessels have been used for thousands of years. Silver was always considered a precious metal, precisely for its many qualities and antibacterial effects were very early recognised. It was no surprise that novel nanotechnology developments included silver as suitable metal for different applications.

Silver nanoparticles are recognised as suitable for many applications where their properties of high electrical conductivity and low sintering temperatures, can be utilized. Two good reviews of different methods to produce silver nanoparticles are prepared by Haider and Kang (2015) and Wei et al. (2015). There are different methods to produce Ag NPs and basically these can be divided based on four approaches: chemical, physical, photochemical and biological methods, such as chemical reduction, laser ablation, photo reduction and green synthesis by using bioorganic plants, for example algae. Chemical methods might be considered as standard, such as reduction of $AgNO_3$. Photochemical methods often use UV light to support chemical process. Irradiation assisted chemical methods are also often applied. Physical methods usually involve evaporation, condensation or other high energy processes and their costs are often very high, making this approach less applied in general. Application of synthesis processes that involve plants or other bioorganic sources is novel direction of research in chemistry of nanoparticles synthesis (Dhand et al. 2016; Joseph et al. 2016; Liang et al. 2015; Gurunathan et al. 2013a, b; Vivekanandhan et al. 2012; Rajeshkumar et al. 2012). Dhand et al. (2016) produced crystalline Ag NPs with sizes between 20–30 nm, and the cubic, face centered crystal lattices. Average size of 5 nm of Ag NPs was obtained by Gurunathan et al. (2013a) and the same group produced anisotropic nanoparticles of 10–50 nm size by using another synthesis route (Gurunathan et al. 2013b). Another green process provided 10–25 nm spherical Ag NPs (Vivekanandhan et al. 2012). Rajeshkumar et al. (2012) used marine algae and their green synthesis produced spherical Ag NPs of 96 nm average size. There are also various attempts to fabricate core-shell nanoparticle systems, where the shell is made of metal (e.g. silver or gold) and the core is another material (e.g. SiO_2) (Khurana et al. 2014). These systems are investigated as possible drug or antibiotics carrier for targeted drug delivery systems. It is proven that the synthesis method essentially determine the size and shape of nanoparticles, as well as their further behaviour, in general. And the influence of bioorganic media adds to more complexity in those relations, but the way the particles are produced is the first essential step to define them. For example, green synthesis of Ag NPs involving hyaluronic acid (HA) to target CD44 + cell receptors which are related to several cancer types had a focus of synthesis to target tumor cells and showed promising results (Liang et al. 2015).

3 Toxicity of Silver Nanoparticles

Medicine is one of the vital areas where it is expected that nanoparticles will add a new dimension to many existing treatments and diagnostics and open up new solutions. However, toxicity is unavoidable issue closely connected with any application of nano or micro particles. The underlining influence is the nano size of particles that enable them to penetrate the cell membrane, unlike any other traditionally known treatments so far. The medical indications or exact consequences of such interactions are however largely unknown so far, especially if long term effects are considered, such as genetic or immune response. Some of the recognised toxic effects of metal NPs are tissue inflammation, oxidative stress, immune system stimulation, changes of the oxygen metabolism towards increase of free radicals, abnormal functioning or death of the cell, etc. Toxicity of silver nanoparticles is established and many researchers have been investigated this aspect of Ag NPs applications, trying to determine responsible mechanisms for cell deaths induced by Ag NPs (Foldbjerg et al. 2015; Castiglioni et al. 2015; Gaillet and Rouanet 2015; Soenen et al. 2015; Murphy et al. 2015; Durán et al. 2015; Rizzello and Pompa 2014; Kim and Ryu 2012; Kendall and Holgate 2012; Yang et al. 2012; Lim et al. 2012; Park et al. 2010; Marambio-Jones and Hoek 2010; Ahamed et al. 2010; Kim et al. 2009; Miura and Shinohara 2009; Arora et al. 2008; Wadhera and Fung 2005).

Externally induced nanoparticles of any kind stimulate immune cells of the living organism through their chemical or physical properties (Kononenko et al. 2015). Immune system of any living organism represents the defense system because numerous response mechanisms are available to protect the body when foreign molecules are recognised (Gamucci et al. 2014). The immune system usually remembers these 'attacks' and specific antibodies are produced in the body to act against possible repeated reactions. Antigens or antibody generators are all those substances that can provoke the reaction of the immune system. Experiments with metallic NPs showed that immune system observes them in the same way as pathogens and phagocytes try to destroy those, as confirmed for nanoparticles of Co, TiO_2, and ZnO (Petrarca et al. 2015). Phagocytes are those cells such as white blood cells that ingest, engulfs or absorb harmful material (e.g. microorganisms, foreign particles, bacteria, dead cells) found in the tissue or blood. Quite different activation of immune response is obtained for nanoparticles than microparticles or agglomerated particles, by activating one or the other signalling pathway for releasing mechanisms or producing specific anti-inflammatory reaction. Different sizes of gold nanoparticles (Au NPs) have been successfully used for targeted killing of the tumor cells of breast carcinoma because they can be easily navigated to the center of carcinoma cells (Moon et al. 2012). It is proven that nanoparticles can help in efficient immunotherapy. However, even with Au NPs which are not cytotoxic and show no biological activity, the total immune response is still under consideration. Protective cells in the body (white blood cells) rapidly act to any possible harmful material, while adaptive immune cells (T-cell subsets and B-cells) act slowly but they can produce memory cells and remember mechanisms related to certain pathogens.

White blood cells that act against harmful particles (macrophages, monocytes, neutrophils, natural killer cells, or NK cells, granulocytes) can be killed by NPs by necrosis or apoptosis, depending on their sizes, concentrations, chemical reactions, etc. (Castiglioni et al. 2015; Kim and Ryu 2012; Buerki-Thurnherr et al. 2013; Wilhelmi et al. 2013; Moon et al. 2012; Vandebriel and De 2012). Cell death by apoptosis is considered to be a defense mechanism of the body and a part of the normal healthy process, while cell death by necrosis is always triggered by external factor (e.g. infection) and is considered as harmful and abnormal. Some processes that can trigger apoptosis are shrinkage of cell, nuclear fragmentation, etc., while necrosis can be provoked by membrane disruption, metabolic collapse, rupture of cell leading to inflammation, etc. Cytotoxic effects of Ag NPs have been studied through different cell conditions they provoked and by measuring the production of oxidative stress genes or the release of IL-8 or IL-1β signalling proteins, or heat shock protein-70 (Hsp70) and other biochemical indicators (Petrarca et al. 2015; Lim et al. 2012; Yang et al. 2012; Kim et al. 2009). Oxidative stress appears to be one of the main causes of cytoxicity and is related to the occurrence of imbalance between the production of free radicals and the inability of the body to act against it through detoxification by antioxidants. Interleukin 8 (IL-8) and interleukin 1 (IL-1) belong to a family of proteins or small cytokines produced and stored by different cells (macrophages, monocytes, epithelial cells, endothelial cells, etc.) and secreted by the cells to indicate some harmful influence and provoke immune system reaction. Heat shock proteins are triggered by abrupt heat changes or some other physical or chemical stress and related Hsp70 gene has evolved as a consequence.

Large doses or concentrations of Ag NPs are cytotoxic, but there is also a proven effect of size dependent influence of nanoparticles, in general. Small Ag NPs (e.g. <10 nm) provoked rather immediate different cytotoxic effects, while large particles or agglomerated particles (>100 nm) sometime did not produce any recognised effects on cells, by measuring via previously listed signalling indicators (Foldbjerg et al. 2015; Rizzello and Pompa 2014; Yang et al. 2012; Marambio-Jones and Hoek 2010). Yang et al. (2012) observed cell reactions induced by 5 and 28 nm sizes of silver nanoparticles and reported that 5 nm Ag NPs exhibited more cytotoxic behaviour than those of 28 nm size, due to higher production of H_2O_2 in the process, whereas both induced production of both hydrogen peroxide and superoxide. Another study also proved the difference between 5 and 100 nm silver nanoparticles where 100 nm particles did not induce cytotoxic indicators unlike 5 nm particles, such as IL-8 and stress genes (Lim et al. 2012). Arora et al. (2008) studied the safe concentration levels for silver nanoparticles for two cell types and with range of nanoparticle sizes of 7–20 nm, and indicated 0.78 μg/mL as the low level onset of apoptosis and 12.5 μg/mL as the necrotic concentration. They recommended 1.56–6.25 μg/mL range of concentrations of Ag NPs, aimed for antimicrobial wound healing applications.

Consequently, there is a range of sizes and concentrations related to Ag NPs that should be safer for use. Kim and Ryu (2012) presented a very good review of Ag NPs tested sizes and concentrations from aspects of their cell toxic effects, especially exhibited for small nanoparticles (2–5 nm). However, there is some evidence that the behaviour of small particles (<100 nm) at nanoscale can be predicted by observing their behaviour when they are of larger sizes (>500 nm) (Maynard et al. 2011). But for both small and larger nanoparticles, direct effects, such as inflammation, oxidative stress, activation of signalling pathways, and carcinogenicity, have been observed (Maynard et al. 2011). The behaviour of cells in contact with a nanoparticles is primarily size dependent. For particle sizes larger than 500 nm, phagocytes take up the particle thus forming the internal vesicle, through a process called phagocytosis, and pinocytosis or macropinocytosis for larger particles and aggregates (0.5–5 μm), as a normal process of the immune system (Zhao et al. 2011). Zhao et al. (2011) made a good review article about different cellular uptake processes related to different nanoparticles. Shape of nanoparticles is also very influential factor regarding cytotoxicity (Soenen et al. 2015; Pal et al. 2007). Spherical nanoparticles with diameter of 30 nm are less toxic then silver nanowires with diameter of 100–160 nm and length of 1.5–25 μm (Soenen et al. 2015). Accordingly, different production routes have been tried for nanosilver, in order to obtain wanted sizes of Ag NPs, which is not easy to do. Spherical and ellipsoid Ag nanoparticles with sizes in the range 20–30 nm, were obtained by Dhand et al. (2016), by using so called 'green' synthesis. It is well-known fact that production routes strongly influence properties of any final material and even slight post-processing can result in quite different properties of the surface (Mitrovic et al. 2014).

Essentially the question of silver degradation in cell culture media determines the process of its effects and toxicity via corresponding chemical reactions during degradation. Dissolution rate is increased with temperature and medium acidity (Liu and Hurt 2010). Many authors proved that in colloidal suspension such as cell culture media, silver nanoparticles appeared into three forms: elemental silver $(Ag^0)_n$, Ag^+ ions and its complexes, and Ag^+ ionic silver adsorbed on the surface of Ag nanoparticles. Within cell culture media, that is, in human body environment, Ag NPs degrade under the influence of oxygen presence with higher dissolution rate than in pure water (Zook et al. 2011; Liu and Hurt 2010).

The most significant influence of Ag NPs originates from Ag ions effects, which might be different than those of nano and micro particles, or agglomerated nanoparticles. Ag^+ ions are produced by the oxidation of Ag NPs in the water, thus indicating that the size of the particles, pH, presence of oxygen, temperature, etc., becomes important parameters, since they all have the significant influence on ions release rate. Ag^+ ions favour production of free radicals and different biologically toxic reactive oxygen species (ROS), due to their reaction with oxygen and those can induce significant cell damage (e.g. cell apoptosis) if increased above certain levels. Superoxide (contains superoxide anion O_2^-), peroxides (contains peroxide anion O_2^{2-}), hydroxyl radical (OH^-) and singlet oxygen (high energy oxygen O_2, having the singlet state with a total quantum spin of 0) are all highly reactive and, as such, are grouped as reactive oxygen species (ROS). ROS are produced during the regular normal cell function, but any abnormal increase due to some external influence can cause serious damage. Oxidative stress refers to the imbalance state of cells where cells cannot repair or detoxify damage resulting from increased presence of any oxidizing species, whereas especially ROS has deleterious effects. Normal redox cell state means that the cells regularly detoxify reactive intermediates made as a product of ROS influence within their systemic manifestation. When this state is disturbed, various toxic effects can occur, and damage of all cell components, such as DNA, proteins and lipids or breaking the normal cell signalling mechanisms.

Surface chemistry of the Ag NPs is a very complex process. Immediately after the nanoparticles entered the biological environment, the coating is formed made of proteins, usually large number of them. The layer of adsorbed protein on the surface of nanoparticle, or "protein corona" can be formed around a single molecule by for example human serum albumin, human blood plasma or hyaluronic-binding protein, or as a complex protein corona (e.g. proteins from yeast extract) (Durán et al. 2015). In any case, this coating has a dynamic composition, changing in time, depending on the presence of proteins in a specific site (Durán et al. 2015; Foldbjerg et al. 2015; Kendall and Holgate 2012). Accordingly this process also affects its influence on the cell, via the influence on the proteins as cell receptors. But it also changes the particle in contact with the surrounding tissue, making the large difference between in vitro and in vivo testing of the nanoparticles influences, because it becomes quite different particle when in the biological environment. Pure silver is not soluble in the water nor is reactive with water and Ag NPs show the same behaviour, but coupled with oxidising processes, production of ROS and protein mediation, Ag ions are produced, through surface oxidation of Ag NPs. Contrary to the pure silver, Ag ions are highly reactive. According to Lui and Hurt (2010), presence of Ag NPs in aqueous environment results in several silver forms and the common ones are: Ag solids, dissolved silver nanoparticles in a forms of ionic silver and silver complexes which are soluble, and ionic silver adsorbed on the surface of Ag nanoparticles. They showed that ion release rate is increased with temperature increase to 37 °C and with decrease of pH, with minor influence of sea salts. But Ag NPs in vivo behaviour can be profoundly different, due to previously mentioned proteins effects.

In vivo behaviour of Ag NPs is still under investigation and represents a challenge due to many complexities of the chemical processes in the living tissue. In vitro investigations are usually very different from in vivo, but should point out the promising directions of research. For example, silver ions were not found in several simulated body fluids after the addition of Ag NPs (Stebounova et al. 2011) and it was detected in minor concentrations in simulated cell culture medium (Jiang et al. 2015) or other Ag NPs solutions (Kim et al. 2009). The major suggested mechanism of Ag NPs cytotoxicity was related to the silver ions intracellular release, and reaction of ions with sulfhydryl (SH-) groups leading to Ag ion complexation, thus forming transition Ag complexes (Jiang et al. 2015). SH- groups are part of natural cell chemistry, such as in proteins, and any disturbance in protein function or natural shape leads to the beginning of incurable diseases. Natural cell system for disease prevention is based on the antioxidant cell reactions. However, Miura and Shinohara (2009) showed that solid elemental silver nanoparticles also produced upregulation of oxidative stress genes (ho-1 and mt-2A), thus proving that not only ionic silver is cytotoxic but also elemental Ag NPs form. One of the latest works provided more accurate answers related to the nanotoxicity of Ag NPs. Wang et al. (2015) used novel analytical techniques, synchrotron radiation microscopy and spectroscopy to observe the behaviour of Ag NPs inside the single cell (human monocyte THP-1) and obtained very precise information related to chemical transformation of silver. Their work showed that the cytotoxicity of silver nanoparticles originates from synergies between Ag^+ ions, Ag-O- and Ag-S- species, produced during the chemical transformation of elemental silver $(Ag^0)_n$. This work proved the presence of Ag^+ ions and it is known that these ions stimulate the reactive oxygen species production due to its reaction with oxygen (superoxide and ROS in general), which leads to cell destruction in several ways, as previously shortly listed. It seems that the main mechanism of cytotoxicity of Ag NPs is oxidative stress induced by silver ions. However, other studies suggested that this is just one part of the silver nanoparticles toxic effects (Foldbjerg et al. 2012). This study compared the effects of particulate Ag NPs and Ag ions on gene expression. They determined that regulation of more than 1000 genes was changed when subjected to Ag nanoparticles in comparison with change of 133 genes regulation when subjected to Ag ions only. In this work, some upregulated genes belong to groups of heat shock protein, metallothionein (intracellular protein with multipurpose function related to cell protective mechanisms against heavy metals, oxidative stress, redox control, homeostasis, etc.), and histone families (proteins related to DNA ordering—compacting of DNA strands). Tests with mice by monitoring of silver concentrations in blood over time, indicated that there is significant difference in elimination of intravenously injected Ag NPs (dose of 120 mg/kg), for male and female mice, whereas in female mice twice longer time was observed (Xue et al. 2012). This study concluded that silver nanoparticles were distributed to all major organs (spleen, liver, kidneys and lungs), but the highest levels were detected in spleen and liver also with the longest time for elimination of Ag NPs presence.

Silver nanoparticles, especially those of less than 50 nm sizes, can also induce DNA and chromosomal damage, as genotoxic effects, which might provoke carcinogenesis, speculated to be the result mainly related to the ROS production. Genotoxicity of Ag NPs has been a subject of many studies but still with many open questions (Foldbjerg et al. 2011, 2015; Jiang et al. 2013; Kim et al. 2011; Hackenberg et al. 2011; Kawata et al. 2009; Asharani et al. 2009a, b; Ahamed et al. 2008). Also, prevention methods of toxic effects of Ag NPs have been studied, usually associated with some antioxidant pre-treatment of nanoparticles, already proven to decrease the level of ROS and cell damage (Castiglioni et al. 2015; Foldbjerg et al. 2011; Kim et al. 2009).

Ag NPs can be added to another drug as a stimulant of the immune system, such as adjuvant in vaccines related to lung problems, where Ag NPs are proven to have positive effect of decreasing inflammation (Mohamud et al. 2014; Park et al. 2010). On the other hand, inhaled Ag NPs provoke rapid increase of reactive oxygen species (ROS) and induce intracellular oxidative stress, related to the harmful inflammation mechanism of airway lung cells (e.g. bronchial epithelial cells, NHBE). Kim et al. (2009) elaborated that the main reason for Ag NPs toxicity is related to the provoked intracellular oxidative stress and not related to the presence of Ag^{+} ions. Arora et al. (2008) proved that higher concentrations of Ag NPs (6.25–50 μg/mL) resulted in changed cell morphologies and they reported some critical values of concentrations above which cell apoptosis occurred.

The current aspect of research is highly focused on mitigation of Ag NPs toxic effects while maintaining its antibacterial effects, by using polymer coatings, functionalization or by embedding nanoparticles into the polymer matrix (Khanna et al. 2015; Hu et al. 2015; Stevanovic 2014). The goal is to find the optimal balance between strong antibacterial properties and side-effect toxicity. One direction is to control the silver effects by controlling its dissolution rate and release of Ag^{+} ions. The important fact is that within the body, silver nanoparticles are always coated with some organic component, such as proteins and it was shown that protein capped Ag NPs are more biocompatible than bare ones (Bhunia et al. 2015). Also, without these organic coatings, Ag NPs would instantly agglomerate. This, in any case, indicates that the behaviour of Ag NPs must be investigated not just as sole interactions of particles with the environment but as the complex behaviour of its colloidal solution. Additional aspect that should be encountered is the question of the rate of agglomeration of silver nanoparticles in living environment, thus lowering overall toxic effects of Ag NPs (Soenen et al. 2015). Also, the gender related results are unexplained, yet proven (Xue et al. 2012). Several researchers stated that toxic effects are highly dependent on the cell type with which Ag NPs come in contact (Liu et al. 2012; Liu and Hurt 2010; Asharani et al. 2009b). There are evidences that in some cases, Ag NPs did not have harmful effects on normal cells, and had highly toxic effects against cancer cells (Liu et al. 2012; Sur et al. 2010; Asharani et al. 2009b).

All previously stated strongly indicate that synergies of different influences must be considered to determine the effect of Ag NPs against the living cells. Physico-chemical properties of the nanoparticles comprise complex relationships and characteristics of many influential factors, starting with the size, shape and biochemical properties of the cells and media in the vicinity of the particle. Even the properties of normal aging process should be considered as well, such as Ostwald ripening (small crystals dissolve in time and reposit themselves onto the larger crystals, thus producing larger nanoparticles), tendency for aggregation, surface state modifications, corrosion and others (Gil et al. 2010). Numerous available data on both beneficial and toxic effects of silver nanoparticles strongly implies that further research is needed, as well as systematic collection, analysis and standardization of related data, methods and characterisation. For example, within lungs, silver nanoparticles can act as hazardous and provoke abnormalities in pulmonary function (Theodorou et al. 2014), but Ag NPs can also result in decreasing inflammation and stimulation positive immune reactions in lungs (Park et al. 2010).

4 Antibacterial Properties of Silver Nanoparticles and Relevant Mechanisms of Action

As previously elaborated and based on existing research results in literature, antibacterial properties of silver is mainly linked to the release of silver ions due to the contact with the organic fluids such as skin moisture, or wound fluids. Silver metal is inert material in both pure air and pure water and it even does not react with oxygen being a noble metal, but it reacts when in contact with sulphur compounds producing silver ions and its ionic form is highly reactive. Silver acts against bacteria through its influence on bacteria cells, and depending on the amount and ionic release rate, it results in bacteria cells death and non-reversible detrimental changes in cell structure and inhibits its further growth. For example, silver sulfadiazine damages the cell membrane and binds to DNA thus inhibiting bacteria growth (Rai et al. 2009). Different silver compounds have been investigated to serve as antibacterial agents, such as silver zeolite coating on ceramics aimed for food and medical materials containers, bathrooms, public spaces, etc. Combinations of silver with UV radiation have been investigated for possible inhibition of viruses (Butkus et al. 2004) or its synergies with different antibiotics (Shahverdi et al. 2007) and both showed good results. Very recent work showed that Ag NPs can be efficiently used in combating well known dangerous dengue virus (Sujitha et al. 2015). However, even with the very promising results so far, the medical products that can be commercially found at the market today are mainly treatments for wounds, burns and chronic ulcers, and majority is based on application of silver sulfadiazine (Bhattacharya and Mukherjee 2008). There is also some evidence that nanosilver can act as antiviral agent (Murphy et al. 2015; Wei et al. 2015).

Along with a lot of debates about nanoparticles use in general, especially their long-term effects, different consumer and medical products containing silver nanoparticles have been approved by relevant official accreditation bodies, such as FDA in USA, EPA in USA, SIAA in Japan and others, depending on the specifics of application (Abou El-Nour et al. 2010). In 2014, Scientific Committee on Emerging and Newly Identified Health Risks (SCENIHR) of European Commission (EC) issued the official document: “Opinion on Nanosilver: safety, health and environmental effects and role in antimicrobial resistance” and expressed the need for more data on cytotoxic effects of silver nanoparticles, especially in longer time. But they also stated that no significant health risks exist related to the use of silver nanoparticles.

Even with years of applications and numerous commercial products on the market, a mechanism of antibacterial action of both silver and silver ions is still under debate. Silver nanoparticles exhibit different forms, corresponding to the previously stated mechanisms related to the Ag NPs toxicity. Accordingly, antibacterial effects can be observed either by the influence of whole Ag NPs or by produced Ag^{+} ions (Le Ouay and Stellacci 2015; Liu and Hurt 2010; Rai et al. 2009). Once again, it seems that the most important effect originates from released silver ions and the presence of whole silver nanoparticles (Ag^{0}) can be observed as a reservoir for producing more Ag^{+} ions beside its interactions with the environment. It is obvious that surface properties of silver nanoparticles, as well as their size and shape determines their interactions with cells, such as membrane penetration, particles aggregation, dissolution, passivation, oxidation, etc. As mentioned previously, Ag NPs stimulates ROS generation and initiate high oxidative stress, which represents the main cytotoxic mechanism. However, the main antibacterial agents are silver ions (Ag^{+}), according to many studies (Le Ouay and Stellacci 2015; Liu and Hurt 2010; Rai et al. 2009; Park et al. 2009). Monovalent silver ions are highly reactive with different molecules which can be found either in the cell vicinity or inside it, such as DNA, organic amines, peptides, phosphates, thiols, and cofactors (e.g. enzymes) and this results in several possible destructive mechanisms for bacteria cells (Le Ouay and Stellacci 2015). Antibiotics are commonly developed to target some specific bacteria and their molecules usually cannot destroy other sources of disease, while silver ions react with all surrounding molecules, resulting in antibacterial reaction against broad spectrum of micro-organisms. Rai et al. (2009) suggested that silver ions penetrate bacteria cells forcing its DNA to change into a condensed state, thus being unable to replicate further.

Influence of silver ions on ROS generation is very important cytotoxic mechanism and some estimates state that more than half of cell processes related to the antibacterial mechanism of Ag NPs is directly linked to ROS species and resulting oxidative stress (Park et al. 2009). On the other hand, another research group studied possible pathways for Ag NPs affecting allergic reactions in pulmonary responses and concluded that presence of silver nanoparticles resulted in decreased intracellular ROS levels in bronchoalveolar lavage fluid, via developed molecular mechanisms (Park et al. 2010). It seems that silver nanoparticles can be tuned to have detrimental effect on bacteria cells, while simultaneously not affecting normal human cells in such way, but there are still many unresolved questions before final conclusions can be made.

Another form of influence comes from the sole silver nanoparticles, primarily due to their extremely high surface areas which is obviously size-dependent property, primarily related to the surface area of particles (Sotirioua et al. 2011; Sadeghi et al. 2012; Molleman and Hiemstra 2015; Soenen et al. 2015; Le Ouay and Stellacci 2015). Smaller particles have larger contact surface, thus producing more effects. Shape of particles is important again due to surface aspects: crystal lattice with more exposed surfaces tends to be more active. It was shown that silver nanoparticles in the shape of plates (Ag-nanoplates) exhibited the best antibacterial properties, in comparison with nanorods (Sadeghi et al. 2012), what is in consistence of previously stated by Soenen et al. (2015). Nanospheres had the least activity, being with the least exposed crystallographic facets. The whole Ag NPs can attach themselves to the cell membrane or go through it to the inside of bacteria cells which contain proteins with sulphur, as well as compounds with phosphorous (e.g. DNA) (Rai et al. 2009). Silver nanoparticles react with both of these molecules resulting in cell death. It is suggested that major pathways of antibacterial effects is related to the reactions of silver soluble complexes: thiols in enzymes or sulphur analogues of alcohols (Liu et al. 2010). Additionally, Ag NPs serves as reservoirs by release silver ions inside the cell provoking yet more detrimental effects.

Silver as noble element does not react with water but can be readily dissolved in water in presence of oxidant, such as airborne oxygen dissolved in water, and this mechanism is called oxidative dissolution. Oxidative dissolution of silver is recognised as a very important mechanism for its antibacterial action, probably due to the increased production of free Ag^+ ions during the process (Le Ouay and Stellacci 2015; Molleman and Hiemstra 2015; Ho et al. 2010, 2011). It has been observed as influenced by the presence of dyoxygen (Ho et al. 2011) and by biologically relevant oxidants, such as strong oxidizer hydrogen peroxide, H_2O_2 (Ho et al. 2010). Oxidation of Ag NPs by O_2 is amine activated and this process is rather different than dissolution of Ag NPs by H_2O_2. Oxidation involving H_2O_2 is size irrelevant and pH level of media has significant role, whereas increase of pH increased release rates. It should be noted that H_2O_2 belongs to ROS species which are also produced by the Ag^+ ions and the complexity of the process is rather obvious. Surface coating of Ag NPs can greatly influence the oxidation process, as expected.

It is shown recently that oxidative dissolution is perhaps the major influential factor to the increased antibacterial action but dissolution kinetics and thermodynamics of nanosilver in water environment and more complex organic fluids is still not completely understood. Understanding of oxidative dissolution of silver can provide mechanisms for control of the silver ions release rate and consequently resulting effects of this major factor of antibacterial activity. If the ions release is done in a controlled manner it would also enable prevention of its toxic effects on healthy cells, as well. Liu et al. (2010) suggested that efficient controlled silver release (accelerate or inhibit) can be achieved through the control of particles sizes, and surface chemistry modifications, such as presence of thiols, citrate ligand, sulfidic coatings or by changing peroxide intermediates. The close environment of the Ag NPs can serve as the influential factor for rate release, such as biological reductants and sulfur-containing biomolecules (Liu et al. 2010). Preoxidation accelerate the release of ions, as well as size reduction. Presence of sodium citrate or free thiols compounds such as cystein, or glutathione inhibit silver ions release and suggested mechanism is either surface binding or prevention of contact with oxygen. In vivo tests showed that silver nanoparticles without any coating were completely prevented to show any antibacterial activity in the presence of protein in media (e.g. from blood, serum or plasma) (Gnanadhas et al. 2013). But when citrate or PVP coating were applied on these Ag NPs they showed high antibacterial activity. Bhunia et al. (2015) showed that protein capped silver nanoparticles are more biocompatible with normal cells than bare particles.

Media containing anions that can interact with Ag^+ cations in such way to form insoluble precipitates, can slow down or completely inhibit antibacterial activity (Le Ouay and Stellacci 2015). The recognised types of such anions that can be found in biological environment originate from: sulfides which can facilitate passivation layer around Ag NPs because Ag_2S has very low solubility; chloride (AgCl) and phosphate (Ag_3PO_4). However, not all the underlining mechanisms are comprehensively understood here. For example, AgCl can contribute to antibacterial activity under some specific conditions, and its complex behaviour is still not yet completely resolved. It is obvious that many complex factors influence the activity of the Ag NPs system in the living environment. Some of the proposed factors which should be considered in order to adapt the antibacterial activity of any nanoparticle system to the living tissue media are: relevant microbial strains, presence and amount of oxidizers, biochemistry of tissue culture media, fluid dynamics (circulation) within the affected zone, physical properties of the system (e.g. temperature), possibility for photoreduction of Ag (+1) due to light, as well as many other feasibility factors starting with storage conditions (stability) and cost (Le Ouay and Stellacci 2015).

Two very important conclusions can be made from these studies: increased oxygen presence and increased media acidity highly facilitates silver ions release.

5 Application Areas

5.1 Application of Ag NPs in Wound Dressings

Medical treatment of wounds and chronic ulcers is perhaps the oldest medical application of silver, thus proving that antibacterial properties of silver have been known for centuries. The first forms of silver used in such purposes were metallic silver and silver nitrate. Eye drops with silver nitrate, as well as wound dressings, were successfully used as medical treatments in 1881, but it was not seriously considered in medicine until 1960s when it started to be used again for the treatment of burns (Rai et al. 2009). In 1968, the silver sulfadiazine cream was developed to treat burns and as antibacterial cream against gram-negative and gram-positive bacteria such as *Escherichia coli*, *Klebsiella species*, *Pseudomonas species*, *Staphylococcus aureus*, etc. Nowadays, the majority of available dressings for wounds and burns are based on use of silver sulfadiazine (Bhattacharya and Mukherjee 2008). There are several commercially available dressings, by different manufacturers: Smith and Nephew Health care Limited, UK, Canada; Motiff Laboratories Pvt. Ltd. Kare Health Specialities, Verna, Goa; Solvay, Brussels, Belgium; Johnson & Johnson, New Brunswick, NJ, USA; Laboratories Urgo, Chenova, France; Activated Charcoal Dressing Hasti Medic India; Argentum Medical, Geneva, IL,

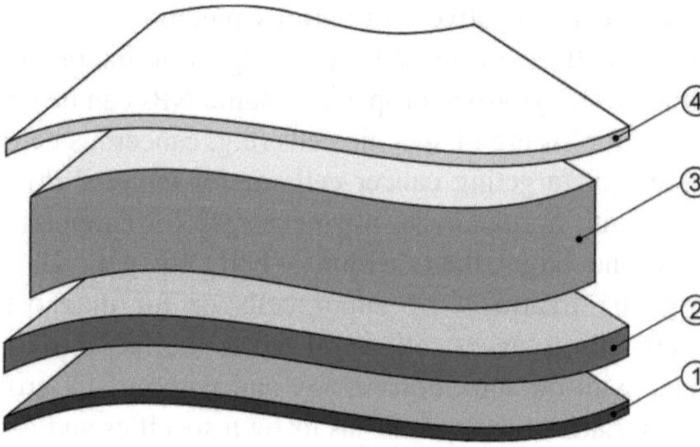

Fig. 2 Different layers in a typical wound dressing: *1* Thin layer in contact with tissue (e.g. Ag NPs impregnated or coated polymer mesh); *2* Thin protection/distribution layer; *3* Thick absorbent foam layer and *4* Thin outer layer—water repellent

USA; Coloplast, Minneapolis, USA; Mölnlycke Health Care, Australia; ConvaTec Australia; Argentum Medical, Chicago, USA and others.

Essentially, wound dressing is used to help the healing of wounds and to prevent further possible complications such as infection. It helps the wound stop bleeding and start healing and absorb excess fluids (blood, plasma). Dressings are in direct contact with the wound, unlike bandage that holds the dressing. There are variety of dressing types at market, such as absorbent dressings, antimicrobial dressings, burn treatments, debridement solutions, compression bandages, film dressings, foam dressings, hydrogel dressings, negative pressure wound therapies, odour control dressings, prevention products (e.g. to prevent pressure ulcers), scar treatment, skin care, skin substitutes, surgical dressings, wound contact layers, etc. Typical wound dressing can have several layers (Fig. 2), whereas two layers represent basic design: polyethylene layer with incorporated treatment agent such as nanocrystalline silver which comes into direct contact with a wound and second layer of polyester gauze over it (Chaloupka et al. 2010).

In comparison with traditional silver dressings, dressings with Ag NPs exhibited significant decrease of healing time (Huang et al. 2007). Good review of benefits obtained by introducing nanocrystalline silver is given by Fong and Wood (2006). There are several research directions in development of improved wound dressing incorporating silver nanoparticles, in order to mitigate possible side effects such as argyria (Wadhera and Fung 2005). One successful solution is combination of nanocrystalline silver with chitosan, which showed significant improvement in healing rates (Lu et al. 2008; Chaloupka et al. 2010).

5.2 *Application of Ag NPs in Drug Delivery*

In general, influence of nanoparticles can be observed from different aspects: immune system response, chemical reactions with surroundings, genetic influence, morphological transformations (inherent property related mainly only to the nanoparticles), effects on cell metabolism, etc. Cytotoxic effects of NPs can be observed as negative and various methods have been investigated to mitigate or eliminate these in case NPs are designed to interact with the healthy tissue or blood. Otherwise, cytotoxic property of some NPs can be observed as beneficial if they are aimed for killing of specific cells (e.g. cancerous cells), or targeted drug delivery. If aimed for targeting cancer cells or for targeted drug delivery, beside biocompatibility with normal cells, it is necessary for nanoparticles to provide a possibility to trace and target them within a body. In general, nanoparticles can be used for targeting treatment of tumor cells or for diagnostics. The main problem with treatment of cancer cells is related to the fact that each tumor has different type of cells, with random morphology and without any order in proliferation, viability or death. Each of these cells has its own specifics and for majority of them, it is still the unknown. Some recent evidence suggests that cancer stem cells (CSC) are involved in majority of cancer cells (Joseph et al. 2016). CSCs are very resistant to known drugs and extremely aggressive in a way that they seem responsible for tumor re-occurrence because they support growth of all other cancer-specific cells. Accordingly, in order to destroy the tumor without its relapse, it is necessary to destroy both CSCs and cancer-specific cells—so called 'dual targeting' (Joseph et al. 2016). CSCs cells accounts for only a small part in the whole tumor but they share many properties of the normal cells, such as being organ-specific, cell differentiation and self-renewal. From the very beginning of the discovery of CSCs many strategies to target these cells have been tried with no success and emerging field of targeted drug delivery by nanocarriers finally showed excellent results (Joseph et al. 2016; Shen et al. 2016; Hong et al. 2015; Malhi and Gu 2015; Hu et al. 2010). Various combinations of nanoparticles and antibodies against specific markers of CSCs are under investigations with ultimate goal to tune the treatment in such way to destroy all tumor cells and leave normal cells unharmed, which is yet unattainable. Combination of nanoparticles and drugs showed promising results in dual targeting of tumor cells. One successful approach was the modification of the surface of the nanocarrier with CSCs surface market antibodies (Malhi and Gu 2015). Also, combinations of standard chemotherapy and nanocarriers with cancer related drugs are investigated because it showed good results (Hu et al. 2010). Recent works showed evidences that efficient inhibition of CSCs by nanoparticles can be done by targeting the following (Shen et al. 2016; Hong et al. 2015):

(1) CSC—specific markers: Aldehyde dehydrogenases; transmembrane receptor CD44 for hyaluronic acid; CD cell surface proteins; stem cell marker CD133,
(2) ABC efflux transporters related to drug resistance,
(3) Inhibition of CSC-related signaling pathways: Wnt/β-catenin (regulation of growth, self-renewal, regeneration), Notch (governing cell differentiation), transforming growth factor-β, and hedgehog signaling (regulation of pattern formation during early development and regeneration),
(4) Tumor microenvironment or CSCs niches.

The investigation of silver nanoparticles for treatment of tumors has started rather recently, and clinical data are scarce or very limited, with research mostly pertaining to in vitro studies or preliminary research in relation to some specific cancer cell lines. For example, it was shown that 100 μg/ml of Ag NPs resulted in necrosis of 40.2% cells of human colon cancer (Arora et al. 2009) but there is no comprehensive approach to all details needed for clinical application. One of the very few published works is the in vivo study of Jacob and Shanmugam (2015) on Ag NPs effect on DAL, Dalton's ascites lymphoma cells in tumor of peritoneal cavity. They realised animal testing on mice and obtained several good results. Firstly, tumor growth were stopped and volume decreased by administering 35 μg/kg of body weight of Ag NPs, with size of particles in the range of 19–42 nm. Life of the animals treated with Ag NPs was prolonged in around 46%, what is a very significant increase. This study did not show complete healing of the control animal, but the results showed significant enhancement of the health and life prolongation. Another in vivo study proved anticancer activity of Ag NPs against multidrug resistant (MDR) cancers (Liu et al. 2012). They conjugated Ag NPs with TAT peptide (Ag NP—TAT) and obtained excellent results in killing animal tumor cells and inhibiting their further growth (malignant tumor cells in mice), as shown in Fig. 3. TAT peptide (trans-activating transcriptional activator) belongs to a group of cell-penetrating peptides (CPPs), which are utilized as carriers, for different molecular cargoes (small chemical molecules or nanoparticles and even large DNA fragments) to the inside of cells where it performs some desired function. Very important finding of that study is efficiency of Ag NP—TAT in comparison with

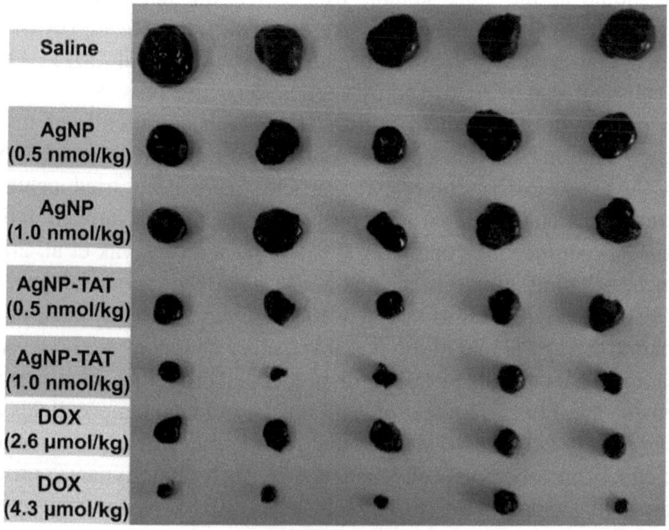

Fig. 3 Representation of in vivo anticancer activity of Ag NPs: comparison of individual tumors removed from animals after different treatments as indicated on the *left*. Reprinted from Biomaterials, 33, Liu J, et al., TAT-modified nanosilver for combating multidrug-resistant cancer, 6155–6161, Copyright (2012), with permission from Elsevier

standard anticancer agent doxorubicin (DOX) which exhibits numerous harmful side-effects unlike Ag NPs (Liu et al. 2012). The same research group studied possible targeting of Ag NPs to cancer cells which have CD44 receptors and showed that modification of Ag NPs with hyaluronic acid (HA) can efficiently enhance targeting and delivery to specific sites (Liang et al. 2015).

Based on numerous in vitro studies, Ag NPs have been recognised as a very suitable anti-cancer agent or supplemental treatment at least, and research in this area has extensively started during the last years. Marambio-Jones and Hoek (2010) presented a good review of existing evidences on toxic effects of Ag NPs on different cancer cells. Depending on the end application, different approaches in modification of nanoparticles' properties have been investigated, by varying production routes, size, surface, electromagnetic properties or nanocarriers (Govindaraju et al. 2015; Gurunathan et al. 2013a, b, 2015; Hu et al. 2015; Petrarca et al. 2015; Liang et al. 2015; Sharma et al. 2014; Pimentel et al. 2013; Sanpui et al. 2011). One of the last published papers indicate that modification of silver nanoparticles with α-tocopherol succinate (α-TOS) produced simultaneous reduction of cytotoxic effects against normal cells, while improving ability to destroy cancerous cells (Hu et al. 2015).

Recent works showed that breast cancer cells treated by silver nanoparticles in different concentrations (1–10 μg/mL) can be inhibited (Gurunathan et al. 2013a, b). Other studies indicated the amount of 36 μg/mL of Ag NPs (Venil et al. 2016) or 7.19 μg/mL (Remya et al. 2015) as adequate quantities to provide cytotoxic effect against breast cancer cells. Remya et al. (2015) used flower extract to synthesize Ag NPs what should be the more suitable from environmental aspects. Some studies indicated that the synthesis method of Ag NPs can influence higher inhibition of breast cancer cells viability than in normal cells (Ortega et al. 2015). Other types of tumor cells can also be destroyed by application of Ag NPs. Silver nanoparticles of 10 nm size produced death of human HL-60 cancerous cells of myeloblastic leukemia, by entering the nucleus and causing cell apoptosis and DNA damage (Govindaraju et al. 2015). Silver nanoparticles have been tried against Mg-63 osteosarcoma cells and inhibited their proliferation and also showed antimicrobial effects against gram negative and gram positive bacteria (Nayak et al. 2016). Even the treatment of retinal diseases, such as neovascularization, can be done by application of silver nanoparticles as targeted drug delivery (Kalishwaralal et al. 2010). Photodynamic therapy is one of the possible cancer treatment methods, which uses the irradiation of cancer cells by laser with predefined properties such as wavelength, power density, etc. Prior to irradiation, cell uptake of metal nanoparticles is realised in some way and in vitro results showed that Ag NPs can efficiently serve to produce cytotoxic effects on cancerous cells mediated by such photodynamic process, even better than gold nanoparticles (El-Hussein et al. 2015). Nanocomposite made of graphene oxide—silver nanoparticles (rGO-Ag) also showed cytotoxic effects and inhibited cell viability of ovarian cancer cells, A2780 and was evaluated as non-toxic for normal cells (Gurunathan et al. 2015). The Ag NPs size was around 20 nm.

Advantage of this approach was in low cost and easy production of large quantities of such nanocomposite, especially because the fabrication process is mediated by widely available plant extract.

The most promising results in using Ag NPs as anti-cancer treatment have been shown by application of different nanocarriers that will provide the targeted release of their content precisely at the wanted location within a body. In such way, possible side-effects of Ag NPs on the normal cells would be ideally eliminated. The question of ideal nanocarrier or nanoparticle coating represents the cutting edge area of research. So far, nanocarriers with gold nanoparticles were mainly investigated as anti-cancer agent, but silver nanoparticles recently have started to be considered as well. Efficient results in application of metallic nanoparticles against tumor cell line were obtained by using nanocarriers made of mesoporous silica nanoparticles (MSNs) with incorporated metallic nanoparticles within pores and cancer-specific compounds (e.g. ligands) on the surface (Giret et al. 2015; Fang et al. 2014). Pores of MSNs can be functionalized in such way that they open upon contact with tumor cells and release their cargo. Opening of the pores can be designed according to changes in media (pH, certain enzymes) or due to changes in temperature, magnetic field, light, etc. In essence, this is the concept of targeted drug delivery, by using nanoparticles in general. Very good review of current state of development in MSNs for drug delivery are given by Giret et al. (2015) who presented benefits, methods to modify the surface of MSNs, methods to introduce drugs into the MSNs and means to release it where planned. Targeting the specific location or specific type of cells is the most difficult thing to achieve. Standard drugs in use today do not have such function and beside healing function, they all have some side-effects on cells, tissues and organs which suffer from exposure to it. Side-effects of antibiotics are well known example. Fang et al. (2014) broadened the targeting function of the MSNs nanocarrier with possibility of the tracing throughout the body until it reaches the targeted cells. They investigated the possibility to incorporate elements that will enable external physical monitoring of the nanocarrier system movement through the body and obtained excellent results with chitosan-polymethacrylic acid (CS-PMAA) based nanocarrier of Ag NPs. Simultaneously it provided targeting of the tumor cells recognised by Tf receptors (HeLa and MRC-5 cells), controlled release of Ag NPs (40–60 nm size) together with DOX anti-cancer drug and surface enhanced Raman scattering (SERS) tracing of the nanocarrier, with strong SERS signals. The complete composite structure was pH sensitive, meaning that the drug release can be controlled by changes of pH values. However, this research was done under in vitro conditions and further in vivo testing is needed. Liang et al. (2015) modified Ag NPs with hyaluronic acid (HA) and obtained efficient targeting of MCF-7 breast cancer cells and SW480 colon adenocarcinoma cells.

Several other nanocarrier options have been pursued, depending on the type of tumor cells and desired functions. Under in vitro conditions, spherical Ag NPs of approximately 3–5 nm diameter, encapsulated in the polyethylene glycol

(PEG) based nanocarrier (altogether around 30–120 nm in dry state and 40–250 nm in wet state), showed excellent features against cervical cancer cell lines, while at the same time being biocompatible towards normal cells (Pimentel et al. 2013). Wang et al. (2012) studied the possibility to use folate receptors in cancer cells to destroy them by targeting it with silver nanoparticles coated by folic acid. Folic acid (FA) binds to the folate receptors which mediate delivery of certain compounds into the cell and accordingly, this nanocarrier should enable efficient targeting of cancer cells where Ag NPs can be then released. Additionally, they bonded standard chemotherapeutic drug doxorubicin (DOX) onto the FA-AgNPs to target its release inside the tumor cell. They obtained excellent results in targeting the FA-AgNPs-DOX to the cancer cells. Drug release was done via changes in pH value of the media inside the cell, which is one reliable method of drug release and was recommended also by other researchers (Fang et al. 2014).

Two well recognised biocompatible materials were investigated as suitable nanocarrier of AgNPs as the anti-cancer agent and both showed very good results. Nanocarrier made of chitosan resulted in a very low quantities of AgNPs which were able to kill cancerous cells, such as 0.33 μg/mL (Sanpui et al. 2011). Significantly larger quantity of 2.4 μg/mL of Ag NPs was needed to destroy human glioblastoma cells with nanocarrier based on alginate-chitosan combination (Ag NPs-Alg-Chi) (Sharma et al. 2014). Hybrid drug delivery systems made of poly (lactic-co-glycolic acid) (PLGA) shell and with Ag NPs as a core and loaded also with some standard drug (e.g. paclitaxel) is a research direction that is vigorously pursued (Pandita et al. 2015). In vitro comparison of coated and uncoated Ag NPs for cytotoxic effects on bladder cancer cells showed that both are deleterious for bladder tumor cells, but this study indicated that extracellular-signal-regulated kinases (ERK 1/2) signal pathways are closely connected with cytotoxicity effects of Ag NPs (Castiglioni et al. 2015).

Silver nanoparticles have also been investigated for possible use in manipulation and stimulation of neural cells, in order to help regeneration of nerves in post-traumatic injuries or degenerative neural diseases and this research direction showed beneficial effects of Ag NPs (Polak et al. 2015). It was shown that several types of nanoparticles can enhance neuron growth and influence electrical activity as one of the well recognised tool in treatment of several neural diseases via application of mechanical tension. If, for example, magnetic nanoparticles could be used in this way, narrowly focused tension of desired cells could be performed. Silver nanoparticles were shown to enhance differentiation of some neural cell types (Polak and Shefi 2015).

It can be seen from the previous text that there is ongoing debate about beneficial or toxic effects of silver nanoparticles in medical use. It is evident that more research is needed, especially in vivo testing. The time will show the best way to use Ag NPs to combat disease and to preserve normal cells, since its antibacterial and anticancer effects are clearly proven, even though exact methods and dosage not yet.

5.3 *Application of Ag NPs in Biosensors and Medical Diagnostics*

Another promising direction of application of silver nanoparticles is using the well known optical properties of silver. Thin silver film has enabled mirrors as we know it, due to its high reflectance and strong interaction with light. Silver nanoparticles have the high ability to absorb and scatter light. Excitation of electrons on the surface of silver by light at specific wavelengths results in collective resonant oscillation of those surface electrons, known as surface plasmon resonance (SPR). Intensities of absorption and scattering of light for these types of materials are thus much higher than of those with non-plasmonic properties. Resonant oscillation enhances the effects otherwise produced by light. It can result in radiating light (scattering) or convert it to heat (absorption). Enhanced scattering means that single molecules can be detected and this phenomenon is used in very precise diagnostics. Silver is one of the best plasmonic materials, together with gold. Changes of Ag NPs shape and size along with changes of refractive index near the surface can modify the absorption and scattering properties of Ag NPs. Round nanoparticles with smaller sizes (e.g. 10 nm) mainly absorb light with narrow peaks at some specific wavelength, while larger particles (e.g. 100 nm) mainly scatter light and have broad peaks shifted towards longer wavelengths. Local refractive index depends on the media/environment where the Ag NPs are located when exposed to light and different media influence different optical intensity (the sum of absorption and scattering) curves.

Plasmonic properties of Ag NPs are used in surface-enhanced Raman spectroscopy (SERS), which can efficiently be applied for medical diagnostics (Coluccio et al. 2016; Langer et al. 2015; Larguinho and Baptista 2015; Li et al. 2015; Lin et al. 2011; Soenen et al. 2015). Very good review of theoretical background related to optical properties of nanoparticles and its use in signalling and biosensors are given in Langer et al. (2015). Raman scattering signals are commonly used for identification of chemical elements due to the unique properties of each chemical element regarding Raman scattering, or energy shifts after exposure to specific light wavelengths. This is the foundation of Raman spectroscopy, widely applied in material science today. The problem with Raman scattering is that the signal is usually very weak and easy to misinterpret. Gold and silver are used to enhance that signal, and silver is the best known material to achieve it, even better than gold (Coluccio et al. 2016; Yang et al. 2013). The problem with application of silver is that the homogeneity of the signal is not steady (Li et al. 2015). But enhancement of the signal obtained by Ag NPs enables detection of single biomolecules, such as blood plasma proteins, urea, DNA and RNA sequences and this technology is very promising for early cancer detection and diagnostics. It can also be used in various different areas of material science, such as drug detection and development of pharmaceuticals, analysis in food quality, environmental analysis, and other.

In area of medical diagnostics, early detection of cancer is one of the pre-conditions of successful treatment and various approaches exist, whereas application of SERS technology showed very beneficial results (Larguinho and Baptista 2015). In general, clinical diagnostics is based on identification of substances specifically linked to the occurrence of some disease, such as biomarker proteins in case of cancer cells. Several studies showed that Ag NPs can be used to prepare SERS substrate for the cancer screening or very sensitive biosensing applications. For example, Li et al. (2015) proposed a method for screening of prostate cancer by using SERS spectroscopy, as non-invasive method. They obtained 98.1% accuracy. For efficient application of Raman scattering for cancer detection, SERS-active nanoparticles are of the utmost importance, such as Ag NPs based SERS substrates. Another research also showed that SERS technology should be considered in this view. New method of blood plasma analysis by using Ag NPs based SERS in combination with membrane electrophoresis (Lin et al. 2011). They obtained unique information within SERS spectra indicating biochemical elements of proteins and showed 100% sensitivity of the technique, in case of gastric cancer screening. In case of esophageal cancer detection, accuracy of this technology was around 90% (Li et al. 2014). It can be concluded that changes in Raman spectra of specific bioorganic substances (e.g. proteins, ligands) can efficiently indicate if the cancer is present. Plasmonic 3D structures with 40 nm silver nanoparticles have been proposed for biosensing of proteins, peptides and other biomolecules within the blood flow (Coluccio et al. 2016).

Silver nanoparticles can assist in other techniques of molecular detection and labeling, such as methods based on fluorescence, dynamic light scattering, micrroarays, and others (Zhu et al. 2015). Fluorescent labeling techniques are widely used and molecules such as proteins, amino acids, peptides or antibodies can be tagged by this technique. Enhancement of fluorescent tagging by Ag NPs is shown by Li et al. (2010). They used Ag NPs in a range of 100–200 nm size and obtained increased sensitivity of several assays, due to the signal increase obtained by fluorescent emission from deposited Ag NPs. Combination of Ag NPs with other nanoparticles (e.g. plasmonic gold nanoparticles) and combination of several techniques can provide efficient sensing of living cells, along with tracking the nanoparticles throughout the tissue (Chen et al. 2016). Developed biosensors based on Ag NPs can also be used in other applications such as detection of toxic metals in water (e.g. arsenic III) (Song et al. 2016). Detection of different metals in environment (water or soil), especially heavy metal ions are of great interest because they exhibit extremely high toxicity from various aspects of living ecosystems, but this is a very complex issue and ongoing research topic (Gumpu et al. 2015).

5.4 Application of Ag NPs in Orthopedics

The provision of focal adhesion sites and antibacterial properties exhibition by nanomaterials have drawn significant attention of biomaterials research community to use them for biomedical engineering applications. The wound-healing ability of silver nanoparticles (Ag NPs) is well known and they also exhibit excellent antibacterial properties. This makes them potential candidate for coating of biomaterials use for orthopedic regeneration, repair and replacement.

Surgical site infections (SSI) are expected to grow in orthopedic implants due to their lack or lower degree of biocompatibility. Such implants are primarily composed of metal alloys or polymer composites. The microbiological properties of these implants are not suitable to stay for prolonged durations in human biological conditions. The cells consider implants foreign materials and unwanted cell attachment to implants result in formation of biofilms that cause SSI. Antibacterial Silver nanoparticles can help in prevention of biofilm formation around the implant.

Due to antimicrobial properties, silver nanotechnology holds strong potential for applications in orthopedic implants. Therefore comprehensive assessment of proposed benefits and potential risks associated with the use of silver nanoparticles in orthopedics is prudent for widespread implementation of this novel technology.

The toxicity and antibacterial properties of Ag NPs are well discussed in Sects. 3 and 4 of this chapter. Succinctly the ions released from silver surface are biologically active and exhibit antimicrobial effects against Gram-positive and Gram-negative bacteria, protozoa, fungi, and viruses (Nair and Laurencin 2008). These ions are able to attach with various bacterial structures including cell wall, plasma membrane and DNA (Chaloupka et al. 2010). The binding of silver ions with cellular components result in destruction of cell wall, interaction with sulfhydryl (SH) groups causing cell function impairment, and interaction with DNA inhibits reproduction (Brennan et al. 2015). The release of ions depends upon the surface area of the silver. Therefore smaller Ag NPs less than 10 nm in size have significantly higher efficacy as greater surface is available for the release of ions (Alt et al. 2004). It has been also under discussion that the direct attachment of Ag NPs with bacterial cell could also induce lysis. However the mechanism of direct attachment and bacterial lysis due to Ag NPs toxicity has not yet properly elucidated (Percival et al. 2005).

Although the survival rate of patients treated with tumor prosthesis is quite satisfactory, about 14–35% of patients induce infections around tumor prosthesis that often leads to amputation (Brennan et al. 2015). Despite observing hygienic precautions, the potential of post-surgery complications is high due to growth of bacteria in biofilms. The prevention or minimizing the risks of periprosthetic infection could be performed by employing antibacterial prosthetics.

As the coating of silver on medical devices proved efficient antibacterial surfaces, an in vivo study on rabbit model showed lower infection rate of 7% by using silver coated titanium megaprosthesis as compared to uncoated group where 47% infection rate was observed. Although the silver concentration was elevated in blood and in organs, no toxicological effects were observed (Gosheger et al. 2004). In a later study by same group, bone sarcoma patients who underwent megaprosthesis placement were observed for five years. Silver coated and pristine titanium megaprosthesis were implanted in two groups of 51 and 74 patients respectively. It was observed that infection rate was higher (17.6%) in titanium group comparing to silver group (5.9%). Consequently, 38.5% of titanium group patients had to undergo amputation (Hardes et al. 2010). This study showed that use of silver-coated megaprothesis not only reduces the infection rate but also less aggressive infection treatment is required after surgery.

The external fixator pins are essential armamentarium in orthopedic surgery for temporary stabilization in emergency or to provide tension band when used with rigid internal fixations. External fixators are also used for long-term fixation in conservative orthopedic treatments. It has been reported that pin tract infection occurred in 42% cases of fixator pin implants (Brennan et al. 2015). Such infections could lead to loosening and damaging of implant and inflammation of bone or bone marrow (osteomyelitis) (Parameswaran et al. 2003). Therefore replacement or repair of implant often required after induction of infection. The huge risks of infection imply development of antimicrobial fixator pins.

In order to examine the efficacy of silver coated fixator pins against infections, a study was performed by incubating external fixator pins with *Staphylococcus epidermidis*. The antimicrobial surface activity of stainless steel pins coated with polymer-containing nanoparticulate silver was compared with control (pristine stainless steel), titanium and copper pins. The incubation of pins with bacteria was performed for 20 h followed by bacteria adhesion test. The silver coated pins showed a 2 log step reduction in S. epidermidis number as compared to control and titanium pins. The copper pins were used as positive control as they showed highest antibacterial activity (Furkert et al. 2011). In another extended study, three biofilm forming bacteria *E. coli, Pseudomonas aeriginosa* and *S. aureus* showed reduced adhesion on silver coated stainless steel fixator pins comparing to control samples (Wassall and Santin 1997). In vivo studies on model animal sheep have been performed to examine the effect of silver coating on stainless steel pins. The study showed reduced infection (from 84 to 62%) and less loosening occurred in silver coated specimen compared to pristine stainless steel pins (Collinge et al. 1994). The authors suggested that reduced glycocalyx protected colonization examined by Scanning Electron Microscopy (SEM) could be extenuation of high antimicrobial surface activity.

Osteomyelitis could evolve into chronic inflammation despite of systematic antibiotic treatment that leads to infection nonunion that may require surgical debridement or amputation. Weak direct current and silver ions have demonstrated satisfactory results to treat osteomyelitis. However the direct use of silver ions in localized osteomyelitis sites could lead to irritation or allergic effects on surgical site and neighbouring tissues. This limits the application of silver iontophoresis for osteomyelitis treatment. IN such cases, specially engineered graft could be use that can induce antimicrobial activity and promote bone regeneration. Composite bone graft made up of silver nanoparticles with poly(lactic-co-glycolic acid) (PLGA) demonstrated high biocompatibility and excellent antibacterial properties. It has been demonstrated that Ag NP mixed PLGA promote osseoincorporation by ongrowth of pre-osteoblasts. Studies show healing of infected femoral defects in 12 weeks by bone morphogenetic protein 2 coupled Ag NP mixed PLGA graft (Zheng et al. 2010).

In the presence of periodontal defects, osseoincorporation require supplement stimulation that can be achieved by additional coating. However these coatings should be antibacterial to avoid osteomyelitis. Hydrooxyapatite (HA) demonstrate high degree of biocompatibility that makes it excellent material for bone grafts and coating of bioimplants. In order to incorporate antibacterial properties, Ag NPs can be incorporate to HA to make Ag-HA bone grafts or implant coatings. The incorporation of Ag into HA coatings can be achieved by a number of techniques including magnetron sputtering, thermal decomposition, electro-chemical deposition, or ion exchange (Brennan et al. 2015).

In vitro studies show significantly reduced *S. epidermidis* and *S. aureus* adhesion on Ag-HA surfaces comparing to control HA and titanium surfaces. Non-toxicity of Ag-HA has also been demonstrated. Therefore Ag-HA provides a multifunctional surface offering both antibacterial and osteoconductive properties (Chen et al. 2006). In vivo studies to evaluate the performance of Ag-HA coatings against Methicillin-resistant *Staphylococcus aureus* (***MRSA***) infection have been performed by using rat model. Significantly reduced number of viable MRSA have been observed on Ag-HA coated titanium implants comparing to control specimen (Shimazaki et al. 2010; Akiyama and Miyamoto 2013).

The addition of antibiotics or silver nanoparticles to bone cements makes them multifunctional material for filling up cavities or creating a mechanical fixation. Studies demonstrated antimicrobial activity by Ag NPs mixed bone cements against *S. epidermidis*, MRSA and MRSE comparing to control bone cement and gentamicin mixed bone cement (Alt et al. 2004). Another study showed reduced formation of biofilms while using Ag NPs loaded acrylic bone cement against *S. aureus* and *S. epidermidis* (Slane et al. 2015).

Traditionally antibiotics specific to particular infection or class of bacteria have been used to fight against orthopedic infections. The broad spectrum of Ag NPs antimicrobial activity makes them propitious substitute to traditional antibiotics for orthopedic prosthesis. The in vitro and in vivo studies on model animal provide promising results of Ag NPs incorporation with orthopedic implants and coatings. However, the toxicity of silver and elevation of silver level in blood after implant requires further research to fully elucidate the harmful effects. A main focus should be development of efficient techniques for incorporation of silver nanoparticles with coating or implant materials so that silver ions could be released in a controlled non-harmful fashion, thus achieving not only antibacterial property but also retaining biocompatibility of the implant (Table 1).

5.5 Application of Ag NPs in Textile

Functional clothes are used for human protection in different areas where hazardous workplaces exist, such as presence of bacteria and viruses in hospitals, or other exposed sites. Functionalization of textile to produce clothes with antibacterial properties, or self-cleaning, UV protection, heat resistant and/or other functions has

Table 1 Recent studies on antimicrobial activity of silver nanoparticles in orthopedic applications

Reference	Bacteria	Substrate/coating	Ag NPs incorporation technique	Biocompatibility test
Li et al. (2016)	*Staphylococcus aureus*	Titanium foils	Facile one-step	Bacteria adhesion, osteoblast cellular activity
Tian et al. (2016)	*Escherichia coli* and *Staphylococcus aureus*	Hydroxyapatite coatings with oriented block arrays on Ti–6Al–4V titanium alloy	Facile Ag mirror reaction	Bacteria adhesion, osteoinductivity
Fu et al. (2016)	*E. coli*	Hydroxyapatite coating on titanium	Electrochemical reduction	Bacteria adhesion
Farhadian et al. (2016)	*Streptococcus mutans colony forming units*	Acrylic baseplates of retainers	Helium-driven chemical-vapor transport and the sputtering method	*S. mutans* colony-forming units count
Jia et al. (2016)	*Staphylococcus aureus*	TiO_2 coating on titanium dics	Mussel-inspired immobilization	Bacteria growth, osteoblast adherence, spreading, proliferation, and differentiation
Actis et al. (2015)	*S. aureus* and methicillin-resistant *S. aureus (MRSA)*	Silver nanoparticles of different geometries	Chemical reduction	Bacterial adhesion, human fetal osteoblast (hFOB) viability
Jia et al. 2015	osteoblast-like MG63 cells	Ti6Al4V scaffolds with 3D hierarchical porosity	Mussel-inspired strategy	Cytocompatibility

been a subject of research for quite a while. Usually one property is the focus of functionalization, but combination of several would lead to superior clothes. Antibacterial properties of textile have considered application of Ag NPs and this is rather new area of research (Balakumaran et al. 2016; Emam et al. 2016; Rehan et al. 2015; Tang et al. 2013; Perera et al. 2013; Freeman et al. 2012).

From the health point of view, the goal of this area of research is to develop such fabrics that ideally would be able to disinfect themselves and minimize possible health infections, through prevention of attachment of bacteria or microbes on the clothes surface or to delay such biofilm deposition. Another wanted property is UV-protection what is especially necessary with all the ongoing climate changes and increase of UV radiation and associated health problems.

Commercial application has started, but it is not widely present yet and still represents wide research area. Antibacterial socks to prevent bad odour, antibacterial gloves or surgical clothes to prevent bacterial infections are just a few cases of existing clothes (Haider and Kang 2015; Abou El-Nour et al. 2010; Rai et al. 2009; Duran et al. 2007). It is proven that incorporation of Ag NPs into the medical clothes can show anti-bacterial properties (Duran et al. 2007). But application of Ag NPs in such manner is still associated with a few unresolved problems. Two main problems can be distinguished related to the incorporation of nanoparticles into the fabrics or functionalization of textile with Ag NPs. Firstly, Ag NPs synthesis process for textile is associated with several problems and the second issue is durability of such deposited Ag NPs after washing. There are also other problems related to design, textile colour, or environmental protection. Some recent results showed that direct deposition of Ag NPs into the cotton fabrics provided antibacterial properties and satisfying quantity of Ag NPs within the clothes after 15–20 washings (Balakumaran et al. 2016; Emam et al. 2016; Perera et al. 2013) and 5 washing in case of wool fabrics (Boroumand et al. 2015). Other authors showed that sliver nanoparticles were mainly distributed over the textile surface with stable function during the six months period (Raza et al. 2015). Direct solventless techniques of Ag NPs deposition have been investigated to avoid environmental pollution occurring in case of the separate synthesis process. Duran et al. (2007) also studied biological synthesis in order to prevent environmental pollution and obtained promising results. Green technology should represent the right solution as shown by other researchers as well (Balakumaran et al. 2016; Boroumand et al. 2015).

Silver nanoparticles of 20–90 nm sizes were incorporated into cotton fabric and nanoparticles of 70–150 nm sizes into wool fabric with addition of specific stabilizers and the produced fabrics simultaneously exhibited three new properties: antibacterial effects, UV protection and vibrant colors (Rehan et al. 2015). New functional properties utilized nanosilver properties: antibacterial action and plasmonic properties and provided multifunctionality to the clothes. Ag NPs provided brilliant colours of the fabrics, in both cases for cotton and wool, and the exhibited colours can be changed by changing the deposition process of particles, which influence size and shape of particles. Increase of Ag NPs density influenced darker colours. Influence on the fabrics colour is based on plasmonic properties of silver nanoparticles and in a rather wide range of wavelengths. This study obtained mainly yellow cotton colour and reddish-yellow or reddish-brown wool colour after treatment, whereas both fabrics were originally white. Very interesting results were obtained regarding UV protection. Namely, cotton treated by Ag NPs exhibited 90% reduction in UV transmittance and 25% reduction in case of wool. This is explained by the ability of Ag NPs to reflect UV radiation. Considering the fact that UV radiation from sun is already found to be dangerous for people, this finding is very important for further research on development of UV protective clothes. Plasmonic properties of Ag NPs can be used in many end applications even in case of solar cells improvement, where their ability of high light scattering over broad wavelengths is used to enhance light trapping performance of the silicon germanium solar cells (Zi et al. 2016). Deposition of silver nanoparticles can produce different colours of fabrics and it was also shown in case of silk fibres (Tang et al. 2013). This was important study because colouring was performed at room temperature unlike traditional silk dying. Different colours were obtained by mixture of different morphologies of Ag NPs in various ratios. Different sizes of anisotropic silver nanoparticles exhibit different localized surface plasmon resonances (LSPRs) and accordingly produce various colours of silk fibres.

It seems that medical field again represents the area which could benefit the most from using Ag NPs in medical clothes, which are exposed to a high level of bacterial and microorganism contents. However, the study by Freeman et al. (2012) showed that surgical scrub suits impregnated by silver nanoparticles did not prevent from bacteria deposition on it, after 4 and 8 h of wearing the cloths. Initially, it exhibits the antibacterial properties, but after a while it behaves as ordinary scrub suit.

5.6 Application of Ag NPs in Water Disinfection Systems and Environment Related Issues

As previously mentioned, antibacterial properties of silver have been used from ancient times for storage of water and its disinfection. Silver cups and vessels for drinking water, together with food silverware were probably the first widely known use of silver. It was rather natural to try using silver for the disinfection of waste water today, as well as for the water purification systems. Filters for disinfection of drinking water, based on silver are in commercial applications for some time and significant research efforts have been devoted to development of such systems especially if lack of drinking water in certain world areas is considered (Praveena and Aris 2015). Also, wastewater treatment by the use of silver has been studied (Lovatel et al. 2015; Siripattanakul-Ratpukdi and Fürhacker 2014). For example, combinations of polymer matrix material with dispersed silicate particles with Ag NPs attached (MMT–ALG–AgNPs) as rather inexpensive hybrid material, showed good potential for waste water treatment (Lovatel et al. 2015). Especially, heavy metal ions (arsenic, cadmium, mercury, lead, chromium, silver, zinc, copper) are highly toxic for the environment and their detection in water or other ecosystem is vital for further disinfection, but standard and low cost approach still does not exist (Gumpu et al. 2015).

However, in cases of using Ag NPs for treatment of water, there are very realistic concerns regarding toxicity of silver nanoparticles on the living organisms when disposed to the water and soil (Zhang et al. 2016; He et al. 2016; Schaumann et al. 2015a, b; Gambardella et al. 2015a, b; Cornelis et al. 2014; Walters et al. 2014; Siripattanakul-Ratpukdi and Fürhacker 2014; Yu et al. 2013; Myrzakhanova et al. 2013; Fabrega et al. 2011; Liu and Hurt 2010; Blaser et al. 2008). The behaviour of silver nanoparticles when released into the natural environment is largely unknown, but even the limited study results indicate their toxicity on living organisms. Not only the amount from disinfectant systems, but also silver coming from modern textile, plastics and other commercial products, is released to the water and soil ecosystems (Benn and Westerhoff 2008). However, related data mainly does not exist and risk assessment cannot be done yet, but strong need is present to investigate the fate and behaviour of silver nanoparticles released to the environment, along with other engineering particles, especially in relation to its influence on the living ecosystems, water, soil and especially on agricultural fields. A scarce data exists regarding silver originating from textile. Benn and Westerhoff (2008) conducted a study with six commercially available socks impregnated with silver nanoparticles and showed that maximum of 650 μg silver in both ionic and colloid forms are released during washing in 500 mL of distilled water (from initial 1360 μg of Ag per gram of textile). They also showed that these quantities can be decreased to the non-dangerous levels within usual procedures in plants for treatment of waste water. But other influences should be considered as well, such as concentration of oxygen and generation of ROS species since these affects antibacterial properties, as well as other harmful consequences on normal living cells (Zhang et al. 2016; Park et al. 2009). On the other hand, some studies pointed that concentration of even ng/L of Ag NPs in water influence the changes in water organisms (fish, invertebrates and prokaryotes) and there is the urgent need to establish valid interrelationships between factors in such complex systems (Schaumann et al. 2015b; Olasagasti et al. 2014; Fabrega et al. 2011). Studies in ecotoxicology area indicated that 100 mg/L of silver nanoparticles (1–20 nm sizes) killed all zebraFish embryos exposed to it (Myrzakhanova et al. 2013). Very good review of possible pathways of released engineered nanoparticles in the environment is given by Cornelis et al. (2014). Influence of Ag NPs on the living organisms in the water ecosystem (plants, invertebrates and fish) are very harmful and ecotoxicity of Ag NPs must not be overlooked in attempts to use it for different products or treatments that might release these nanoparticles into the soil or water (Walters et al. 2014; Gambardella et al. 2015a, b; Olasagasti et al. 2014). Modifications on molecular level are especially emphasized as very harmful, such as gene modifications by Ag nanoparticles (Olasagasti et al. 2014) or its effects on immune system (Myrzakhanova et al. 2013) that surely results in much more harmful consequences on the whole ecosystem in general. Another issue in this area is related to the determinants of the environment pollution or reliable biomarkers that can provide reliable quantifiable values of ecotoxicity levels in such complex systems as the natural environment. One list of such biomarkers is given in Walters et al. (2014). Biochemical markers as indicators of environmental damage are complex and depend on number of influential factors (Gambardella et al. 2015a).

Regarding the treatment of waste water, very few studies exist related to the evaluation of wastewater quantities and silver nanoparticles effects (Schlich et al. 2013; Del Real et al. 2016). One study indicated around 5 t/hectar of land of sewage sludge per every 3 years as the main source of Ag NPs released into environment which represents the soil and water contamination (Schlich et al. 2013). Contamination of soil by Ag NPs has been related to the inhibition of microbial activity what influenced the change in C and N activity, and especially influenced changes of nitrification potential of agricultural soil (He et al. 2016). Very recent study indicated up to 400 mg/kg concentration of silver in the wastewater sludge as the realistic amount and showed that main silver species in sludge was silver sulphide, Ag_2S (Del Real et al. 2016). They proved that silver nanoparticles mainly aggregated as heteroaggregates with other compounds in soil, ranging up to 3 μm in size. It is assumed that the large amount of silver in environment ends up as silver sulphide, but there is also limited data on toxicity of this compound, even though it is reported as less toxic than pure Ag NPs. Research conducted with different focuses (beneficial vs toxic Ag NPs effects) clearly shows the need of one synergetic approach in future research.

Studies showed that filters for drinking water release some quantities of silver nanoparticles into the drinking water what should be considered by those who drink it and who are mainly unaware of that (Gatti and Montanari 2015). There are no available data on the exact quantities or their health related consequences. However, some case studies related to the existing commercially applied nanosilver (e.g. in shampoo or skin creams) proved to be only the marketing story and only 18% of products tested in this case study analysis could be classified as nanoproducts really containing nanoparticles which were advertised (Gatti and Montanari 2015). Very recent survey on commercial nanoproducts stated that for the majority of products (around 71%) manufacturers do not show supporting information or sufficient details that would prove that nanomaterials are really applied (Vance et al. 2015). Some case studies in lab environments, such as exposure of aquatic species to silver nanoparticles showed evident harmful influence on their immune system (Gatti and Montanari 2015; Gambardella et al. 2015a, b; Yu et al. 2013; Myrzakhanova et al. 2013). These studies showed that plants, algae and other marine life forms exposed to different nanoparticles all exhibited some form of harmful structural modifications, starting with colour changes. Dose dependency is emphasized, but the shape, size and other structural properties of Ag NPs probably influence it as well, and further research is needed.

Silver cations can form several complexes with DNA and RNA (Arakawa et al. 2001). On the other hand, silver and some silver compounds are known to be photo reactive (e.g. silver nitrate, $AgNO_3$). Some researchers have studied possibilities to use synergies of this silver property with UV radiation to enhance disinfectant effects of silver and to include killing of viruses in drinking water beside microbes

(Kim et al. 2008; Butkus et al. 2004). The results look promising, indicating that only a small amount of added silver (e.g. 0.05 mg/L) in combination with UV or even visible light can inhibit some type of viruses, but further research is needed. Enhancement of Ag NPs antibacterial properties when combined with UV radiation is proven, but the exact mechanism of activation is still not known.

Silver nanoparticles in combination with magnetic materials showed suitable properties to be used as disinfectant of waste water whereas magnetic properties of such particles would enable their removal from the water after disinfection is finished and their reuse, thus efficiently preventing further environment pollution by engineered nanoparticles (Haider and Kang 2015; Zhang et al. 2011; Chudasama et al. 2009). Fe_3O_4 nanoparticles were coated with silica (SiO_2) and these core-shell nanoparticles were coated further with Ag nanoparticles, making a composite magnetic nanoparticles Fe_3O_4–SiO_2–Ag for waste water disinfection (Zhang et al. 2011). They reported that concentrations of around 15–30 mg/L and 250–500 mg/L are suitable to inhibit and eliminate infectious agents in water, depending on the observed microbes and bacteria, which are still rather high concentrations considering previously stated aspects of ecotoxicity.. Similar core-shell system (Fe_3O_4–Ag) and its superparamagnetic properties in order to remove them from the solution after treatment have been investigated by Chudasama et al. (2009). They reported concentrations of 80–360 mg/L as effective doses.

Generally, there are no exact guidelines or legal limitations on allowed quantities of any nanoparticles in the drinking water or in wider frames, in water and soil ecosystem, except a few technical recommendations recently published by the World Health Organization, WHO (Fewtrell 2014), www.who.int and US Environmental Protection Agency (USEPA https://www.epa.gov). National regulations are also mainly lacking, except for a few such as regulation for drinking water in Germany (Trinkwasserverordnung) that sets 0.01 mg/L of silver as allowed maximum (Chernousova and Epple 2013). A few specified limits on silver use, advises following concentrations of silver as tolerable from aspects of health risks (WHO (Fewtrell 2014); USEPA https://www.epa.gov):

- in general up to 0.1 mg/L concentration of silver should be tolerated well,
- child of 10 kg: 0.2 mg/L for 1–10 days intake,
- cosmetic effects: 0.005 mg/kg per day,
- concentration in drinking water: 0.2 mg/L.

The lack of regulations that would legally force purification of drinking water from nanoparticles is a consequence of lack of health related data and consistent conclusions when effect of nanoparticles on human health is observed. This will surely change in future. The main reason is due to the fact that numerous research results related to Ag NPs in general are not consistent and cannot provide general exact recommendations. Existing data are obtained by different methods, different cells and living organisms have been studied, different concentrations and external environment and exposures, as well as wide array of silver forms, from nanostructures to micro and macro size, different shapes and chemical compositions of compounds. There is a strong need for standardised approach. There is a need to adopt reliable methods for detection and determination of different factors related to the presence of Ag NPs in a soil or water (concentrations, species), as well as analytical tools appropriate for investigations of complex environmental samples. Some recommendations for mitigation of toxic effects of Ag NPs on living organisms, upon its release into the environment, are given in Siripattanakul-Ratpukdi and Fürhacker (2014).

5.7 Application of Ag NPs in Food Industry

In general, food industry is searching for better solutions by using nanoparticles in several areas (Martirosyan and Schneider 2014). In agriculture, improvement of pesticides is important, as well as targeted genetic engineering and preservation, germination and crops yield all expecting significant results from nanotechnology. Phenomena of 'smart food' is imposed by the evolving market and different nanoencapsulation of nanoparticles to provide better flavours, taste, colour or anti-caking is focus of many research groups in this area. Nutrition in different aspects is expected to be highly improved by novel nanoparticles and their effects on nutrient delivery, fortification of minerals and vitamins, purification of water, or some biosensing properties. These three general areas in food industry is at the beginning of nanoparticles application in general and Ag NPs are not yet the subject of extensive research here, even though some attempts exist. The main sector within this industry where silver nanoparticles have been investigated, is food packaging and detection of contaminants, so called 'smart packaging'.

Based on the existing research results, silver nanoparticles could be used in food industry for different purposes such as antibacterial packaging to increase the life and quality of products (Singh et al. 2016; Azlin-Hasim et al. 2016; Haider and Kang 2015; Kuorwel et al. 2015; Martirosyan and Schneider 2014; Duncan 2011), biosensors to detect various contaminants (e.g. pathogens) or monitor lifetime (Stephen Inbaraj and Chen 2016; Berekaa 2015; Bülbül et al. 2015; Li and Sheng 2014), but also as additives in dietary supplements (e.g. agents to obtain clear fruit juices). So far, the main use of Ag NPs in food industry is related to antimicrobial packaging, either as the packing material itself or in a form of thin films. All existing studies are related to polymer—silver combinations (polymer nanocomposites or thin films). There are also attempts to use Ag NPs for biosensing applications in food, but that research is mainly related to other nanoparticles (e.g. gold).

In general, nanoparticles application in food packaging is studied aiming to prevent food contamination and/or environmental pollution, to prolong the shelf life and to monitor food behaviour during storage (Duncan 2011). Many data is lacking, especially related to health risks of food packing with nanoparticles incorporated within the material. These types of commercial products need to be fully investigated, even though that cannot be stated for all present products nowadays (e.g. beverage packaging with oxygen protection based on polymers). Silver is stable and has long shelf life and it is very easy to incorporate Ag NPs into different polymers which are the main material for food packaging. Mechanical properties of chitosan films, namely strength, are increased by addition of small amounts of Ag NPs (5% wt), but it resulted in small decrease of elongation to break at LDPE polymer (Kuorwel et al. 2015). But the main reason to use Ag NPs for food packages is their proven antimicrobial effects. Polymer nanocomposites with incorporated silver nanoparticles have been studied, exhibiting good prevention of contamination by bacteria, fungi or different other pathogens to enter food (Singh et al. 2016; Haider and Kang 2015; Duncan 2011). Another benefit is the sustained longer release of Ag NPs into the food thus preventing microbial activities in longer time. Beside nanocomposites containing Ag NPs, thin films and coatings on food packages have been studied. Coatings with incorporated Ag NPs are used to prevent biofilm growth and bacteria/microbes development inside the food containers such as cans. Several researchers investigated thin edible films containing Ag NPs directly on the food surface as the method to preserve fruits and vegetables and proved longer storage time and less ripening (Azlin-Hasim et al. 2016; Duncan 2011).

However, the resulting effects on health if and when Ag NPs migrate into the food from the packaging materials are not well studied or not studied at all, even though there are evidences that they can provoke harmful effects depending on the particle sizes, or amounts (Kuorwel et al. 2015; Gaillet and Rouanet 2015; Martirosyan and Schneider 2014; Cushen et al. 2012). Risk assessment of nanoparticles migration from packaging to food is needed and only very limited data exist (Hannon et al. 2015). For example, it is proven that silver nanoparticles from orange juice container migrate into the juice after some time, or from cellulose pads containing Ag NPs into the meat exudates (not the meat itself) (Martirosyan and Schneider 2014). The conditions under which the leaching of Ag NPs or its ions or some other resulting Ag NPs compound into the food will occur is not well understood or studied. It is very difficult to measure and precisely quantify migration levels of nanoparticles because it is commonly in very low amounts and also in a very complex environment. A few studies are reviewed in Kuorwel et al. (2015). It could be elaborated that pH, temperature, storage life, pressure, humidity or exposure to light (especially natural light) would have the influence but it is largely unknown. For example, it is known that polymer exposure to natural light results in its degradation through photo-oxidation what might cause migration of nanoparticles into the food, but it is assumed that food packages are not exposed to natural light for longer time (Hannon et al. 2015). There are some studies related to the reusing of food packages (lunch boxes or bags) indicating that with repeated use of the packaging, migration of nanoparticles is likely to occur (Hannon et al. 2015). Also, the extents of Ag leaching from different polymer matrices have been studied but there are no general recommendations. EFSA (European Food Safety Authority, https://www.efsa.europa.eu) limits the amounts and size of silver nanoparticles in food or related products (Cushen et al. 2012). Very good review of currently existing regulations and legislations related to the safety of nanotechnologies and nanoparticles is given in Wacker et al. (2016) and earlier review in Cushen et al. (2012). But real experimental data for particular cases of food packaging and used nanoparticles are missing, even though it is needed to support nanoparticles use in food sector.

Important area of research is related to the biosensors for detection of contaminants in the food by using nanoparticles. These are mainly based on gold nanoparticles or carbon nanotubes, but silver also has been studied. SERS efficiency and optical characteristics of Ag NPs can be used for the development of food sensors. Usually it is the change of solution colour that indicates the presence of some particular compound that needs detection. Ag NPs are usually associated with a detection of metal ions, proteins, melamine or pesticides (Bülbül et al. 2015). Less attention is given to the silver nanoparticles as sensing agents due to their high affinity towards Ag ions release and surface oxidation, both related to highly unwanted further effects in the surroundings. But very good optical properties and lower cost than gold nanoparticles are main reasons to investigate silver in this way. Metal contaminants in food originating from processing and existing in very small amounts can be detected by aggregation of Ag NPs and functionalized Ag NPs has been used for detection of pesticides at their concentrations of 24 μg/mL (Bülbül et al. 2015). It was shown that Ag NPs incorporated as microspheres in polystyrene (PS-NH2/Ag NPs) can provide detection of melamine in milk powder, with accuracy of 1.9×10^{-8} mol/L (Li and Sheng 2014). Detection of E.coli bacteria strain with accuracy of 5–10^9 cells/mL was achieved with Ag NPs conjugated with rabbit IgG antibody, by using UV-VIS spectroscopy as a method of monitoring the levels of bacteria in a meat (Stephen Inbaraj and Chen 2016).

Apart from Ag NPs, silver zeolite has gained much attention. Ag-zeolite incorporates silver ions and shows sustained antibacterial activity over longer time than zeolite which is also well known for its antibacterial properties and widely used in agricultural fields (Berekaa 2015). Also, combinations of Ag NPs with nanoclay or ZnO nanoparticles are studied, usually for antibacterial effects (Kuorwel et al. 2015). Another possible application of silver nanoparticles is in agriculture for improvement of germination and crop yield. One recent study showed that Ag NPs in colloidal solution used for treatment of oilseeds resulted in 10–20% improved germination and grain crops especially showed better effect in relation to using AG NPs for treatment of seeds (Fedorenko et al. 2015). However, authors emphasized that lacking of safety data and health risks in application of such seeds improvement method still makes it hard to evaluate whether it is safe for use or not.

6 Conclusions

Extremely broad possibilities to apply silver nanoparticles in diverse sectors for different promised benefits, are the key of extensive studies devoted to them. From medical devices, drug delivery systems, dietary supplements, food packaging and containers, wound dressings, biosensors, textile, agriculture, purification systems, home appliances, cosmetics to semiconductor products and optical devices and many more emerging nowadays, nanosilver has proven value, especially in relation to its antibacterial effects. Application of Ag NPs due to their optical properties showed great potential, especially in biosensing. Wide availability and low cost of manufacturing and easiness of processing makes it unavoidable noble element for many purposes. However, major concerns have been raised related to toxicity of Ag NPs. From first enthusiastic applications in medical devices it has been proven that there are high health risks, depending on the amount, size and shape of nanoparticles, as well as many other still undetermined factors affecting its toxicity and final resulting effects in the human body. Also, environmental concerns need to be taken into account with the increase of nanosilver application in products such as water purification systems or construction materials such as paints or tiles and even refrigerators, because all of these products are in mass production and even small amounts of harmful materials migrating to water or soil from one product are not negligible for eco-system in general. There is an urgent need to study in detail many aspects of using Ag NPs in medicine, but also in other indirectly and directly health related sectors, such as textile, cosmetics or water treatment systems, food industry and agriculture. Antibacterial properties are rather well determined but mechanisms that can provoke unwanted harmful effects on the cells and organs are yet under investigations, especially of possible genetic effects. If both superb antibacterial properties, and high cell toxicity are combined, one area might be emphasized as highly promising: targeted cancer treatment. Some recent results showed very promising future in this area of research.

Acknowledgements This work has been supported by a research grant from Science Foundation Ireland (SFI) under Grant Number 12/IA/1576 and part of the project III41017—Virtual Human Osteoarticular System and its Application in Preclinical and Clinical Practice which is sponsored by the Ministry of Education, Science and Technological Development of the Republic of Serbia for the period of 2011–2016.

References

Abou El-Nour KMM, Eftaiha A, Al-Warthan A et al (2010) Synthesis and applications of silver nanoparticles. Arab J Chem 3:135–140

Actis L, Srinivasan A, Lopez-Ribot JL et al (2015) Effect of silver nanoparticle geometry on methicillin susceptible and resistant Staphylococcus aureus, and osteoblast viability. J Mater Sci Mater Med 26:215

Ahamed M, Karns M, Goodson M et al (2008) DNA damage response to different surface chemistry of silver nanoparticles in mammalian cells. Toxicol Appl Pharmacol 233:404–410

Ahamed M, AlSalhi MS, Siddiqui MKJ (2010) Silver nanoparticle applications and human health. Clin Chim Acta 411:1841–1848

Akiyama T, Miyamoto H (2013) Silver oxide-containing hydroxyapatite coating has in vivo antibacterial activity in the rat tibia. J Orthop Res 31(8):1195–1200

Alt V, Bechert T, Steinrücke P et al (2004) An in vitro assessment of the antibacterial properties and cytotoxicity of nanoparticulate silver bone cement. Biomaterials 18:4383–4391

Arakawa H, Neault JF, Tajmir-Riahi HA (2001) Silver(I) complexes with DNA and RNA studied by fourier transform infrared spectroscopy and capillary electrophoresis. Biophys J 81:1580–1587

Arora S, Jain J, Rajwade JM et al (2008) Cellular responses induced by silver nanoparticles: in vitro studies. Toxicol Lett 179:93–100
Arora S, Jain J, Rajwade JM et al (2009) Interactions of silver nanoparticleswith primary mouse fibroblasts and liver cells. Toxicol Appl Pharmacol 236:310–318
Asharani PV, Mun GLK, Hande MP et al (2009a) Cytotoxicity and genotoxicity of silver nanoparticles in human cells. ACS Nano 3:279–290
Asharani PV, Hande MP, Valiyaveettil S (2009b) Anti-proliferative activity of silver nanoparticles. BMC Cell Biol 10:65
Azlin-Hasim S, Cruz-Romero MC, Cummins E et al (2016) The potential use of a layer-by-layer strategy to develop LDPE antimicrobial films coated with silver nanoparticles for packaging applications. J Colloid Interface Sci 461:239–248
Balakumaran MD, Ramachandran R, Jagadeeswari S et al (2016) In vitro biological properties and characterization of nanosilver coated cotton fabrics—an application for antimicrobial textile finishing. Int Biodeterior Biodegradation 107:48–55
Berekaa MM (2015) Nanotechnology in food industry; advances in food processing, packaging and food safety. Int J Curr Microbiol App Sci 4(5):345–357
Benn TM, Westerhoff P (2008) Nanoparticle silver released into water from commercially available sock fabrics. Environ Sci Technol 42(11):4133–4139
Bhattacharya R, Mukherjee P (2008) Biological properties of "naked" metal nanoparticles. Adv Drug Deliv Rev 60:1289–1306
Bhunia AK, Samanta PK, Aich D (2015) Biocompatibility study of protein capped and uncapped silver nanoparticles on human hemoglobin. J Phys D: Appl Phys 48:235305
Blaser SA, Scheringer M, MacLeod M et al (2008) Estimation of cumulative aquatic exposure and risk due to silver: contribution of nano-functionalized plastics and textiles. Sci Total Environ 390:396–409
Boroumand MN, Montazer M, Simon F et al (2015) Novel method for synthesis of silver nanoparticles and their application on wool. Appl Surf Sci 346:477–483
Brennan SA, Ní Fhoghlú C, Devitt BM et al (2015) Silver nanoparticles and their orthopaedic applications. Bone Joint J 97-B:582–589
Buerki-Thurnherr T, Xiao L, Diener L et al (2013) In vitro mechanistic study towards a better understanding of ZnO nanoparticle toxicity. Nanotoxicology 7:402–416
Butkus MA, Labare MP, Starke JA (2004) Use of aqueous silver to enhance inactivation of coliphageMS-2 by UV disinfection. Appl Environ Microbiol 70:2848–2853
Bülbül G, Hayat A, Andreescu S (2015) Portable nanoparticle-based sensors for food safety assessment. Sensors 15:30736–30758
Buzea C, Pacheco Blandino II, Robbie K (2007) Nanomaterials and nanoparticles: sources and toxicity. Biointerphases 2(4):17–172
Castiglioni S, Cazzaniga A, Perrotta C et al (2015) Silver nanoparticles-induced cytotoxicity requires ERK activation in human bladder carcinoma cells. Toxicol Lett 237:237–243
Chaloupka K, Malam Y, Seifalian A (2010) Nanosilver as a new generation of nanoproduct in biomedical applications. Trends Biotechnol 28(11):580–588
Chen P, Wang Z, Zong S et al (2016) pH-sensitive nanocarrier based on gold/silver core–shell nanoparticles decorated multi-walled carbon manotubes for tracing drug release in living cells. Biosens Bioelectron 75:446–451
Chen W, Liu Y, Courtney HS et al (2006) In vitro anti-bacterial and biological properties of magnetron co-sputtered silver-containing hydroxyapatite coating. Biomaterials 27:5512–5517
Chernousova S, Epple M (2013) Silver as antibacterial agent: ion, nanoparticle, and metal. Angew Chem 52:1636–1653
Chudasama B, Vala AK, Andhariya N et al (2009) Enhanced antibacterial activity of bifunctional Fe3O4-Ag core-shell nanostructures. Nano Res 2(12):955–965
Collinge C, Goll G, Seligson D et al (1994) Pin tract infections: silver vs uncoated pins. Orthopedics 17(5):445–448
Coluccio ML, Francardi M, Gentile F et al (2016) Plasmonic 3D-structures based on silver decorated nanotips for biological sensing. Opt Lasers Eng 76:45–51

Cornelis G, Hund-Rinke K, Kuhlbusch T et al (2014) Fate and bioavailability of engineered nanoparticles in soils: a review. Crit Rev Environ Sci Technol 44:2720–2764
Cushen MK, Kerry J, Morris M et al (2012) Nanotechnologies in the food industry—recent developments, risks and regulation. Trends Food Sci Technol 24:30–46
Del Real AEP, Castillo-Michel H, Kaegi R et al (2016) Fate of Ag-NPs in sewage sludge after application on agricultural soils. Environ Sci Technol 50:1759–1768
Dhand V, Soumya L, Bharadwaj S et al (2016) Green synthesis of silver nanoparticles using coffea arabica seed extract and its antibacterial activity. Mater Sci Eng, C 58:36–43
Duncan TV (2011) Applications of nanotechnology in food packaging and food safety: barrier materials, antimicrobials and sensors. J Colloid Interface Sci 363:1–24
Durán N, Silveira CP, Durán M et al (2015) Silver nanoparticle protein corona and toxicity: a mini-review. J Nanobiotechnol 13:55
Duran N, Marcarto PD, De Souza GIH et al (2007) Antibacterial effect of silver nanoparticles produced by fungal process on textile fabrics and their effluent treatment. J Biomed Nanotechnol 3:203–208
El-Hussein A, Mfouo-Tynga I, Abdel-Harith M et al (2015) Comparative study between the photodynamic ability of gold and silver nanoparticles in mediating cell death in breast and lung cancer cell lines. J Photochem Photobiol B 153:67–75
Emam HE, Saleh NH, Nagy KS et al (2016) Instantly AgNPs deposition through facile solventless technique for poly-functional cotton fabrics. Int J Biol Macromol 84:308–318
Fabrega J, Luoma SN, Tyler CR et al (2011) Silver nanoparticles: behaviour and effects in the aquatic environment. Environ Int 37:517–531
Farhadian N, Usefi Mashoof R, Khanizadeh S et al (2016) Streptococcus mutans counts in patients wearing removable retainers with silver nanoparticles vs those wearing conventional retainers: a randomized clinical trial. Am J Orthod Dentofac Orthop 149:155–160
Fang W, Wang Z, Zong S et al (2014) PH-controllable drug carrier with SERS activity for targeting cancer cells. Biosens Bioelectron 57:10–15
Fedorenko VF, Buklagin DS, Golubev IG et al (2015) Review of Russian nanoagents for crops treatment. Nanotechnol Russ 10:318–324
Fewtrell L (2014) Silver: water disinfection and toxicity. Centre for Research into Environment and Health, World Health Organization, WHO, www.who.int
Foldbjerg R, Dang DA, Autrup H (2011) Cytotoxicity and genotoxicity of silver nanoparticles in the human lung cancer cell line, A549. Arch Toxicol 85:743–750
Foldbjerg R, Irving ES, Hayashi Y et al (2012) Global gene expression profiling of human lung epithelial cells after exposure to nanosilver. Toxicol Sci 130:145–157
Foldbjerg R, Jiang X, Micləuş T et al (2015) Silver nanoparticles—wolves in sheep's clothing? Toxicol Res 4:563–575
Fong J, Wood F (2006) Nanocrystalline silver dressings in wound management: a review. Int J Nanomed 1:441–449
Freeman AI, Halladay LJ, Cripps P (2012) The effect of silver impregnation of surgical scrub suits on surface bacterial contamination. Vet J 192:489–493
Fu C, Zhang X, Savino K et al (2016) Antimicrobial silver-hydroxyapatite composite coatings through two-stage electrochemical synthesis. Surf Coat Technol 301:13–19
Furkert F, Sörensen J, Arnoldi J (2011) Antimicrobial efficacy of surface-coated external fixation pins. Curr Microbiol 62(6):1743–1751
Gaillet S, Rouanet JM (2015) Silver nanoparticles: their potential toxic effects after oral exposure and underlying mechanisms—a review. Food Chem Toxicol 77:58–63
Gambardella C, Ferrando S, Gatti AM et al (2015a) Review: morphofunctional and biochemical markers of stress in sea urchin life stages exposed to engineered nanoparticles. Environmental toxicology. In Press, doi:10.1002/tox.22159
Gambardella C, Costa E, Piazza V et al (2015b) Effect of silver nanoparticles on marine organisms belonging to different trophic levels. Mar Environ Res 111:41–49
Gamucci O, Bertero A, Gagliardi M et al (2014) Biomedical nanoparticles: overview of their surface immune-compatibility. Coatings 4:139–159

Gatti A, Montanari S (2015) Chapter 8: food, drugs and nanoparticles. In: Case studies in nanotoxicology and particle toxicology. Elsevier Inc. pp 163–194

Gil PR, Oberderster G, Elder A et al (2010) Correlating physico-chemical with toxicological properties of nanoparticles: the present and the future. ACS Nano 4:5527–5531

Giret S, Wong Chi Man M, Carcel C (2015) Mesoporous-silica-functionalized nanoparticles for drug delivery. Chem Eur J 21:13850–13865

Gosheger G, Hardes J, Ahrens H et al (2004) Silver-coated megaendoprostheses in a rabbit model —an analysis of the infection rate and toxicological side effects. Biomaterials 25:5547–5556

Govindaraju K, Krishnamoorthy K, Alsagaby SA et al (2015) Green synthesis of silver nanoparticles for selective toxicity towards cancer cells. IET Nanobiotechnol 9:325–330

Gnanadhas DP, Ben Thomas M, Thomas R et al (2013) Interaction of silver nanoparticles with serum proteins affects their antimicrobial activity in vivo. Antimicrob Agents Chemother 57:4945–4955

Gumpu MB, Sethuraman S, Krishnan UM et al (2015) A review on detection of heavy metal ions in water—an electrochemical approach. Sens Actuators B 213:515–533

Gurunathan S, Raman J, Abd Malek SN et al (2013a) Green synthesis of silver nanoparticles using Ganoderma neo-japonicum Imazeki: a potential cytotoxic agent against breast cancer cells. Int J Nanomed 8:4399–4413

Gurunathan S, Han JW, Dayem AA et al (2013b) Green synthesis of anisotropic silver nanoparticles and its potential cytotoxicity in human breast cancer cells (MCF-7). J Ind Eng Chem 19:1600–1605

Gurunathan S, Han JW, Park JH et al (2015) Reduced graphene oxide-silver nanoparticle nanocomposite: a potential anticancer nanotherapy. Int J Nanomed 10:6257–6276

Hackenberg S, Scherzed A, Kessler M et al (2011) Silver nanoparticles: evaluation of DNA damage, toxicity and functional impairment in human mesenchymal stem cells. Toxicol Lett 201:27–33

Haider A, Kang IK (2015) Preparation of silver nanoparticles and their industrial and biomedical applications: a comprehensive review. Adv Mater Sci Eng 2015:165257

Hannon JC, Kerry J, Cruz-Romero M et al (2015) Advances and challenges for the use of engineered nanoparticles in food contact materials. Trends Food Sci Technol 43:43–62

Hardes J, von Eiff C, Streitbuerger A et al (2010) Reduction of periprosthetic infection with silver-coated megaprostheses in patients with bone sarcoma. J Surg Oncol 101:389–395

He S, Feng Y, Ni J et al (2016) Different responses of soil microbial metabolic activity to silver and iron oxide nanoparticles. Chemosphere 147:195–202

Ho CM, Yau SKW, Lok CN et al (2010) Oxidative dissolution of silver nanoparticles by biologically relevant oxidants: a kinetic and mechanistic study. Chem Asian J 5:285–293

Ho CM, Wong CK, Yau SKW et al (2011) Oxidative dissolution of silver nanoparticles by dioxygen: a kinetic and mechanistic study. Chem Asian J 6:2506–2511

Hong IS, Jang GB, Lee HY et al (2015) Targeting cancer stem cells by using the nanoparticles. Int J Nanomed 10:251–260

Hu CMJ, Aryal S, Zhang L (2010) Nanoparticle-assisted combination therapies for effective cancer treatment. Ther Deliv 1:323–334

Hu G, Cai Y, Tu Z et al (2015) Reducing the cytotoxicity while improving the anti-cancer activity of silver nanoparticles through α-tocopherol succinate modification. RSC Adv 5:82050–82055

Huang Y, Li X, Liao Z et al (2007) A randomized comparative trial between acticoat and SD-Ag in the treatment of residual burn wounds, including safety analysis. Burns 33:161–166

Jacob JA, Shanmugam A (2015) Silver nanoparticles provoke apoptosis of Dalton's ascites lymphoma in vivo by mitochondria dependent and independent pathways. Colloids Surf B 136:1011–1016

Jia Z, Xiu P, Li M et al (2016) Bioinspired anchoring AgNPs onto micro-nanoporous TiO2 orthopedic coatings: trap-killing of bacteria, surface-regulated osteoblast functions and host responses. Biomaterials 75:203–222

Jiang X, Foldbjerg R, Miclaus T et al (2013) Multi-platform genotoxicity analysis of silver nanoparticles in the model cell line CHO-K1. Toxicol Lett 222:55–63

Jia Z, Li M, Xiu P et al (2015) A novel cytocompatible, hierarchical porous Ti6Al4V scaffold with immobilized silver nanoparticles. Mater Lett 157:143–146
Jiang X, Miclaus T, Wang L et al (2015) Fast intracellular dissolution and persistent cellular uptake of silver nanoparticles in CHO-K1 cells: implication for cytotoxicity. Nanotoxicology 9:181–189
Joseph MM, George SK, Sreelekha TT (2016) Bridging 'green' with nanoparticles: biosynthesis approaches for cancer management and targeting of cancer stem cells. Curr Nanosci 12:47–62
Kalishwaralal K, BarathManiKanth S, Pandian SRK et al (2010) Silver nano—a trove for retinal therapies. J Control Release 145:76–90
Kawata K, Osawa M, Okabe S (2009) In vitro toxicity of silver nanoparticles at non-cytotoxic doses to HepG2 human hepatoma cells. Environ Sci Technol 43:6046–6051
Kendall M, Holgate S (2012) Health impact and toxicological effects of nanomaterials in the lung. Respirology 17:743–758
Khanna P, Ong C, Bay BH et al (2015) Nanotoxicity: an interplay of oxidative stress, inflammation and cell death. Nanomaterials 5:1163–1180
Khurana P, Thatai S, Kumar D (2014) Chapter 4: fabrication, properties of nanoshells with controllable surface charge and its applications. In: Tiwari A, Nordin AN (eds) Advanced biomaterials and biodevices. Scrivener Publishing LLC, Wiley, pp 121–146
Kim JY, Lee C, Cho M et al (2008) Enhanced inactivation of E. coli and MS 2 phage by silver ions combined with UV-A and visible light irradiation. Water Res 42:356–362
Kim S, Choi JE, Choi J et al (2009) Oxidative stress-dependent toxicity of silver nano-particles in human hepatoma cells. Toxicol In Vitro 23:1076–1084
Kim HR, Kim MJ, Lee SY et al (2011) Genotoxic effects of silver nanoparticles stimulated by oxidative stress in human normal bronchial epithelial (BEAS-2B) cells. Mutat Res 726: 129–135
Kim S, Ryu DY (2012) Silver nanoparticle-induced oxidative stress, genotoxicity and apoptosis in cultured cells and animal tissues. J Appl Toxicol 33:78–89
Kononenko V, Narat M, Drobne D (2015) Nanoparticle interaction with the immune system. Arh Hig Rada Toksikol 66:97–108
Kumar A, Vemula PK, Ajayan PM et al (2008) Silver-nanoparticle-embedded antimicrobial paints based on vegetable oil. Nat Mater 7:236–241
Kuorwel KK, Cran MJ, Orbell JD et al (2015) Review of mechanical properties, migration, and potential applications in active food packaging systems containing nanoclays and nanosilver. Compr Rev Food Sci Food Safety 14:411–430
Langer J, Novikov SM, Liz-Marzánl LM (2015) Sensing using plasmonic nanostructures and nanoparticles. Nanotechnology 26:322001
Larguinho M, Baptista PV (2015) Gold and silver nanoparticles for clinical diagnostics—from genomics to proteomics. J Proteomics 75:2811–2823
Le Ouay B, Stellacci F (2015) Antibacterial activity of silver nanoparticles: a surface science insight. Nano Today 10:339–354
Lewis G (2016) Properties of nanofiller-loaded poly (methyl methacrylate) bone cement composites for orthopedic applications: a review. J Biomed Mater Res—Part B Appl Biomater (in press). doi:10.1002/jbm.b.33643
Li P, Tong Z, Huo L et al (2016) Antibacterial and biological properties of biofunctionalized nanocomposites on titanium for implant application. J Biomater Appl 31(2):205–214
Li S, Guo Z, Liu Z (2015) Surface-enhanced Raman spectroscopy + support vector machine: a new noninvasive method for prostate cancer screening? Expert Rev Anticancer Ther 15:5–7
Li D, Feng S, Huang H et al (2014) Label-free detection of blood plasma using silver nanoparticle based surface-enhanced Raman spectroscopy for esophageal cancer screening. J Nanosci Nanotechnol 10:478–484
Li Z, Sheng C (2014) Nanosensors for food safety. J Nanosci Nanotechnol 14:905–912
Li J, Wang Z, Gryczynski I, Mandecki W (2010) Silver nanoparticle-enhanced fluorescence in microtransponder-based immuno- and DNA hybridization assays. Anal Bioanal Chem 398:1993–2001

Liang J, Zeng F, Zhang M et al (2015) Green synthesis of hyaluronic acid-based silver nanoparticles and their enhanced delivery to CD44+ cancer cells. RSC Adv 5:43733
Lim DH, Jang J, Kim S et al (2012) The effects of sub-lethal concentrations of silver nanoparticles on inflammatory and stress genes in human macrophages using cDNA microarray analysis. Biomaterials 33:4690–4699
Lin J, Chen R, Feng S et al (2011) A novel blood plasma analysis technique combining membrane electrophoresis with silver nanoparticle-based SERS spectroscopy for potential applications in noninvasive cancer detection. Nanomed Nanotechnol Biol Med 7:655–663
Liu J, Hurt RH (2010) Ion release kinetics and particle persistence in aqueous nano-silver colloids. Environ Sci Technol 44:2169–2175
Liu J, Sonshine DA, Shervani S et al (2010) Controlled release of biologically active silver from nanosilver surfaces. ACS Nano 4:6903–6913
Liu J, Zhao Y, Guo Q et al (2012) TAT-modified nanosilver for combating multidrug-resistant cancer. Biomaterials 33:6155–6161
Lovatel RH, Nevesa RM, Oliveira GR et al (2015) Disinfection of biologically treated industrial wastewater using montmorillonite/alginate/nanosilver hybrids. J Water Process Eng 7:273–279
Lu S, Gao W, Gu HY (2008) Construction, application and biosafety of silver nanocrystalline chitosan wound dressing. Burns 34:623–628
Malhi S, Gu X (2015) Nanocarrier-mediated drugs targeting cancer stem cells: an emerging delivery approach. Expert Opin Drug Deliv 12:1177–1201
Marambio-Jones C, Hoek EMV (2010) A review of the antibacterial effects of silver nanomaterials and potential implications for human health and the environment. J Nanopart Res 12: 1531–1551
Martirosyan A, Schneider Y-J (2014) Engineered nanomaterials in food: implications for food safety and consumer health. Int J Environ Res Public Health 11:5720–5750
Maynard AD, Warheit DB, Philbert MA (2011) The new toxicology of sophisticated materials: nanotoxicology and beyond. Toxicol Sci 120:109–129
Miola M, Fucale G, Maina G et al (2017) Composites bone cements with different viscosities loaded with a bioactive and antibacterial glass. Journal
Mitrovic S, Adamovic D, Zivic F et al (2014) Friction and wear behavior of shot peened surfaces of 36CrNiMo4 and 36NiCrMo16 alloyed steels under dry and lubricated contact conditions. Appl Surf Sci 290:223–232
Miura N, Shinohara Y (2009) Cytotoxic effect and apoptosis induction by silver nanoparticles in HeLa cells. Biochem Biophys Res Commun 390:733–737
Molleman B, Hiemstra T (2015) Surface structure of silver nanoparticles as a model for understanding the oxidative dissolution of silver ions. Langmuir 49:13361–13372
Mohamud R, Xiang SD, Selomulya C et al (2014) The effects of engineered nanoparticles on pulmonary immune homeostasis. Drug Metab Rev 46:176–190
Moon JJ, Huang B, Irvine DJ (2012) Engineering nano- and microparticles to tune immunity. Adv Mater 24:3724–3746
Murphy M, Ting K, Zhang X et al (2015) Current development of silver nanoparticle preparation, investigation, and application in the field of medicine. J Nanomater 2015:696918
Myrzakhanova M, Gambardella C, Falugi C et al (2013) Effects of nanosilver exposure on cholinesterase activities, CD41, and CDF/LIF-like expression in zebraFish (Danio rerio) larvae. Biomed Res Int 2013:205183
Nair L, Laurencin C (2008) Nanofibers and nanoparticles for orthopaedic surgery applications. J Bone Jt. Surg Am 90:128–131
Nayak D, Ashe S, Rauta PR et al (2016) Bark extract mediated green synthesis of silver nanoparticles: evaluation of antimicrobial activity and antiproliferative response against osteosarcoma. Mater Sci Eng C 58:44–52
Olasagasti M, Gatti A, Capitani F et al (2014) Toxic effects of colloidal nanosilver in zebrafish embryos. J Appl Toxicol 34:562–575
Ortega FG, Fernández-Baldo MA, Fernández JG et al (2015) Study of antitumor activity in breast cell lines using silver nanoparticles produced by yeast. Int J Nanomed 10:2021–2031

Pal S, Tak YK, Song JM (2007) Does the antibacterial activity of silver nanoparticles depend on the shape of the nanoparticle? A study of the gram-negative bacterium Escherichia coli. Appl Environ Microbiol 27:1712–1720

Pandita D, Kumar S, Lather V (2015) Hybrid poly(lactic-co-glycolic acid) nanoparticles: design and delivery prospectives. Drug Discovery Today 20:95–104

Parameswaran AD, Roberts CS, Seligson D et al (2003) Pin tract infection with contemporary external fixation: how much of a problem? J Orthop Trauma 17:503–507

Park HJ, Kim JY, Kim J et al (2009) Silver-ion-mediated reactive oxygen species generation affecting bactericidal activity. Water Res 43:1027–1032

Park HS, Kim KH, Jang S et al (2010) Attenuation of allergic airway inflammation and hyperresponsiveness in a murine model of asthma by silver nanoparticles. Int J Nanomedicine 5:505–515

Percival S, Bowler P, Russell D (2005) Bacterial resistance to silver in wound care. J Hosp Infect 60(1):1–7

Perera S, Bhushan B, Bandara R et al (2013) Morphological, antimicrobial, durability, and physical properties of untreated and treated textiles using silver-nanoparticles. Colloids Surf A: Physicochem Eng Aspects 436:975–989

Petrarca C, Clemente E, Amato V et al (2015) Engineered metal based nanoparticles and innate immunity. Clin Mol Allergy 13:13. doi:10.1186/s12948-015-0020-1

Pimentel RC, Martínez ESM, García AM et al (2013) Silver nanoparticles nanocarriers, synthesis and toxic effect on cervical cancer cell lines. BioNanoScience 3:198–207

Polak P, Shefi O (2015) Nanometric agents in the service of neuroscience: manipulation of neuronal growth and activity using nanoparticles. Nanomed Nanotechnol Biol Med 11: 1467–1479

Praveena SM, Aris AZ (2015) Application of low-cost materials coated with silver nanoparticle as water filter in Escherichia coli removal. Water Qual Exposure Health 7(4):617–625

Rai M, Yadav A, Gade A (2009) Silver nanoparticles as a new generation of antimicrobials. Biotechnol Adv 27:76–83

Rajeshkumar S, Kannas C, Annadurai G (2012) Green synthesis of silver nanoparticles using marine brown algae turbinaria conoides and its antibacterial activity. Int J Pharm Bio Sci 3:502–510

Raza ZA, Rehman A, Mohsin M et al (2015) Development of antibacterial cellulosic fabric via clean impregnation of silver nanoparticles. J Clean Prod 101:377–386

Rehan M, Mashaly IIM, Mowafi S et al (2015) Multi-functional textile design using in-situ Ag NPs incorporation into natural fabric matrix. Dyes Pigm 118:9–17

Remya RR, Rajasree SRR, Aranganathan L et al (2015) An investigation on cytotoxic effect of bioactive AgNPs synthesized using Cassia fistula flower extract on breast cancer cell MCF-7. Biotechnol Rep 8:110–115

Rizzello L, Pompa PP (2014) Nanosilver-based antibacterial drugs and devices: mechanisms, methodological drawbacks, and guidelines. Chem Soc Rev 43:1501–1518

Sadeghi B, Garmaroudi FS, Hashemi M et al (2012) Comparison of the anti-bacterial activity on the nanosilver shapes: nanoparticles, nanorods and nanoplates. Adv Powder Technol 23:22–26

Sanpui P, Chattopadhyay A, Ghosh SS (2011) Induction of apoptosis in cancer cells at low silver nanoparticle concentrations using chitosan nanocarrier. ACS Appl Mater Interfaces 3:218–228

Schaumann GE, Baumann T, Lang F et al (2015a) Engineered nanoparticles in soils and waters. Sci Total Environ 535:1–2

Schaumann G, Philippe A, Bundschuh M et al (2015b) Understanding the fate and biological effects of Ag- and TiO2-nanoparticles in the environment: the quest for advanced analytics and interdisciplinary concepts. Sci Total Environ 535:3–19

Schlich K, Klawonn T, Terytze K et al (2013) Hazard assessment of a silver nanoparticle in soil applied via sewage sludge. Environ Sci Europe 25:17

Shahverdi AR, Fakhimi A, Shahverdi HR et al (2007) Synthesis and effect of silver nanoparticles on the antibacterial activity of different antibiotics against Staphylococcus aureus and Escherichia coli. Nanomed Nanotechnol Biol Med 3:168–171

Sharma S, Chockalingam S, Sanpui P et al (2014) Silver nanoparticles impregnated alginate-chitosan-blended nanocarrier induces apoptosis in human glioblastoma cells. Adv Healthc Mater 3:106–114

Shimazaki T, Miyamoto H, Ando Y et al (2010) In vivo antibacterial and silver-releasing properties of novel thermal sprayed silver-containing hydroxyapatite coating. J Biomed Mater Res B Appl Biomater 92:386–389

Shen S, Xia JX, Wang J (2016) Nanomedicine-mediated cancer stem cell therapy. Biomaterials 74:1–18

Singh PK, Jairath G, Ahlawat SS (2016) Nanotechnology: a future tool to improve quality and safety in meat industry. J Food Sci Technol 53:1739. doi:10.1007/s13197-015-2090-y

Siripattanakul-Ratpukdi S, Fürhacker M (2014) Review: issues of silver nanoparticles in engineered environmental treatment systems. Water Air Soil Pollut 225:1939

Slane J, Vivanco J, Rose W et al (2015) Mechanical, material, and antimicrobial properties of acrylic bone cement impregnated with silver nanoparticles. Mater Sci Eng C Mater Biol Appl 48:188–196

Soenen SJ, Parak WJ, Rejman J et al (2015) (Intra)Cellular stability of inorganic nanoparticles: effects on cytotoxicity, particle functionality, and biomedical applications. Chem Rev 115:2109–2135

Song L, Mao K, Zhou X et al (2016) A novel biosensor based on Au@Ag core–shell nanoparticles for SERS detection of arsenic(III). Talanta 146:285–290

Sotirioua GA, Teleki A, Camenzind A et al (2011) Nanosilver on nanostructured silica: antibacterial activity and Ag surface area. Chem Eng J 170(2011):547–554

Stebounova LV, Guio E, Grassian VH (2011) Silver nanoparticles in simulated biological media: a study of aggregation, sedimentation, and dissolution. J Nanopart Res 13:233–244

Stephen Inbaraj B, Chen BH (2016) Nanomaterial-based sensors for detection of foodborne bacterial pathogens and toxins as well as pork adulteration in meat products. J Food Drug Anal 24:15–28

Stevanovic M (2014) Chapter 10: assembly of polymers/metal nanoparticles and their applications as medical devices. In: Tiwari A, Nordin AN (eds) Advanced biomaterials and biodevices. Scrivener Publishing LLC, Wiley, pp 121–146

Sujitha V, Murugan K, Paulpandi M et al (2015) Green-synthesized silver nanoparticles as a novel control tool against dengue virus (DEN-2) and its primary vector Aedes aegypti. Parasitol Res 114:3315–3325

Sur I, Cam D, Kahraman M et al (2010) Interaction of multi-functional silver nanoparticles with living cells. Nanotechnology 21:175104

Tang B, Li J, Hou X et al (2013) Colorful and antibacterial silk fiber from anisotropic silver nanoparticles. Ind Eng Chem Res 52:4556–4563

Theodorou IG, Ryan MP, Tetley TD et al (2014) Inhalation of silver nanomaterials—seeing the risks. Int J Mol Sci 15:23936–23974

Tian B, Chen W, Yu D et al (2016) Fabrication of silver nanoparticle-doped hydroxyapatite coatings with oriented block arrays for enhancing bactericidal effect and osteoinductivity. J Mech Behav Biomed Mater 61:345–359

Vance ME, Kuiken T, Vejerano EP et al (2015) Nanotechnology in the real world: redeveloping the nanomaterial consumer products inventory. Beilstein J Nanotechnol 6:1769–1780

Vandebriel RJ, De JWH (2012) A review of mammalian toxicity of ZnO nanoparticles. Nanotechnol Sci Appl 5:61–71

Venil CK, Sathishkumar P, Malathi M et al (2016) Synthesis of flexirubin-mediated silver nanoparticles using chryseobacterium artocarpi CECT 8497 and investigation of its anticancer activity. Mater Sci Eng, C 59:228–234

Vivekanandhan S, Christensen L, Misra M et al (2012) Green process for impregnation of silver nanoparticles into microcrystalline cellulose and their antimicrobial bionanocomposite films. J Biomater Nanobiotechnol 3:371–376

Wacker MG, Proykova A, Santos GML (2016) Dealing with nanosafety around the globe—regulation vs. innovation. Int J Pharm 509:95–106

Wadhera A, Fung M (2005) Systemic argyria associated with ingestion of colloidal silver. Dermatol Online J 11:12
Walters CR, Pool EJ, Somerset VS (2014) Ecotoxicity of silver nanomaterials in the aquatic environment: a review of literature and gaps in nano-toxicological research. J Environ Sci Health Part A Toxic/Hazard Subst Environ Eng 49:1588–1601
Wang Y, Newell BB, Irudayaraj J (2012) Folic acid protected silver nanocarriers for targeted drug delivery. J Biomed Nanotechnol 8:751–759
Wang L, Zhang T, Li P et al (2015) Use of synchrotron radiation-analytical techniques to reveal chemical origin of silver-nanoparticle cytotoxicity. ACS Nano 9:6532–6547
Wassall M, Santin M (1997) Adhesion of bacteria to stainless steel and silver-coated orthopedic external fixation pins. J Biomed Mater Res 36(3):325–330
Wei L, Lu J, Xu H et al (2015) Silver nanoparticles: synthesis, properties, and therapeutic applications. Drug Discov Today 20:595–601
Wilhelmi V, Fischer U, Weighardt H et al (2013) Zinc oxide nanoparticles induce necrosis and apoptosis in macrophages in a p47phox- and Nrf2-independent manner. PLoS ONE 8:e65704
Xue Y, Zhang S, Huang Y et al (2012) Acute toxic effects and gender-related biokinetics of silver nanoparticles following an intravenous injection in mice. J Appl Toxicol 32:890–899
Yang EJ, Kim S, Kim JS et al (2012) Inflammasome formation and IL-1β release by human blood monocytes in response to silver nanoparticles. Biomaterials 33:6858–6867
Yang J, Palla M, Bosco FG et al (2013) Surface-enhanced Raman spectroscopy based quantitative bioassay on aptamer-functionalized nanopillars using large-area Raman mapping. ACS Nano 7:5350–5359
Yu SJ, Yin YG, Liu JF (2013) Silver nanoparticles in the environment. Environ Sci Processes Impacts 15:78–92
Zhang C, Hu Z, Deng B (2016) Silver nanoparticles in aquatic environments: physiochemical behavior and antimicrobial mechanisms. Water Res 88:403–427
Zhang X, Niu H, Yan J et al (2011) Immobilizing silver nanoparticles onto the surface of magnetic silica composite to prepare magnetic disinfectant with enhanced stability and antibacterial activity. Colloids Surf, A 375:186–192
Zhao F, Zhao Y, Liu Y et al (2011) Cellular uptake, intracellular trafficking, and cytotoxicity of nanomaterials. Small 7:1322–1337
Zheng Z, Yin W, Zara JN et al (2010) The use of BMP-2 coupled—nanosilver-PLGA composite grafts to induce bone repair in grossly infected segmental defects. Biomaterials 31:9293–9300
Zhu X, Li J, He H et al (2015) Application of nanomaterials in the bioanalytical detection of disease-related genes. Biosens Bioelectron 74:113–133
Zi W, Ren W, Xiao F et al (2016) Ag nanoparticle enhanced light trapping in hydrogenated amorphous silicon germanium solar cells on flexible stainless steel substrate. Sol Energy Mater Sol Cells 144:63–67
Zivic F, Babic M, Grujovic N, Mitrovic S, Adamovic D (2013) Influence of loose PMMA bone cement particles on the corrosion assisted wear of the orthopaedic AISI 316LVM stainless steel during reciprocating sliding. Wear 300:65–77
Zivic F, Babic M, Grujovic N et al (2012) Effect of vacuum-treatment on deformation properties of PMMA bone cement. J Mech Behav Biomed Mater 5:129–138
Zook JM, Long SE, Cleveland D et al (2011) Measuring silver nanoparticle dissolution in complex biological and environmental matrices using UV-visible absorbance. Anal Bioanal Chem 401:1993–2002

Nanotechnology in Ophthalmology

Nenad Petrovic, Mirjana Janicijevic Petrovic, Suncica Sreckovic, Svetlana Jovanovic, Dusan Todorovic and Tatjana Sarenac Vulovic

Abstract Nanotechnology has already been used for the purpose of the measurement as well as treatment of great number of ophthalmological diseases. The applications of Nanotechnology to Ophthalmology refer to the following: delivery of drugs, peptides and genes, biomedical imaging and diagnostics of distinct disorders of the eye, study of pathogenesis of the eye disease. Regenerative nanomedicine is considered to be a prospective approach for treating retinal degenerations and the second cranial nerve regeneration. Our manuscript included possible different application for ophthalmological drug using nanotechnology. Also, we described the importance of nanotechnology for imaging, diagnostic and monitoring of ocular diseases. Nanobiotechnology for glaucoma treatment, as well as nanoparticles for gene delivery therapy of retinal degenerative disease, is most attractive approach for the ophthalmological treatment.

Keywords Nanotechnology · Ophthalmology · Nanoparticles · Nanocarriers · Ocular drug delivery

N. Petrovic (✉) · M.J. Petrovic · S. Sreckovic · S. Jovanovic · D. Todorovic · T.S. Vulovic
Department of Ophthalmology, Faculty of Medical Sciences, University of Kragujevac, Svetozara Markovica 69, Kragujevac, Serbia
e-mail: nenadpet@yahoo.com

M.J. Petrovic
e-mail: mira.andreja@yahoo.com

S. Sreckovic
e-mail: suncesun@yahoo.com

S. Jovanovic
e-mail: drsvetlanajovanovic@yahoo.com

D. Todorovic
e-mail: drdusantodorovic@yahoo.com

T.S. Vulovic
e-mail: tvoja.tanja@yahoo.com

D. Brabazon et al. (eds.), *Commercialization of Nanotechnologies–A Case Study Approach*, DOI 10.1007/978-3-319-56979-6_11

1 Introduction

Nanotechnology has already been used for the purpose of the measurement as well as treatment of great number of ophthalmological diseases. The applications of Nanotechnology to Ophthalmology refer to the following: delivery of drugs, peptides and genes, biomedical imaging and diagnostics of distinct disorders of the eye, study of pathogenesis of the eye disease. Regenerative nanomedicine is considered to be a prospective approach for treating retinal degenerations and the second cranial nerve regeneration.

2 Absorption of Drugs into the Eye

Generally, the major problem in ocular therapeutics is to preserve an effective drug concentration at the site of action for a proper time period, to gain adequately pharmacological response (Bucolo et al. 2008). The active compound topical instilling is the first chosen procedure for delivery in ocular therapy. Almost 90% of all ophthalmic drug formulations are used as eye drop (Bourlais et al. 1998).

Nevertheless, the bioavailability of an instilled compound is mostly low. This unbioavailability can be explained by the specific characteristics of the ocular surface and of the anterior segment. Those facts obstruct the efficient transit to the inner part of the eye. Ocular surface, which includes tree layer tear film, corneal epithelium with tight junctions and desmosomes, with blood-aqueous barrier of ciliary epithelium non-pigmented layer restrict the passing of molecules from the blood to the inner part of the eye, and eventually retinal pigment epithelium (RPE) along with the retinal vessels endothelium that forms the blood-retina barrier, which restrains the passing of molecules from the blood to the retina and vitreous cavity (Urtti 2006). Moreover, specific physiological processes are the important factors in the low efficiency of conventional drug formulations. For example, blinking and tear drainage via the lachrymal drainage system function to minimize the residence time of the molecules that are topically applied. Due to these anatomical and physiological limitations after topical application, an important fraction of the administered drug is lost by distinct mechanisms, which leads to a very low ocular bioavailability (Gref et al. 1994; Fresta et al. 1999). Less than 5% of topically administered drugs get to intraocular tissues (Keister et al. 1991). The majority of the instilled dose is frequently systemically absorbed by the naso-lacrimal duct and conjunctiva.

In order to increase the success of the treatment of pathologies that affect the posterior eye segment, different routes other than topical, i.e. the systemic route or intravitreal injections, have also been taken into consideration for the purpose of the augmentation of the bioavailability of drugs in the internal eye structure. Considering the systemic administration, the major hindrance is that only 1–2% of the administered drug gets to the vitreous cavity (Fresta et al. 2001). Consequently, this creates the frequent administration of high amounts of drug causing significant systemic undesirable effects. Currently, the best manner for treating the most diseases of the posterior eye segment, like intraocular inflammation, is by injecting drugs into the vitreous. Frequent intraocular injections are necessary to treat retinal disorders. These injections cause potential undesirable side effects, the increased risk of infections, and inadequate acceptance by the patient. More frequent side effects connected with repeated intravitreal injections involve the higher risk of cataract progression, retinal detachment, intraocular pressure elevation, and endophthalmitis.

Taking everything into consideration, for the efficient treatment of pathologies that affect the posterior eye segment, amending the corneal and/or conjunctival permeability has become one of the main issues in ocular drug delivery.

3 Ocular Drug Delivery Nanocarriers

Ophthalmic drug delivery might make improvement from the features of nanotechnology-based drug delivery systems (Raju and Goldberg 2008). Nanocarriers are employed to intensify ocular drug delivery (Vandervoort and Ludwig 2007). There are various types of drug delivery technologies that are created for the purpose of the drug delivery systems. There has been a progression in the structure of nanocarriers. Nanosystems that could be applied in ophthalmology differ in their structure and they involve the following (Diebold and Calonge 2010):

Nanoparticles (*NPs*) refer to the first generation of colloidal systems made of polymerized matrices. Nanoparticles could be defined as the small polymeric colloidal particles containing a therapeutic agent that is dispersed in polymer matrix or encapsulated in polymer. Accordingly, they might occur as nanospheres or nanocapsules, Fig. 1.

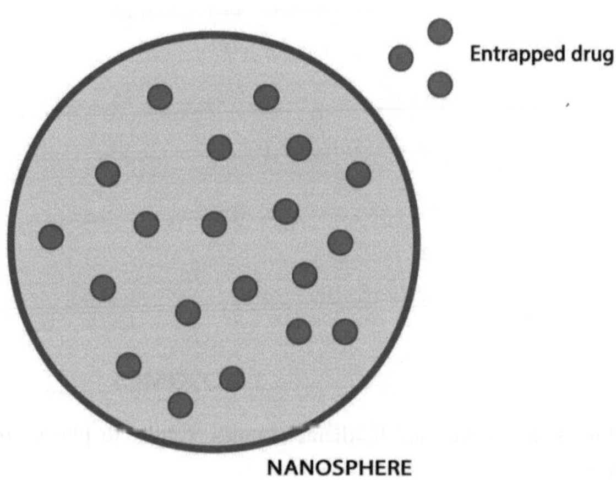

Fig. 1 Nanoparticles (NPs) refer to the first generation of colloidal systems made of polymerized matrices

Polymeric micelles refer to self-assembled block co-polymers that create an outer hydrophilic layer in aqueous solution and an inner hydrophobic core. The micellar core could be filled with a water-insoluble therapeutic agent, Fig. 2.

Liposomes are lipidicmembranes, similar to plasma membrane surrounding an aqueous core. The diameter of liposomes is ~200-nm diameter micelles. A variant of liposomes are niosomes, consisting of non-ionic surfactants, Fig. 3.

Nanoemulsions and Nanosuspensions refer to the mixtures made of different percentage of water and oil.

Dendrimers occur as synthesized, monodispersed symmetric macromolecules. They are created around a small molecule containing an internal cavity enclosed by numerous reactive end groups. Dendrimers, highly branched polymers, are

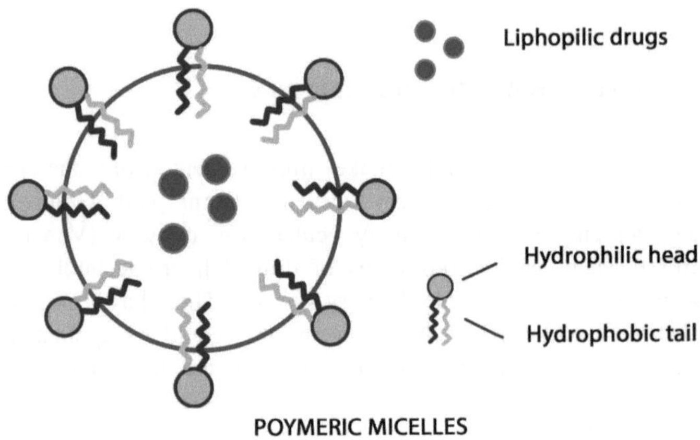

Fig. 2 Polymeric micelles refer to self-assembled block co-polymers that create an outer hydrophilic layer in aqueous solution and an inner hydrophobic core

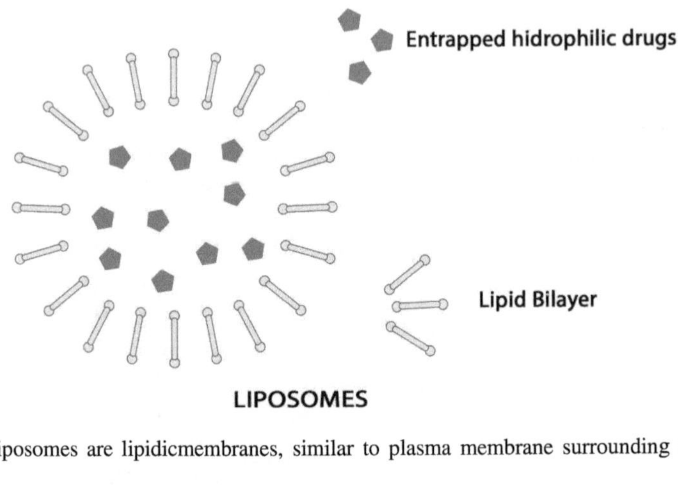

Fig. 3 Liposomes are lipidicmembranes, similar to plasma membrane surrounding an aqueous core

characterized by strictly controllable nano-scale scaffolding and nanocontainer features, which in some way simulate the features of macromolecules such as DNA and ribonucleic acid (RNA) (Tomalia et al. 2007). The diameter of dendrimer varies widely (Hahn et al. 2002).

Nanotubes and fullerenes are made of carbon-based nanomaterials.

Quantum dots are nanocrystals with fluorescent properties covered with other materials, and thus they can be used for imaging purposes.

Ferrofluids refer to the colloidal solutions of iron oxide magnetic nanoparticles enclosed by a polymeric layer, which could be additionally covered with antibodies.

Nanoparticleloaded contact lenses.

3.1 Nanocarriers-Based Topical Drug Application to the Eye

An ideal topical drug delivery system should possess certain desirable features including satisfactory corneal and conjunctival penetration, extended duration of the precorneal residence time, easy instillation, non-irritative and capable of decreasing lachrymation and reflex blinking, and adequate rheological features. It is currently possible to design nanocarriers with specific delivery requirements for ocular administration. The usage of nanocarriers creates significant possibilities for topical ocular drug delivery, mostly due to their ability to ensure the encapsulated molecule while facilitating its transport to the different compartments of the eye (De Campos et al. 2001, 2004; Bourges et al. 2003a, b; Kayser et al. 2005; Sanchez and Alonso 2006; Badawi et al. 2008; De la Fuente et al. 2008a, b). Those carriers have the ability to deliver the loaded therapeutic molecule in a safe manner while hindering it from the damage or deactivation. Advantageous sides of the usage of nanoparticles involve ameliorated topical passage of huge, poorly water/soluble molecules such as immunesuppressive drugs for immune-related diseases that affect vision. Other huge and unstable molecules, such as nucleic acids, delivered by nanoparticles, provide prosperous outcomes for gene transfer therapy in the cases of severe retinal conditions.

The loaded agents might function in a more efficient manner minimizing side effects in comparison to the same agents that are administered without the nanocarrier. These properties are promising for drug treatment of chronic conditions such as glaucoma (Chu et al. 2002), iridocyclitis (Kassem et al. 2007), or retinal venous occlusion causing retinal edema (Bondì et al. 2007). They are also useful for the treatment of intraocular tumors, and other cell proliferation conditions (Marano et al. 2005). These conditions refer to ocular neovascularization and proliferative vitreoretinopathy.

3.2 Ocular Drug Delivery Nanocarriers Concerning the Anterior Eye Segment

Concerning the bioavailability of an instilled conventional drug onto the cornea and conjunctiva it is commonly low. Only a few seconds after instillation, the most part of it is lost owing to the tear drainage and blinking (Bayens and Gurny 1997). Thus, there is a short pre-corneal residence time, usually less than 2 min and a nonproductive absorption thorough the well vascularised conjunctiva and the nasolacrimal drainage system (Jarvinen et al. 1995).

For those reasons, intensive research in recent decades has focused on increasing the corneal penetration of topically applied drugs (Schoenwald 1990; Sasaki 1999; Edwards 2001; Mannermaa 2006). Nanoparticles (NPs) were regarded to provide the prospect of an easier delivery and transport across tissues.

In 1986, it was tried for the first time with the usage of NPs with progesterone, a highly lipophilic molecule (Li et al. 1986). Later, Losa and associates used poly (isobutyl-cyanoacrylate) (PIBCA) and poly-e-caprolactone (PECL) nanocapsules for the ocular delivery of metipranolol employed for treating glaucoma (Losa et al. 1993). To improve the corneal penetration of Cyclosporine (CsA) applied topically, Juberías and associates studied PECL nanocapsules loaded with 1% cyclosporine in a rat model of the rejection after keratoplasty (Juberías et al. 1998).

Topical application of steroidal and nonsteroidal anti-inflammatory drugs on the eye is a common treatment used to treat the inflammatory reaction. Nevertheless, since they are almost entirely insoluble in water and since they are excluded from intraocular sites by the various blood-ocular barriers in the efficient manner. Their capability to get to intraocular tissues is very poor. Pignatello and all, developed ibuprofen-loaded NPs made of inert polymeric resins (Eudragit RS100) optimized as a pharmaceutical preparation (Pignatello et al. 2002). The controlled release profile of IBU from nanoparticles was shown in vitro dissolution tests, while in vivo effectiveness was estimated on the rabbit eye after inducing an ocular trauma. Drug levels in the aqueous humor proved to be higher as well after utilization of the nanosuspension. IBU-loaded nanosuspensions showed no toxicity on ocular tissues.

3.3 Polylactide Nanoparticles Concerning the Topical Drug Application to the Eye

Later, the ameliorated formulations employing polymers, biocompatibility and biodegradability, for instance poly-D-lactic acid (PLA) and its copolymer glycolic acid (PLGA) were elaborated. Poly(lactic/glycol) acid nanoparticles incorporating flurbiprofen have been prepared by the solvent displacement technique using poloxamer 188 as a stabilizer to amend the accessibility of the drug for the purpose of the inflammation prevention after ocular surgery (Vega et al. 2006). In vivo rabbit model of ocular inflammation flurbiprofen-loaded PLGA NPs efficiently minimized the inflammation. When compared to profitable flurbiprofen formulations, flurbiprofen-loaded NP minimized more efficiently the inflammation based on the clinical signs. The commercial eye drop showed a suppression of inflammation, with minimal inflammation reached after 90 min. A comparable nanoparticle formulation demonstrated a higher suppression, which increased throughout the 150 min observation time of the study. In vivo studies in rabbits showed that the formulations caused no toxicity or irritation on ocular tissues.

3.4 Chitosan Nanoparticles for Topical Drug Application to the Eye

Applications using chitosan (CS)-based nanosystems have been developed recently. CS is a natural polysaccharide possessing significant properties. These properties include biocompatibility and biodegradability, mucoadhesiveness, and the capability to shortly magnify the permeability of mucosal barriers. These properties have made it quite useful in the development of drug delivery systems for transmucosal administration (Alonso and Sánchez 2003; Paolicelli et al. 2009).

Chitosan (CS) nanoparticles application concerning the ocular drug delivery was examined by observing their interaction with the conjunctiva in vivo and their toxicity level in conjunctival cells (De Campos et al. 2004). The spectrofluorimetry and confocal microscopy was used for studying the in vivo interaction of fluorescent CS (CS-fl) nanoparticles with the rabbits' ocular surface. In vivo studies indicated prominently higher amounts of CS-fl in cornea and conjunctiva for CS-fl nanoparticles than without it. The stated amounts were invariable for up to 24 h. Confocal studies suggest that nanoparticles penetrate into the corneal and conjunctival epithelia. These examinations point out that CS nanoparticle have a prominent potentiality concerning the ocular drug delivery. According to the histopathological analysis and dealing with the ocular surface structures CS NPs proved to be well-tolerated by, inducing no damage or inflammation.

NPs consisted of CS and carbopol, a polymer of acrylic acid, joined features of both polymers. These features refer to the stability in aqueous solution, small size, better biocompatibility, and positive charge to facilitate interaction with the negatively charged biological membranes. CS/carbopol NPs could be loaded with pilocarpine. Pilocarpine is a parasympathomimetic drug used for treating the open-angle glaucoma. Kao and associates compared the efficacy of pilocarpine-loaded NPs against liposomes, gel, and eye-drops formulations of pilocarpine in normal rabbits. Pilocarpine causes miosis, while the effect of the CS/carbopol NPs occurs up to 24 h, which is significantly longer than the others (Kao et al. 2006).

Nanoparticles could be useful in dealing with the corneal allograft rejection. Yuan and associates produced rapamycin-loaded chitosan/polylactic acid nanoparticles. They observed that these nanoparticles prolong median allograft survival by 17% in rabbits in comparison to aqueous rapamycin eye drops (Yuan et al. 2008).

CS/sodium alginate NPs, another CS-based nanosystem, has been confirmed lately as a possible reservoir for topical delivery of gatifloxacin. It stands as a potent fourth-generation fluoroquinolone showing low intrastromal penetration which is mostly used in the case of microbial keratitis (Motwani et al. 2008).

Regarding the NPs delivery of drugs some questions could be raised. These questions include the depth of tissue penetration, the drug concentration and duration in the target and surrounding tissues. The surface features of the nanocarriers have an impact on the interaction with the ocular surface structures. For example, CS-coated PECL nanocapsules penetrate into the corneal epithelium in vivo in a more efficient manner than uncoated PECL or polyethylene glycol-coated PECL nanocapsules (Dal Pizzol et al. 2016). As mentioned before, there are the exceptional biological features of CS, especially mucoadhesiveness and the ability to transiently enhance the permeability of mucosal barriers. In this sense, Yenice and associates used hyaluronic acid (HA)-coated PECL nanospheres for corneal Cyclosporin (CsA) delivery, and they showed that concentrations in rabbit cornea were 10–15-fold higher than that achieved when CsA was administered as solution in castor oil (Yenice et al. 2008). Diebold and associates joined liposomes and CS NPs to form a new nanosystem "liposome chitosan nanoparticles" (LCS-NPs), and the theoretical advantages of these complexes refer to the process of combining the liposome biocompatibility with biological membranes and the mentioned features of CS NPs (Diebold et al. 2007). Recently, de la Fuente and associates have started to work with NPs composed of hyaluronic acid and chitosan (HA-CS NPs), and this new development is capable of encapsulating macromolecules of hydrophilic and hydrophobic nature. Hydrophilic nature is shown in the protein bovine serum albumin, while hydrophobic nature in the polypeptide CsA (De la Fuente et al. 2008a, b). HA-CS NPs could carry plasmid DNA and might be appropriate regarding the gene delivery as seen in the case of ocular surface-derived cell lines.

3.5 Nanobiotechnology to Treatment of Glaucoma

Topical application of drugs is the leading way of treating glaucoma nowadays. Nevertheless, there is a low percentage of the traditional glaucoma medicines that can enter the eye. It is shown in previous studies with nanoparticles high penetration rates and patient comfort as well.

Zimmer and associates tested different pharmaceutical aspects of pilocarpine-loaded PBCA NPs versus a standard pilocarpine solution in an elevated IOP rabbit model. The NPs proved to be better in minimizing IOP and pertaining miosis, particularly at lower drug concentrations. The NPs induced maximum reduction of IOP at 2–3 h, whereas the drug solution maximum response was at 1–2 h (Zimmer et al. 1994). Epinephrine-loaded poly-N-isopropylacrylamide NPs was also examined in rabbits in order to estimate IOP-lowering effect. In comparison to conventional eye-drops the NPs turned out to have six times more long-lasting effect (Hsiue et al. 2002). Finally, Wadhawa and associates reported a significant IOP reduction in the rabbits' eyes exposed to CS-HA NPs loaded with timolol compared to standard timolol eye-drops or blank NPs (Wadhawa et al. 2010).

It is considered that the accumulating of extracellular matrix (ECM) materials in the trabecular meshwork (TM) induces intraocular pressure (IOP) elevation, the open angle glaucoma risk factor. Matrix metalloproteinase-3 (MMP-3) is the proteinase that could efficiently lower ECM elements. The MMP-3 delivery to the TM is a prospective way concerning IOP reduction. Turturro and associates tested the feasibility of using polymeric microparticles to achieve a slow and sustained release of active MMP-3 to cultured human TM cells. Active MMP-3 was subsequently enclosed in PLGA microparticles. The release of MMP-3 from the microparticles led to maintained decrease of fibronectin up to 10 days. This suggests that controlled and sustained release of active MMP-3 to TM cells via encapsulation into PLGA microparticles can represent a possible treatment of glaucoma (Turturro et al. 2013).

It should be also pointed out that though not strictly NPs but a nanodevice, the development of a prototype for a nanodrainage system to be inserted into the sclera represents an alternative way for humour aqueous outflow in the surgery of glaucoma (Pan et al. 2006).

3.6 Nanoparticleloaded Contact Lenses for Ophthalmic Drug Delivery

The drug loss and the side effects that are common when the drugs are administered as eye-drops could be decreased by using soft contact lenses concerning the ophthalmic drug delivery. Polymeric lens materials can be loaded with nanoparticulate eye medication for ophthalmic drug delivery applications. The main goal is to enclose the ophthalmic drug formulations in nanoparticles, and disseminate these drug-laden particles in the lens. When inserted into the eye, the lens is constantly releasing the drug into the tear film between the air and the lens and the tear film between the cornea and the lens. In that way the drug delivery is provided for the prolonged period. The drug molecules have a much longer residence time of several days in the postlens tear film, compared to approximately 2–5 min in the case of topical application of drugs in the form of eye drops. These lenses reduce drug wastage and side-effects and provide controlled drug release for over 5–10 days (Gulsen and Chauhan 2005).

Core-shell nanospheres filled with an antibiotic could be inserted into a conventional contact lens. A copolymer formed by combining pullulan and polycaprolactone (PCL) was employed to produce core-shell nanospheres that encapsulated ciprofloxacin. The lenses induced efficient anti-bactericidal activity for a few days (Garhwal et al. 2012).

4 Ocular Drug Delivery Nanocarriers Concerning the Posterior Eye Segment

During the last decade much attention has been paid to optimizing drug delivery systems for the posterior eye segment. Alternative options to intravitreal injections, involving scleral implants and devices, transdermal patches, and various iontophoretic devices containing hydrogel reservoirs, have been explored with variable results (Del Amo and Urtti 2008). Several kinds of NPs carrying diverse active molecules, including genetic material, are currently in pre-clinical studies that employ the above mentioned approaches (Bejjani et al. 2005; Normand et al. 2005; Irache et al. 2005; Farjo et al. 2006; Paasonen et al. 2007). The aim is to make use of the ability of vitreous to emit and keep molecules within the tissues with which it is in direct contact as retina for instance, and to exploit it as a biological reservoir when the NPs are put inside. This provides the retaining therapeutic levels of the drug for several months after an intravitreal injection.

There were conducted prospective studies described in the literature on the usage of intravitreally injected NPs. Ganciclovir-loaded albumin NPs represent the significant instance. Ganciclovir is considered to be the common treatment for retinitis caused by cytomegalovirus. In vitro experiments it is shown that albumin NPs released ganciclovir in a sustained manner (Merodio et al. 2001), with a significant improvement of drug uptake by cytomegalovirus infected human cells (Merodio et al. 2002).

Nanoparticles were also examined for intraocular drug delivery in order to enable controlled drug release, to provide the protection of the drug from enzymatic degradation, and to regulate the course of the drug to the place of action. Experimental animals have been used for the purpose of the kinetics of PLA nanoparticle localization within the intraocular tissues and their ability to release encapsulated substance. Intravitreous injection of PLA nanoparticles proves to induce the transretinal movement, mainly concentrated in the retinal pigmented epithelial (RPE) cells. RPE cells are capable of taking up various types of NPs. Bourges and all, tested in rats, intravitreal PLA NPs loaded with fluorochromes and showed a preferential localization at the RPE cells after 24 h. The most significant result was the fact that RPE cells maintained the NPs, which constantly released the fluorochrome for months after the single injection. For up to four months after the injection, fluorescence pouring out from the NPs was noticed in distant parts of retinal tissue, ganglion cells, and the outer segments of the rod (Bourges et al. 2006).

Encapsulated rhodamine dye pours out from the nanoparticles. Then it stains the neuroretina and the RPE cells. The results support the notion that particular aiming at these tissues is possible. Additionally, the occurrence of the nanoparticles within RPE cells 4 months after a single injection indicates that a stable and ceaseless delivery of drugs could be accomplished.

The discovery that RPE cells have the ability to take up different kinds of NPs opens the possibility of using them to treat retinal disorders associated with the process of ageing or the dystrophies of photoreceptor. The goal will be to mark these cells with particular molecules or genetic material able to oppose or cease the processes causing these diseases. PLGA nanoparticles are capable of delivering genes to RPE cells in vitro and in vivo with adequate effectiveness, quickness, and in a safe manner (Bejjani et al. 2007).

C-60 fullerene-based antioxidants (Dugan et al. 1997, 2001) might even be used to treat conditions characterized by oxidative damage e.g. diabetic retinopathy, ROP, AMD and other conditions (Shen et al. 2007; Zarbin 2004; Yorek 2003; Komeima et al. 2007). Additionally, nanoparticles could be employed to provide the delivery of growth factors to cells (Ferreira et al. 2008a, b). Intravitreal nanoparticle-mediated fibroblast growth factor delivery enables a maintained protection of the retina in rats (Sakai et al. 2007). The neuroprotective effects of PLGA nanospheres to encapsulate pigment epithelium-derived factor (PEDF) have been evaluated in the induced retinal ischemic injury (Li et al. 2006). Intravitreal glial cell-derived neurotrophic factor-loaded biodegradable (poly) lactic-co-glycolic acid (PLGA) microspheres allow the maintained protection of ganglion cell in the glaucoma rodent models (Jiang et al. 2007). Intravitreal injection of the naked peptides demonstrated a 44% reduction of cell death of the retinal ganglion cells (RGCs) after 48 h. Injection of the encapsulated peptide showed a very similar protective effect that lasted for at least 7 days. The authors attributed the extended effect to the slow release of PEDF from PLGA particles and do the protection of the peptide against degradation and rapid clearance. The US Food and Drug Administration (FDA) authorized the PLGA usage in humans.

Besides size, grafting polymers with polyethylene glycol (PEG) represent another method of controlling particle distribution. A hydrophobic polymer, cyanoacrylate-co-hexadecyl cyanoacrylate, was coupled to hydrophilic PEG chains in order to produce tamoxifen-loaded nanoparticles (De Kozak et al. 2004). Intraocular injection in rats resulted in a significant inhibition of experimentally induced autoimmune uveoretinitis, whereas injection of the free drug did not alter the disease.

The aim of these therapeutic procedures is to reduce the number of intravitreal injections by the use of NPs and to improve the intraocular usefulness of topically applied drugs as well. Recently, Kassem and associates studied different nanosuspensions prepared by high-pressure homogenization of the three insoluble glucocorticoids: hydrocortisone, prednisolone, and dexamethasone (Elbary et al. 2008). They observed magnified absorption and prominent prolongment of the glucocorticoid action. As an interesting fact, intravenously injected PLA NPs encapsulating betamethasone phosphate efficiently hold the inflammation in check in the case of a rat model of experimental autoimmune uveoretinitis (Sakai et al. 2006).

5 Use of Dendrimers in Ophthalmology

One of the main motives of using dendrimers in ophthalmology is to overcome the limitation of agents targeted to a single molecule or receptor, Fig. 4. Dendrimers enable polyvalent medicines, larger molecules where several ligands can fasten to several receptors in order to get the desired biological response. Dendrimers have been examined as means for regulated drug delivery in relation to cancer therapy, pilocarpine, and vascular endothelial growth factor (VEGF) inhibition (Marano et al. 2005; Vandamme and Brobeck 2005). The use of dendrimers in drug delivery to the eye is also being explored to target multiple pathologies.

Some of the strategies for the treatment of eye disorders involve prevention of neovascularization. For instance, Marano and associates made use of a lipophilic amino acid dendrimer in order to deliver an anti-VEGF oligonucleotide into rats' eyes with choroidal new vessels induced by the laser (CNVs) (Marano et al. 2005). The dendrimer-oligonucleotide conjugate prohibited CNV progression for half of the year up to 95%. The eyes injected only with oligonucleotide indicated no treatment success in comparison to vehicle-injected controls conducted at these particular times. The large proportion of the retina absorbed intravitreally injected dendrimer-oligonucleotide conjugate. It entered all retinal cell layers to the retinal pigment epithelium. The dendrimer-oligonucleotide conjugate showed satisfying tolerance in vivo, demonstrating remarkable biodistribution and no noticeable augmentation in the antigens related to inflammation.

Ideta and associates made use of the dendrimer porphyrin encapsulated by a polymeric micelle to treat laser-induced CNVs in rodents (Ideta et al. 2005). They observed a prominent intensification of photodynamic therapy (PDT) efficiency with less light energy needed for CNV occlusion. The usage of such dendrimers

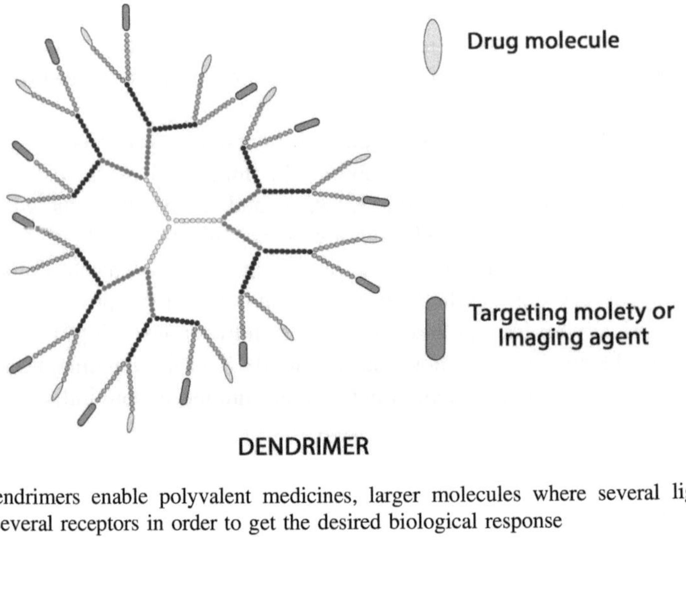

Fig. 4 Dendrimers enable polyvalent medicines, larger molecules where several ligands can fasten to several receptors in order to get the desired biological response

might be successful for the delivery of therapeutic oligonucleotides when treating angiogenic eye diseases.

Dendrimers can be used to prevent scar formation following eye surgery. Another usage could be to interrupt the process of inflammation and angiogenesis in the posterior chamber of the eye. Synthetically engineered dendrimers can be tailored to have defined immunomodulatory and antiangiogenic properties; they can be used synergistically to prevent scar tissue formation (Shaunak et al. 2004).

6 Other Applications of Nanotechnology to Ophthalmology Imaging, Diagnostics and Monitoring

Nanoparticles (dendrimer, gold nanoparticle, quantum dot, superparamagnetic iron oxide) can be used for biomedical imaging. Polyamidoamine dendrimer prototypes might be efficient as marked diagnostic magnetic resonance imaging (MRI) contrast agents. Gold nanoparticles have been utilized both to improve the identification of tumor by computed tomography (CT) and to conduct colorimetric biosensing. Superparamagnetic iron oxide (SPIO) nanoparticles can be visualized by using MRI, and they have been approved for use by the FDA as MRI contrast agents (Reimer et al. 2003), Fig. 5.

Quantum dots (Qdots) refer to the light-emitting nanocrystals (2–10 nm). Their visualization is performed by employing optical imaging involving optical coherence tomography (OCT). Regarding ocular tissues, Yamamoto and associates have shown that quantum dots can be used for imaging the vitreous (Yamamoto et al. 2007).

The process of coupling SPIO nanoparticles or Qdots to antibodies that identify molecules as the elements of the complement system or molecular constituents connected with age-related macular degeneration caused alterations in Bruch membrane. That could serve as a vehicle to image the biochemical and/or structural

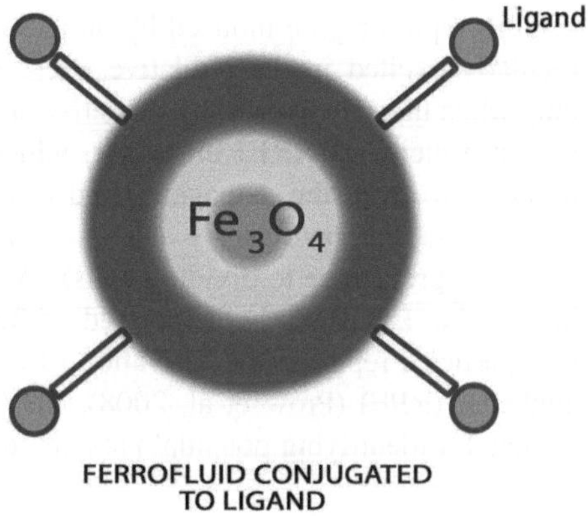

Fig. 5 Nanoparticles (dendrimer, gold nanoparticle, quantum dot, superparamagnetic iron oxide) can be used for biomedical imaging. Superparamagnetic iron oxide (SPIO) nanoparticles can be visualized by using MRI, and they have been approved for use by the FDA as MRI contrast agents

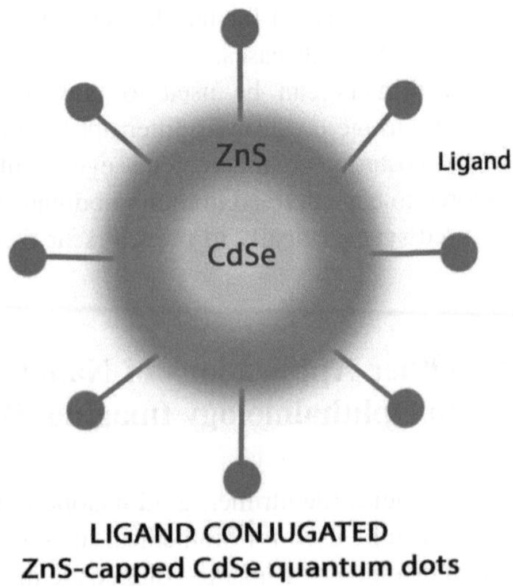

Fig. 6 The process of coupling SPIO nanoparticles or Qdots to antibodies that identify molecules as the elements of the complement system or molecular constituents connected with age-related macular degeneration caused alterations in Bruch membrane. That could serve as a vehicle to image the biochemical and/or structural changes accompanying AMD

changes accompanying AMD (Takeda et al. 2009), Fig. 6. This ability might be essential for the improvement of AMD treatments that are aimed at early molecular alterations and early stages of the disease.

Other inorganic NPs made of noble metals, such as gold (Hayashi et al. 2009) or silver (Gurunathan et al. 2009), are generating much interest due to their small size, about 20 nm, and their great potential of crossing the blood-retina boundary (Kim et al. 2009).

Nanodevices can permit continuous monitoring and the treatment of critical biological variables such as intraocular pressure, oxygen tension, or redox potential (Wei et al. 2007; Helmchen and Denke 2005). The prospect for future treatment of retinopathies based on the nanoparticle drug delivery systems involving biomolecular control biosensors that have the ability to identify oxidative stress proteins has been shown (Prow et al. 2006a, b). Particularly, Prow and associates have generated a biosensor DNA fastened to a magnetic nanoparticle (Prow et al. 2006a, b). The foundation of the biosensor is an enhanced green fluorescent protein (EGFP) reporter gene induced by an antioxidant response element (ARE). ARE is regularly excited in the oxidative stress environment. Genes' expression occurs only when they are excited by oxidative stress proteins that fasten to the antioxidant response element (ARE) biosensor, which could be used further for therapeutic purposes to avert the damage related to oxidative stress that could induce retinopathies and the optic nerve damage. This nanoparticle enters endothelial cells (showing preference to dividing cells). When the cells are exposed to hyperoxia, it incites the EFGP to be expressed. After subretinal injection, these biosensor nanoparticles reported the activation of the ARE in diabetic rat retinal pigmented epithelial (RPE) (Prow et al. 2008). The antioxidant biosensor might represent a vehicle for identifying potential patients for therapy (e.g. babies with retinopathy of prematurity (ROP) who will need laser photocoagulation or other treatment) at the moment before clinical manifestations of severe disease are obvious.

Nanowires could be defined as the structures possessing a diameter of less than 100 nm and the length that is not definite. Semiconductor nanowires exhibit peculiar electronic features and sizes similar to biological structures included in cellular communication. That consequently makes them suitable nanostructures for producing active communications with biological systems (Cohen-Karni et al. 2009). These techniques should be applicable to measuring functions such as electrical activity and contractility in neurons. Micro- and nano-electromechanical systems (M/NEMS)-based engineering permits the construction of small diagnostic devices. For the monitoring of intraocular pressure the microfabricated implantable wireless passive IOP sensor with the surface area of 4×2 mm^2 could be inserted into the anterior chamber (Dresher and Irazoqui 2007).

7 Regenerative Medicine

"Regenerative nanomedicine" refers to a recently formed subgroup of nanomedicine which involves nanoparticles that possess gene transcription factors and other modulating molecules that allow the reprogramming of cells in vivo.

Nanotechnology can be used to create an extracellular matrix (ECM). The ECM can alter cell behavior. The ECM includes both cell adhesion sites and growth factors. Transmembrane proteins called integrins, whose role is to function as an intermediary between cells and ECM proteins including laminin and fibronectin. Integrin binding to ECM ligands results in integrin clustering, formation of focal adhesion complexes, modification of cytoskeleton organization, and activation of intracellular signaling that has an impact on cell adhesion and proliferation (Hynes 2002). For instance, nanotopography could function as the guidance for the processes of keratocyte shape, alignment and migration. These properties could be changed to enhance biointegration of prosthetic devices like artificial corneas (Pot et al. 2010).

Biocompatible/biodegradable scaffolds possessing the adequate nanoscale properties may make improvements to cell survival and differentiation. Nanofiber scaffolds could be employed both to constitute niches for stem cell self-renewal and to function as substrates to transplant sheets of cells (Ferreira et al. 2008a, b). Ellis-Behnke and associates, for example, reported that a designed self-assembling peptide nanofiber scaffold (SAPNS) promoted axonal regeneration through the hamsters' optic tract (Ellis-Behnke et al. 2006). In the process, the regenerated axons connected again to target tissues and supported the recovery of vision. Although the technique is still far away from the usage in humans, the prospects of application are promising.

Regenerative nanomedicine represents a prospective procedure in the treatment of retinal degenerations. Microstructured polymer scaffolds improve retinal progenitor cell survival and differentiation. They also have potentiality regarding the process of spatially targeted (e.g. submacular) cell delivery (Tao et al. 2007; Neeley et al. 2008; Sundback et al. 2005). Beeley and associates have conducted a procedure to create microscale thin films (<10 μm) in polycaprolactone (PCL), which could be defined as a biocompatible, slowly releasing polymer that might be perfect for insertion into the subretinal space (Beeley et al. 2005). Microstructures and nanostructures could be used to enable the regulated release of trophic factors that improve regeneration and integration features.

8 Nanosurgery

The future utilization of nanotechnology for the purpose of producing surgical nanoinstruments is at its very early stage. Kim and Lieber generated nanotweezers for the purpose of interrogation and nanostructures' manipulation (Kim and Lieber 1999). Bhisitkul and Keller created micromechanical forceps (Bhisitkul and Keller 2005). The future usage of nanosurgical devices could be in the conduction of the procedures including internal recanalization of retinal artery and vein occlusions, complex epiretinal membranes' dissection, reparation of retinal breaks, and the separated optic nerves' reanastomosis.

9 Nanoparticles for Gene Delivery/Therapy of Retinal Degenerative Disorders

The eye is an excellent candidate for gene therapy for two main reasons: it has immunity, and its genetic-based diseases are well apprehended.

Genetic-based therapies could be elaborated by employing distinct types of nucleic acids.

Regarding the pharmaceutical aspect, the delivery of genetic material is a challenging feature. What can be considered as an efficient approach to some extent recently has been the utilization of viral vectors including adenovirus, retrovirus, and mainly recombinant adeno-associated virus (rAAV) as gene carriers (Snyder 1999). What is confirmed by various studies is that some viral vectors could successfully deliver transgenes to ocular tissues while the others could not. It should be mentioned as well that the application of viral vectors endangers the safety of patients.

Nanoparticles could be a new method for non-viral gene transfer enhancement. Gene delivery employing non-viral carrier systems has potentiality of being safer and more efficient. Nanoparticles have been explored as a means for gene delivery in the diagnosis and treatment of ocular disease (Prow et al. 2006a, b; Farjo et al. 2007; Mo et al. 2007; Cai et al. 2008). There are a number of studies that have begun to employ distinct approaches, involving the following: naked DNA delivery by electroporation and the creation of lipoplexes by DNA condensation with cationic lipids (Kachi et al. 2005; Bloquel et al. 2006; Johnson et al. 2008), HA-chitosan oligomers (CSO) NPs loaded with a model plasmid encoding for alkaline phosphatase, DNA condensed with polycationic polymers (Farjo et al. 2006), preparation of single molecule of pEGFP compacted with PEG-substituted polysysine (Ding et al. 2009).

Compacted DNA nanoparticles (Copernicus) have been examined as a system capable of nonviral gene transfer to ocular tissues. Compacted DNA nanoparticles could be targeted to different tissues in the eye by varying the injection site: intravitreal injection could target the cornea, trabecular meshwork, and inner layer of the retina; subretinal injection most commonly affect the outer retina and retinal pigment epithelium (RPE) (Farjo et al. 2006). Nanoparticles could transfect nearly all eye cell types and create powerful levels of gene expression that depend on the dose. Preliminary results demonstrate that DNA nanoparticles can correct vision defects in a mouse model of retinitis pigmentosa by the delivery of normal copies of genes into rods and cones layer of the retina. In the clinical trials conducted in humans, the compacted DNA nanoparticles exhibited certain features (Cai et al. 2009). They proved to be safe and efficient, without theoretical restriction on plasmid size, and without inducing immune responses. They also could be highly concentrated. DNA nanoparticles might have capacity for successful treating of more complicated eye disorders including diabetic retinopathy, macular degeneration, and different conditions that affect ganglion cells and the optic nerve.

Albumin-derived nanoparticles used for the transport of plasmids composed of genes for the Flt receptor, which fastens free VEGF, enter keratocyte cytoplasm and enable the maintained restraint of the corneal neovascularization caused by injury (Jani et al. 2007).

A final example of potential gene therapy treatment of retinal diseases is the recent work of Park and associates (Park et al. 2009). They encapsulated in PLGA NPs an expression plasmid for a natural angiogenic inhibitor, plasminogen kringle 5 (K5) that inhibits ischemia induced neovascularization in the case of oxygen-induced retinopathy (OIR) in a rat model. Since the animals with OIR were supplied with K5-loaded NPs intravitreally, it is shown high K5 expression in the inner retina that lasted for four weeks. Additionally, in comparison to fellow control eyes retinal vascular leakage and retina neovascularisation were minimized in those K5-NP injected eyes.

In the process of gene delivery nanoscale-engineered cell substrata (e.g. nanowires) and carbon nanotubes could be applied as well (Yoo et al. 2013).

10 Nanoparticle Safety

Biocompatibility is required concerning not only ocular, but any biomedical usage of NPs as a therapeutic agent. When using NPs, risks exist by both organic and inorganic NPs including aggregation, tissue accumulation, and the adsorption of plasma proteins onto the surface. NP aggregation has the capacity to obstruct the cell metabolism or even to diminish tissue function. For example, the accumulation of topically applied NPs onto the cornea and conjunctiva might obstruct the lachrymal drainage punctum and worsen the process of the tear film recycling.

Similarly to carbon nanotubes, possible cytotoxic activity of NPs could induce changes of cell membranes like membrane disruption (Panessa-Warren et al. 2009). Prow has published lately a very thorough report concerning the toxicity level of nanomaterials in the eye. He made a summary of relevant studies conducted since 1996. It included thirty different types of nanoparticles and other nanocarriers examined for the purpose of ocular administrations (Prow 2010).

He also took into consideration the possible inflammatory and immunosuppressive features depicted for various types of NPs. It is known that some NPs are antigenic themselves. The antigenicity depends on particle size, especially those of ultra-small size (25 nm or smaller), (Dobrovolskaia and McNeil 2007; Zolnik et al. 2010).

Allergic or hypersensitivity reactions can be induced or aggravated in animal models and humans by dendrimers (Toyama et al. 2008), carbon nanotubes (Nygaard et al. 2009), lipid-based NPs (Szebeni et al. 2007), titanium dioxide NPs (Yanagisawa et al. 2009), and polystyrene NPs (Yanagisawa et al. 2010).

The strict estimation of the possible immunotoxic effects is the most significant issue regarding the development of NP-based novel therapies nowadays. Taking this into consideration, the standardization of procedures to examine immunological features and the issues involving toxicology is very significant.

11 Conclusion

Nanotechnology, as one of the most important discoveries of the XX century, can take a great role in ophthalmology. The wide spectrum of the ophthalmological diseases has limited options for the treatment, because of the eye anatomy, physiology, histology, immunity, as well as of the available treatment options. The nanotechnology development can provoke revolution in the treatment of different ophthalmological diseases.

References

Alonso MJ, Sánchez A (2003) The potential of chitosan in ocular drug delivery. J Pharm Pharmacol 55:1451–1463

Badawi AA, El-Laithy HM, El Qidra RK et al (2008) Chitosan based nanocarriers for indomethacin ocular delivery. Arch Pharm Res 31:1040–1049

Bayens R, Gurny R (1997) Chemical and physical parameters of tears relevant for the design of ocular drug delivery formulations. Pharm Acta Helv 72:191–202

Beeley NR, Rossi JV, Mello-Filho PA et al (2005) Fabrication, implantation, elution, and retrieval of a steroid-loaded polycaprolactone subretinal implant. J Biomed Mater Res A 73:437–444
Bejjani R, BenEzra D, Cohen H et al (2005) Nanoparticles for gene delivery to retinal pigment epithelial cells. Mol Vis 11:124–132
Bejjani RA, Andrieu C, Bloquel C et al (2007) Electrically assisted ocular gene therapy. Surv Ophthalmol 52(2):196–208
Bhisitkul RB, Keller CG (2005) Development of microelectromechanical systems (MEMS) forceps for intraocular surgery. Br J Ophthalmol 89:1586–1588
Bloquel C, Bourges JL, Touchard E et al (2006) Non-viral ocular gene therapy: potential ocular therapeutic avenues. Adv Drug Deliv Rev 58:1224–1242
Bourges JL, Gautier SE, Delie F et al (2003a) Ocular drug delivery targeting the retina and retinal pigment epithelium using polylactide nanoparticles. Invest Ophthalmol Vis Sci 44:3562–3569
Bourges JL, Bloquel C, Thomas A et al (2003b) Intraocular implants for extended drug delivery: therapeutic applications. Adv Drug Deliv Rev 58(11):1182–1202
Bondì ML, Craparo EF, Giammona G et al (2007) Nanostructured lipid carriers-containing anticancer compounds: preparation, characterization, and cytotoxicity studies. Drug Deliv 14:61–67
Bourlais CL, Acar L, Zia H et al (1998) Ophthalmic drug delivery systems—recent advances. Prog Retin Eye Res 17(1):33–58
Bucolo C, Maltese A, Drago F (2008) When nanotechnology meets the ocular surface. Expert Rev Ophthalmol 3:325–332
Cai X, Conley S, Naash M (2008) Nanoparticle applications in ocular gene therapy. Vision Res 48:319–324
Cai X, Nash Z, Conley SM et al (2009) A partial structural and functional rescue of a retinitis pigmentosa model with compacted DNA nanoparticles. PLoS ONE 4:e5290
Chu TC, He Q, Potter DE (2002) Biodegradable calcium phosphate nanoparticles as a new vehicle for delivery of a potential ocular hypotensive agent. J Ocul Pharmacol Ther 18:507–514
Cohen-Karni T, Timko BP, Weiss LE et al (2009) Flexible electrical recording from cells using nanowire transistor arrays. Proc Natl Acad Sci USA 106:7309–7313
Dal Pizzol C, O'Reilly A, Winter E et al (2016) Application of response surface methodology for the technological improvement of solid lipid nanoparticles. J Nanosci Nanotechnol 16(2): 47–1238
De Campos AM, Diebold Y, Carvalho EL et al (2004) Chitosan nanoparticles as new ocular drug delivery systems: in vitro stability, in vivo fate, and cellular toxicity. Pharm Res 21:803–810
De Campos AM, Sanchez A, Alonso MJ (2001) Chitosan nanoparticles: a new vehicle for the improvement of the delivery of drugs to the ocular surface. Application to cyclosporin A. Int J Pharm 224:159–168
De Kozak Y, Andrieux K, Villarroya H et al (2004) Intraocular injection of tamoxifen-loaded nanoparticles: a new treatment of experimental autoimmune uveoretinitis. Eur J Immunol 34:12–3702
Del Amo EM, Urtti A (2008) Current and future ophthalmic drug delivery systems. A shift to the posterior segment. Drug Discov Today 13:135–143
De la Fuente M, Csaba N, Garcia-Fuentes M et al (2008a) Nanoparticles as protein and gene carriers to mucosal surfaces. Nanomed 3:845–857
De la Fuente M, Seijo B, Alonso MJ (2008b) Novel hyaluronan-based nanocarriers for transmucosal delivery of macromolecules. Macromol Biosci 8:441–450
Diebold Y, Calonge M (2010) Applications of nanoparticles in ophthalmology. Prog Retin Eye Res 29:596–609
Diebold Y, Jarrín M, Sáez V et al (2007) Ocular drug delivery by liposomeechitosan nanoparticle complexes (LCS-NP). Biomaterials 28:1553–1564
Ding XQ, Quiambao AB, Fitzgerald JB et al (2009) Ocular delivery of compacted DNA-nanoparticles does not elicit toxicity in the mouse retina. PLoS ONE 4:e7410
Dobrovolskaia MA, McNeil SE (2007) Immunological properties of engineered nanomaterials. Nat Nanotechnol 2:469–478

Dresher RP, Irazoqui PP (2007) A compact nanopower low output impedance CMOS operational amplifier for wireless intraocular pressure recordings. Conf Proc IEEE Eng Med Biol Soc 2007:6056–6059

Dugan LL, Turetsky DM, Du C et al (1997) Carboxyfullerenes as neuroprotective agents. Proc Natl Acad Sci USA 94:9434–9439

Dugan LL, Lovett EG, Quick KL et al (2001) Fullerene-based antioxidants and neurodegenerative disorders. Parkinsonism Relat Disord 7:243–246

Edwards A, Prausnitz MR (2001) Predicted permeability of the cornea to topical drugs. Pharm Res 18:1497–1508

Elbary AA, Kassem MA, Abou Samra MM et al (2008) Formulation and hypoglycemic activity of pioglitazone-cyclodextrin inclusion complexes. Drug Discov Ther 2(2):94–107

Ellis-Behnke RG, Liang YX, You SW et al (2006) Nano neuro knitting: peptide nanofiber scaffold for brain repair and axon regeneration with functional return of vision. Proc Natl Acad Sci USA 103:5054–5059

Farjo R, Skaggs J, Quiambao AB (2006) Efficient nonviral ocular gene transfer with compacted DNA nanoparticles. PLoS ONE 1:e38

Farjo R, Skaggs J, Quiambao AB et al (2007) Effect of Rds abundance on cone outer segment morphogenesis, photoreceptor **gene** expression, and outer limiting membrane integrity. J Comp Neurol 504(6):30–619

Ferreira L, Karp JM, Nobre L et al (2008a) New opportunities: the use of nanotechnologies to manipulate and track stem cells. Cell Stem Cell 3:136–146

Ferreira L, Squier T, Park H et al (2008b) Human embryoid bodies containing nano- and microparticulate delivery vehicles. Adv Mater 20:2285–2291

Fresta M, Panico AM, Bucolo C et al (1999) Characterization and in-vivo ocular absorption of liposome encapsulated acyclovir. J Pharm Pharmacol 51:565–576

Fresta M, Fontana G, Bucolo C et al (2001) Ocular tolerability and in vivo bioavailability of poly (ethylene glycol) (PEG)-coated polyethyl-2-cyanoacrylate nanosphere-encapsulated acyclovir. J Pharm Sci 90(3):97–288

Garhwal R, Shady SF, Ellis EJ et al (2012) Sustained ocular delivery of ciprofloxacin using nanospheres and conventional contact lens materials. Invest Ophthalmol Vis Sci 53:1341–1352

Gref R, Minamitake Y, Peracchia MT et al (1994) Biodegradable long-circulating polymeric nanospheres. Science 263:1600–1603

Gulsen D, Chauhan A (2005) Dispersion of microemulsion drops in HEMA hydrogel: a potential ophthalmic drug delivery vehicle. Int J Pharm 292:95–117

Gurunathan S, Lee KJ, Kalishwaralal K et al (2009) Antiangiogenic properties of silver nanoparticles. Biomaterials 30:6341–6350

Hahn U, Gorka M, Vogtle F et al (2002) Light-harvesting dendrimers: efficient intra- and intermolecular energy-transfer processes in a species containing 65 chromophoric groups of four different types. Angew Chem Int Ed Engl 41(3595–3598):3514

Hayashi A, Naseri A, Pennesi ME et al (2009) Subretinal delivery of immunoglobulin G with gold nanoparticles in the rabbit eye. Jpn J Ophthalmol 53:249–256

Helmchen F, Denk W (2005) Deep tissue two-photon microscopy. Nat Methods 2:932–940

Hsiue GH, Hsu SH, Yang CC et al (2002) Preparation of controlled release ophthalmic drops, for glaucoma therapy using thermosensitive poly-N-isopropylacrylamide. Biomaterials 23: 457–462

Hynes RO (2002) Integrins: bidirectional, allosteric signaling machines. Cell 110:673–687

Ideta R, Tasaka F, Jang WD et al (2005) Nanotechnology-based photodynamic therapy for neovascular disease using a supramolecular nanocarrier loaded with a dendritic photosensitizer. Nano Lett 5:2426–2431

Irache JM, Merodio M, Arnedo A et al (2005) Albumin nanoparticles for the intravitreal delivery of anticytomegaloviral drugs. Mini Rev Med Chem 5:293–305

Jani PD, Singh N, Jenkins C et al (2007) Nanoparticles sustain expression of Flt intraceptors in the cornea and inhibit injury-induced corneal angiogenesis. Invest Ophthalmol Vis Sci 48: 2030–2036

Jarvinen K, Jarvinen T, Urtti A (1995) Ocular absorption following topical delivery. Adv Drug Deliv Rev 16:3–19
Jiang C, Moore MJ, Zhang X et al (2007) Intravitreal injections of GDNF-loaded biodegradable microspheres are neuroprotective in a rat model of glaucoma. Mol Vis 13:1783–1792
Johnson LN, Cashman SM, Kumar-Singh R (2008) Cell-penetrating peptide for enhanced delivery of nucleic acids and drugs to ocular tissues including retina and cornea. Mol Ther 16:14–107
Juberías JR, Calonge M, Gómez S et al (1998) Efficacy of topical cyclosporine-loaded nanocapsules on keratoplasty rejection model in the rat. Curr Eye Res 17:39–46
Kachi S, Oshima Y, Esumi N et al (2005) Nonviral ocular gene transfer. Gene Ther 12:843–851
Kao HJ, Lin HR, Lo YL et al (2006) Characterization of pilocarpine-loaded chitosan/Carbopol nanoparticles. J Pharmacokinet Pharmacodyn 58:179–186
Kassem MA, Abdel Rahman AA, Ghorab MM et al (2007) Nanosuspension as an ophthalmic delivery system for certain glucocorticoid drugs. Int J Pharm 340:126–133
Kayser O, Lemke A, Hernandez-Trejo N (2005) The impact of nanobiotechnology on the development of new drug delivery systems. Curr Pharm Biotechnol 6:3–5
Keister JC, Cooper ER, Missel PJ et al (1991) Limits on optimizing ocular drug delivery. J Pharm Sci 80:50–53
Kim JH, Kim JH, Kim KW et al (2009) Intravenously administered gold nanoparticles pass through the blood-retinal barrier depending on the particle size, and induce no retinal toxicity. Nanotechnology 20:505101
Kim P, Lieber CM (1999) Nanotube nanotweezers. Science 286:2148–2150
Komeima K, Rogers BS, Campochiaro PA (2007) Antioxidants slow photoreceptor cell death in mouse models of retinitis pigmentosa. J Cell Physiol 213:809–815
Li H, Tran VV, Hu Y et al (2006) A PEDF N-terminal peptide protects the retina from ischemic injury when delivered in PLGA nanospheres. Exp Eye Res 83:33–824
Li VH, Wood RW, Kreuter J et al (1986) Ocular drug delivery of progesterone using nanoparticles. J Microencapsul 3:213–218
Losa C, Marchal-Heussler L, Orallo F (1993) Design of new formulations for topical ocular administration: polymeric nanocapsules containing metipranolol. Pharm Res 10:80–87
Mannermaa E, Vellonen KS, Urtti A (2006) Drugtransport in corneal epithelium and blood-retina barrier: emerging role of transporters in ocular pharmacokinetics. Adv Drug Deliv Rev 58:1136–1163
Marano RJ, Toth I, Wimmer N et al (2005) Dendrimer delivery of an anti-VEGF oligonucleotide into the eye: a long-term study into inhibition of laser-induced CNV, distribution, uptake and toxicity. Gene Ther 12:1544–1550
Merodio M, Arnedo A, Renedo MJ et al (2001) Ganciclovir-loaded nanoparticles: characterization and in vitro release properties. Eur J Pharm Sci 12:251–259
Merodio M, Espuelas MS, Mirshahi M et al (2002) Efficacy of ganciclovir-loaded nanoparticles in human cytomegalovirus (HCMV)-infected cells. J Drug Target 10:231–238
Mo Y, Barnett ME, Takemoto D et al (2007) Human serum albumin nanoparticles for efficient delivery of Cu, Zn superoxide dismutase gene. Mol Vis 13:746–757
Motwani SK, Chopra S, Talegaonkar S et al (2008) Chitosan-sodium alginate nanoparticles as submicroscopic reservoirs for ocular delivery: formulation, optimisation and in vitro characterisation. Eur J Pharm Biopharm 68:513–525
Neeley WL, Redenti S, Klassen H et al (2008) A microfabricated scaffold for retinal progenitor cell grafting. Biomaterials 29:418–426
Normand N, Valamanesh F, Savoldelli M (2005) VP22 light controlled delivery of oligonucleotides to ocular cells in vitro and in vivo. Mol Vis 11:184–191
Nygaard UC, Hansen JS, Samuelsen M et al (2009) Single-walled and multi-walled carbon nanotubes promote allergic immune responses in mice. Toxicol Sci 109:113–123
Paasonen L, Laaksonen T, Johans C et al (2007) Gold nanoparticles enable selective light-induced contents release from liposomes. J Control Release 122:86–93
Pan T, Brown JD, Ziaie B (2006) An artificial nano-drainage implant (ANDI) for glaucoma treatment. Conf Proc IEEE Eng Med Biol Soc 1:3174–3177

Panessa-Warren BJ, Maye MM, Warren JB et al (2009) Single walled carbon nanotube reactivity and cytotoxicity following extended aqueous exposure. Environ Pollut 157:1140–1151
Paolicelli P, de la Fuente M, Sánchez A et al (2009) Chitosan nanoparticles for drug delivery to the eye. Expert Opin Drug Deliv 6:239–253
Park K, Chen Y, Hu Y et al (2009) Nanoparticle-mediated expression of an angiogenic inhibitor ameliorates ischemia-induced retinal neovascularisation and diabetes-induced retinal vascular leakage. Diabetes 58:1902–1913
Pignatello R, Bucolo C, Ferrara P et al (2002) Eudragit RS100 nanosuspensions for the ophthalmic controlled delivery of ibuprofen. Eur J Pharm Sci 16:53–61
Pot SA, Liliensiek SJ, Myrna KE et al (2010) Nanoscale topography-induced modulation of fundamental cell behaviors of rabbit corneal keratocytes, fibroblasts, and myofibroblasts. Invest Ophthalmol Vis Sci 51:1373
Prow TW (2010) Toxicity of nanomaterials to the eye. Wiley Interdiscip Rev Nanomed Nanobiotechnol 2:317–333
Prow TW, Bhutto I, Grebe R et al (2008) Nanoparticle-delivered biosensor for reactive oxygen species in diabetes. Vision Res 48:478–485
Prow T, Grebe R, Merges C et al (2006a) Nanoparticle tethered antioxidant response element as a biosensor for oxygen induced toxicity in retinal endothelial cells. Mol Vis 12:616–625
Prow T, Smith JN, Grebe R et al (2006b) Construction, gene delivery, and expression of DNA tethered nanoparticles. Mol Vis 12:606–615
Raju HB, Goldberg JL (2008) Nanotechnology for ocular therapeutics and tissue repair. Expert Rev Ophthalmol 3:431–436
Reimer P, Ferucarbotran Balzer T (2003) (Resovist): a new clinically approved RES-specific contrast agent for contrast-enhanced MRI of the liver: properties, clinical development, and applications. Eur Radiol 13:1266–1276
Sakai T, Kuno N, Takamatsu F et al (2007) Prolonged protective effect of basic fibroblast growth factor-impregnated nanoparticles in Royal College of Surgeons rats. Invest Ophthalmol Vis Sci 48:3381–3387
Sakai T, Kohno H, Ishihara T et al (2006) Treatment of experimental autoimmune uveoretinitis with poly(lactic acid) nanoparticles encapsulating betamethasone phosphate. Exp Eye Res 82:657–662
Sanchez A, Alonso MJ (2006) Nanoparticular carriers for ocular drug delivery. In: Torchilin VP (ed) Nanoparticulates as drug carriers. Imperial College Press, London, UK, pp 649–673
Sasaki H, Yamamura K, Mukai T et al (1999) Enhancement of ocular drug penetration. Crit Rev Ther Drug Carrier Syst 16:85–146
Schoenwald RD (1990) Ocular drug delivery: pharmacokinetic considerations. Clin Pharmacokinet 18:255–269
Shaunak S, Thomas S, Gianasi E et al (2004) Polyvalent dendrimer glucosamine conjugates prevent scar tissue formation. Nat Biotechnol 22:977–984
Shen JK, Dong A, Hackett SF et al (2007) Oxidative damage in age-related macular degeneration. Histol Histopathol 22:1301–1308
Snyder RO (1999) Adeno-associated virus-mediated gene delivery. J Gene Med 1:166–175
Sundback CA, Shyu JY, Wang Y et al (2005) Biocompatibility analysis of poly(glycerol sebacate) as a nerve guide material. Biomaterials 26:5454–5464
Szebeni J, Alving CR, Rosivall L et al (2007) Animal models of complement-mediated hypersensitivity reactions to liposomes and other lipid-based nanoparticles. J Liposome Res 17:107–117
Takeda A, Baffi JZ, Kleinman ME et al (2009) CCR3 is a target for age-related macular degeneration diagnosis and therapy. Nature 460:225–230
Tao S, Young C, Redenti S et al (2007) Survival, migration and differentiation of retinal progenitor cells transplanted on micro-machined poly(methyl methacrylate) scaffolds to the subretinal space. Lab Chip 7:695–701
Tomalia DA, Reyna LA, Svenson S (2007) Dendrimers as multipurpose nanodevices for oncology drug delivery and diagnostic imaging. Biochem Soc Trans 35:61–67

Toyama T, Matsuda H, Ishida I et al (2008) A case of toxic epidermal necrolysis-like dermatitis evolving from contact dermatitis of the hands associated with exposure to dendrimers. Contact Dermatitis 59:122–123

Turturro S, Sunoqrot S, Ying H et al (2013) Sustained release of matrix metalloproteinase-3 to trabecular meshwork cells using biodegradable PLGA microparticles. Mol Pharm 10:32–3023

Urtti A (2006) Challenges and obstacles of ocular pharmacokinetics and drug delivery. Adv Drug Deliv Revs 58:1131–1135

Vandamme TF, Brobeck L (2005) Poly(amidoamine) dendrimers as ophthalmic vehicles for ocular delivery of pilocarpine nitrate and tropicamide. J Control Release 102:23–38

Vandervoort J, Ludwig A (2007) Ocular drug delivery: nanomedicine applications. Nanomed 2:11–21

Vega E, Egea MA, Valls O et al (2006) Flurbiprofen loaded biodegradable nanoparticles for ophthalmic administration. J Pharm Sci 95:2393–2405

Yenice I, Mocan MC, Palaska E et al (2008) Hyaluronic acid coated poly-epsilon-caprolactone nanospheres deliver high concentrations of cyclosporine A into the cornea. Exp Eye Res 87:162–167

Wadhawa S, Paliwal R, Paliwal SR et al (2010) Hyaluronic acid modified chitosan nanoparticles for effective management of glaucoma: development, characterization, and evaluation. J Drug Target 18:292–302

Wei C, Nagai T, Wei W et al (2007) New advances in nanomedicine: diagnosis and preventive medicine. Med Clin North Am 91:871–879

Yamamoto S, Manabe N, Fujioka K et al (2007) Visualizing vitreous using quantum dots as imaging agents. IEEE Trans Nanobioscience 6:94–98

Yanagisawa R, Takano H, Inoue KI et al (2009) Titanium dioxide nanoparticles aggravate atopic dermatitis-like skin lesions in NC/NGA mice. Exp Biol Med (Maywood) 234:314–322

Yanagisawa R, Takano H, Inoue KI et al (2010) Size effects of polystyrene nanoparticles on atopic dermatitis-like skin lesions in NC/NGA mice. Int J Immunopathol Pharmacol 23:131–141

Yoo SM, Kang M, Kang T et al (2013) Electrotriggered, spatioselective, quantitative gene delivery into a single cell nucleus by Au nanowire nanoinjector. Nano Lett 13(6):2431–2435

Yorek MA (2003) The role of oxidative stress in diabetic vascular and neural disease. Free Radic Res 37:471–480

Yuan XB, Yuan YB, Jiang W et al (2008) Preparation of rapamycin-loaded chitosan/PLA nanoparticles for immunosuppression in corneal transplantation. Int J Pharm 349:241–248

Zarbin MA (2004) Current concepts in the pathogenesis of age-related macular degeneration. Arch Ophthalmol 122:598–614

Zimmer A, Mutschler E, Lambrecht G et al (1994) Pharmacokinetic and pharmacodynamic aspects of an ophthalmic pilocarpine nanoparticle-delivery-system. Pharm Res 11:1435–1442

Zolnik BS, González-Fernández A, Sadrieh N et al (2010) Minireview: nanoparticles and the immune system. Endocrinology 151:458–465

Ever-Expanding Application Potentials for Iron-Based Nanomaterials: Catalyses and Biomedicine

Suzana Gotovac Atlagić and Verica Pavlić

Abstract Iron oxide nanoparticles are iron-based nanomaterials that are, to date, successfully used in various areas of industry and everyday life. The sustainability of iron is a factor which leads to wide variety of research of the iron oxide roles in industrial catalysis and biomedicine. An industrial catalysis and biomedical applications are here connected, because these two groups of applications demand very similar preparation of materials: shape of the particles and their size, the particular and, as uniform as possible, particle porosity. Further, commonly used synthesis methods are outlined, in order to be able to select the preferred synthesis method according to final desired application of the nanoparticles.

Keywords Iron nanoparticles · Iron oxide magnetic nanoparticles · Biomedical application · Catalyses

1 Introduction

Iron is a chemical element with extraordinarily diverse range of roles in nature, as well as applications in technology. It is widely available, being the sixth most abundant element in the Universe, the most abundant element on Earth and also the fourth most abundant element in the Earth's crust. Native iron is rarely found on the surface of the Earth because it tends to oxidize, however its oxides form the primary

S.G. Atlagić (✉)
Faculty of Science, University of Banja Luka, Mladena Stojanovića 2,
78000 Banja Luka, Bosnia and Herzegovina
e-mail: suzana.gotovac.a@unibl.rs

V. Pavlić
Institute of Dentistry, Zdrave Korde 4, 78000 Banja Luka,
Bosnia and Herzegovina

V. Pavlić
Faculty of Medicine, Department of Dentistry, University of Banja Luka,
Bulevar Petra Bojovića 1a, 78000 Banja Luka, Bosnia and Herzegovina

D. Brabazon et al. (eds.), *Commercialization of Nanotechnologies–A Case Study Approach*, DOI 10.1007/978-3-319-56979-6_12

ores. While it makes up about 5% of the total Earth's crust, both the Earth's inner and outer core are believed to largely consist of an iron-nickel alloy (constituting 35% of the mass of the Earth as a whole). Most of the iron in the crust is found combined with oxygen as iron oxide minerals such as hematite (Fe_2O_3) and magnetite (Fe_3O_4) (Emsley 2011). It can be said with certainty that its potential in industry and other human activities is still to be explored.

The sustainability is, without a doubt, the word that matches iron and its oxides very well. The replacement of many technologies with iron compounds should significantly increase worldwide, even for developing countries. Excellent example for this could be the field of industrial catalysis. Namely, during the last decades, transition-metal catalysts, especially based on precious metals such as Pd, Rh, Ir, and Pt have been proven to be efficient for a numerous applications (Plietker and Beller 2011). However, they have very high prices that are, at present, from approximately few hundred and even up to few thousand €/g for Pd and Pt, depending on the quality. Cheaper traditional catalytic metals, such as nickel, copper, chromium or cadmium, are mostly toxic for environment accumulating in water and air, effecting the plant development and some of them are causing diseases, including cancer (Midander et al. 2009; Bernhoft 2013; Hausinger 2013; Matos et al. 2013). Obviously, their application is not sustainable in a long run.

On the other hand, iron is being cheap and available. It is not only just that iron has a broad potential in constructions as a bulk material. Iron compounds are irreplaceable in many technologies, and iron is also a mineral that our bodies need for many functions. For example, iron is a part of haemoglobin, a protein which carries oxygen from our lungs throughout our bodies. Haemoglobin helps muscles store and use oxygen (Aldridge 1996). If the living organisms lack iron, they develop iron deficiency-anaemia. Causes of low iron levels include blood loss, poor diet or an inability to absorb enough iron from foods (Smith 2012). Life, in any form, without iron, is impossible (Aisen and Listowsky 1980). Namely, biocatalytic roles of iron include oxygen activation and transport, reduction of ribonucleotides, electron transfers, peroxide decomposition and phosphoester bond hydrolysis. On the other hand, the same redox properties that makes iron so useful in biocatalysis and oxygen transport, present hazards to the organism, as too much iron can damage the body. Some people have an inherited disease called hemochromatosis that causes too much iron to build up in the body (Greener 2013). Also, taking too many iron supplements can cause iron poisoning. In the environmental studies, iron levels in drinking water are usually set to levels above 0.3 mg/l, at which iron stains laundry and plumbing fixtures. There is usually no noticeable taste at iron concentrations below 0.3 mg/l, although turbidity and colour may develop (World Health Organization 2011). No health-based guideline value is proposed for iron and it is important to know that iron is not of health concern at levels causing acceptability problems in drinking-water. Suggested values are expected to prevent negative, cumulative effects of iron in water to humans and animals who consume the water, to living organisms in surface water and the nature surrounding it, in general.

When talking about iron nanoparticles, the term is used mostly for particles of magnetite (Fe_3O_4), maghemite (γ-Fe_2O_3), hematite (α-Fe_2O_3), zero-valent iron nanoparticles etc. In general, many forms of these iron-based nanomaterials are already being synthesized on the industrial scale and are used in various areas of industry and life (Huber 2005). To date, the most commercialized role of the iron at the nanoscale is in the electronic industry. However, in that area, one most often speaks about the nano-layers of the few atoms thickness or embedded nanowires. These are usually magnetic materials based on iron or its alloys with e.g. Pt or Au. Nano-sized, highly magnetic iron needles are used in removable electronic media, such as USB and other forms of electronic memory devices.

The industrial catalysis is an emerging area for iron nanoparticles, due to obvious economic and already mentioned environmental reasons. Catalysts are an inseparable part of the contemporary industrial-level synthesis, and very extensively used particularly in the organic compounds productions such as: petroleum refining, production of fuels, fertilizers, polymers, pharmaceuticals. If the catalysts are used, these processes become more efficient in the sense of energy consumption, and total product output consequently is more environmentally friendly. The industry is in constant quest for new catalysts which would have increased activity, last longer, and, if possible be least toxic for environment. For iron particles (mostly micro particles until recently), catalyses applications included as typical ones the Fischer-Tropsch synthesis or ammonia synthesis. In these processes, iron catalyst is traditionally used but iron nanoparticles have far stronger catalytic activities (Pour et al. 2010). The same trend, from micro iron to nano iron particles, is found in the area of catalytic degradation of toxic molecules in water and these promising applications are also discussed.

Considering all of these facts, the subject of catalytic applications of iron nanoparticles in various new industrial areas and its biomedical applications are emphasized here. An industrial catalysis and biomedical applications are shown together, because these two groups of applications demand very similar preparation of materials: shape of the particles and their size, the particular and as uniform as possible porosity. However, the final desired application is directing the optimal synthesis technique. Present chapter will try to show that these are two areas of applications which could even surpass iron and iron oxide nanoparticles most widespread use-nano electronics.

2 Synthesis Methods

Sixteen pure phases of iron oxides, i.e. oxides, hydroxides or oxy-hydroxides are known to date: $Fe(OH)_3$, $Fe(OH)_2$, $Fe_5HO_8 \cdot 4H_2O$, Fe_3O_4, FeO, five polymorphs of FeOOH and four of Fe_2O_3. Characteristics of these oxide compounds include mostly the trivalent state of the iron, low solubility and brilliant colours (Schwertmann and Cornell 2008). Methods for the synthesis of iron oxides in nano-dimension are still explored widely and focus is, as in most inorganic nano-synthesis, on obtaining the right morphology, crystallinity, overall dimensions uniformity etc. Bellow is the short review of the methods for preparation of various iron and iron oxide nanoparticles.

2.1 Chemical Precipitation

This is the most widely available and applicable method for synthesis of the iron oxide nanoparticles. In general case, the alkali such as NaOH or NH_4OH is added drop by drop to the solution containing the iron salt solution. Iron salt can be e.g. ferric nitrate, ferric or ferro chloride or sulphate. Precipitation occurs and the aging of the solution is necessary. Although large quantities of the nanoparticles can be easily obtained, due to difficulties in regulating the uniformity of the particles, numerous solutions are offered. Typically, the suggestions on regulating the pH, ionic strength, temperature and nature of the salts, as well as, the Fe^{2+}/Fe^{3+} ratio are offered (Thapa et al. 2004; Tominaga et al. 2006).

2.2 Sol-Gel Techniques

Not only iron, but also other metal oxides can be prepared by means of the sol-gel route. Sol-gel route is a chemical, wet method often used for preparation of the inorganic/organic hybrid materials. This group of methods usually offers advantages such as homogeneity of the mixture, possibility for good process control of materials formation and relatively low temperatures. However, aging of the sol-gel (mixture which contains hydrolyzed iron ions) is sometimes energy and time consuming process. For example, Pascal et al. (1999) reported necessity for few days to several weeks of aging for their method, in which fresh ferric solution is poured into the concentrated NaOH or KOH solution (and held so long at 60–80 °C). Also, forced hydrolysis of the ferric chloride solution in the presence of the phosphate and hypophosphate, gave colloidal hematite nanoparticles of narrow size distribution, but demand keeping the mixture at 100 °C for as long as 2–7 days (Ozaki et al. 1984). All of these conditions are significantly influencing the final price and demand the final use to be carefully considered before commercialization.

2.3 Organic Iron Compounds Thermal and Sonochemical Degradation

Thermal and sonochemical degradations of iron compounds are based on similar transformations of the organic molecules containing iron. By applying thermal or sonic energy, the molecules are heated and decomposed resulting in formation of the iron or iron oxide nanoparticles. These methods are known to use iron pentacarbonyl $Fe(CO)_5$ or iron (III) acetylacetonate $Fe(C_5H_7O_2)_3$ as precursor, however, the uniformity of the particles has to be farther developed.

2.4 *Ultrasonic Spray Pyrolysis*

Among a variety of approaches, the utilization of ultrasound for materials synthesis has been extensively examined over many years, and is now positioned as one of the most powerful tools in nanostructured materials synthesis (Bang and Suslick 2010). In that field, an ultrasonic spray pyrolysis (UPS) gives nanoparticles of a wide range of composition and phase content both iron and other metals or their combinations (Ebin and Gurmen 2011; Gurmen et al. 2009). There are many technical solutions, some of them even commercialized. In general, the precursor solution is ultrasonically atomized before the fine-sized droplet mist goes into the furnace. There, the thermal decomposition of the material takes place. USP method also provides the possibility of preparing spherical, non-agglomerated and dense particles. Other advantages can be listed as low cost, narrow particle size distribution, and homogeneous composition. Even if the nano-crystalline iron particles are prepared using the same precursor, such as aqueous solution of iron (II) chloride tetrahydrate ($FeCl_2 \cdot 4H_2O$), particle size, morphology, and crystalline size depend on factors such as solution concentration and the reaction temperature (Fig. 1). In general it could be said that increased temperature with higher reaction temperatures (e.g. increasing from 600–1000 °C) gives particles with smaller sizes and higher crystallinity.

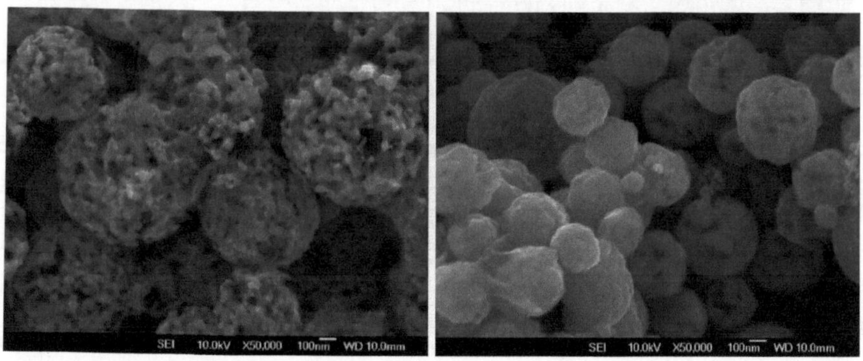

Fig. 1 Iron nanoparticles obtained by USP at 600 °C with 0.2 M iron chloride solution (*left*) and at 1000 °C with 0.08 M solution (*right*). (Reprinted with permission from Ebin and Gurmen 2011)

2.5 *Iron Mining Waste Reuse for Nanoparticle Synthesis*

Recently, scientists from Banja Luka in Bosnia and Herzegovina (B&H) have shown that there is a possibility to synthesize various iron nanoparticles from the waste sludge of the mining industry (Gotovac Atlagić et al. 2014; Stević et al. 2014). To date, nanoparticles with highly nanoporous surface, as well as, by contrast, compact round-shaped, but highly crystalline particles have been synthesized by this group (Fig. 2). B&H possesses many hotspots of this nature, namely the mining industry waste water accumulation lakes, as shown in study supported by JICA (Hydro-Engineering Institute Sarajevo 2014). The method for preparation was chemical precipitation combined with the microemulsion, however, the raw material source is unique in the literature up to date.

As for the iron, the most promising one lies near Prijedor city, with its iron mine of estimated 60 million tons of still unrecovered iron. However, it is important to notice that these results are opening the doors for studying the accumulation lakes of, for example, zincs mining industry, as well. Zinc oxide nanomaterials are still the most widely commercialized metal nanoparticles and would be desirable to find cheap new raw materials. If proven economically acceptable, this technique could help out the resolution of environmental problems that especially China, Brazil and India are facing today. Having many mining hotspots, these countries are even at some disastrous releases of mining waste water into the rivers directly, facing the direct need for treatment and protection of wider water sources. Even more worth the effort is the economical necessity of much more effective use of already excavated ore.

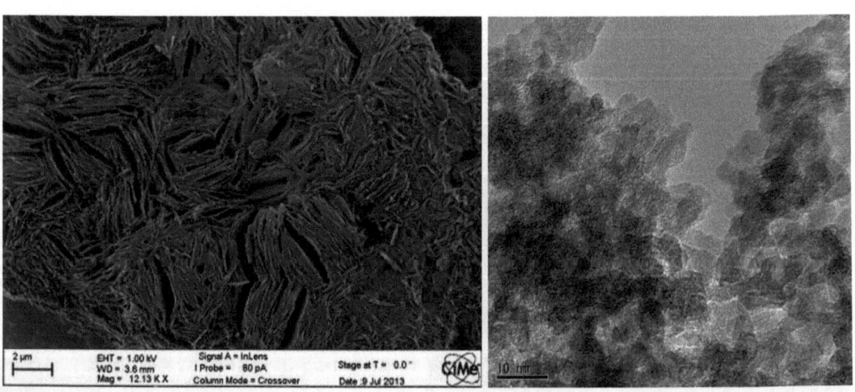

Fig. 2 Amorphous (*left*) and crystalline (*right*) iron nanoparticles synthesized from the iron mining sludge as the precursor (Gotovac Atlagić et al. 2014; Stević et al. 2014)

3 Industrial Catalysys and Industrial Pollution Related Application of Iron Nanoparticles

To start the section on industrial and related catalyses by iron nanoparticles, a representative process of synthesis of ammonia was chosen. With Haber-Bosch process industry effectively separates nitrogen which can then be used in large quantities for production of urea, nitric acid and, extremely importantly-the artificial fertilizers. This process is nearly a hundred years old, discovered in the beginning of the 20th century by Fritz Haber and developed for industrial production by Carl Bosch. The reaction between gaseous nitrogen and hydrogen forming an ammonia molecule is exothermic, carried out at high pressures and temperatures when occurs. However, the Haber with colleagues also circulated the reactant gasses over an iron catalyst contained in a hot tube and then washed out the ammonia at normal temperature (Leigh 2004).

The iron catalyst in this process, most often consists of reduced magnetite ore, the work of which is enhanced with aluminium, potassium, calcium, magnesium or silica oxides, and the reaction temperatures go to 400 °C, with pressure between 150 and 300 Bars. Usually, the Haber-Bosch process to produce ammonia relies on carbon-intensive steam reforming of methane as a hydrogen source (Twigg 1989). Recently, an electrochemical pathway in which ammonia is produced by electrolysis of air and steam in a molten hydroxide suspension of nano-Fe_2O_3 was presented (Licht et al. 2014). In this method, at 200 °C in an electrolyte with a molar ratio of 0.5 NaOH/0.5 KOH, ammonia is produced at 1.2 V under 2 mA/cm^2 of applied current at coulombic efficiency of 35% (35% of the applied current results in the six-electron conversion of N_2 and water to ammonia, and excess H_2 is cogenerated with the ammonia). Other conditions presented there are 250 °C and 25 bar of steam pressure, where the electrolysis voltage necessary for 2 mA/cm^2 current density decreased to 1.0 V. This shows a promising energy-efficiency of such process.

Between the most important industrial catalytic applications of iron nanoparticles, the Fischer-Tropsch synthesis (FTS), a process for the conversion of syngas (H_2/CO) derived from carbon sources such as: coal, peat, biomass, and natural gas into hydrocarbons and oxygenates (Fig. 3). FTS produces a complex multicomponent mixture of linear and branched hydrocarbons and oxygenated compounds. Fuels produced by the FTS are of a high quality, due to a very low aromatic and zero sulfur content (Van Der Laan and Beenackers 1999). Considerable progresses have been made in the past two decades on the development of more active and selective cobalt and iron catalysts and more effective reactor/process technologies. Iron based catalyst systems have remained a preferred choice in commercial FTS plants due to low cost and tendency to yield high amounts of olefines in hydrocarbon distribution (Pour et al. 2010). But, iron is not stable during the FTS; where four different species of iron have been detected including iron carbide (Fe_xC), magnetite (Fe_3O_4), hematite (Fe_2O_3), and metallic phase (Fe) (Niemantsverdriet

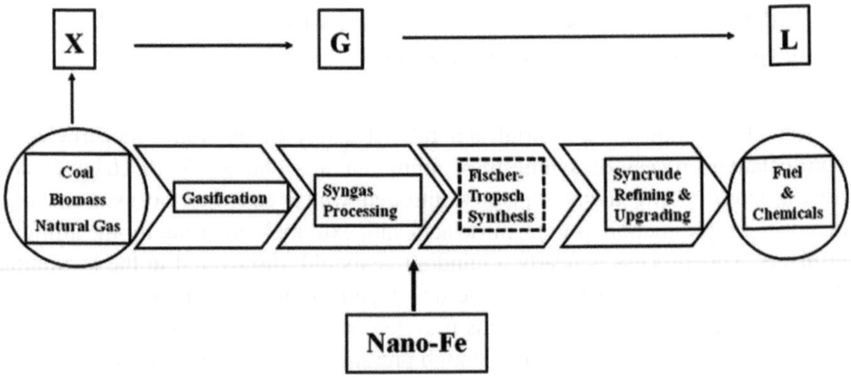

Fig. 3 Simplified scheme of the Fischer-Tropsch synthesis work flow

et al. 1980). It is becoming clear, however, that nano-sizing the iron catalyst is the key to achieving better results in FTS effectiveness.

It could be clearly understood again from the study, in which nanostructured catalysts were prepared in two series, mixing iron with other nanoparticles. There, firstly, Fe_2O_3, CuO and La_2O_3 nanoparticles were prepared separately by microemulsion method and then mixed together (Pour et al. 2010). In another method by same authors, iron and Cu oxides were prepared by coprecipitation microemulsion method. Catalyst evaluation was conducted in a fixed-bed stainless steel reactor and compared with conventional iron catalyst. It was found that the catalytic activity in FTS reaction was increased with decreasing the crystal size of the catalyst from conventional to nano catalysts. These results contribute higher amount of ε-carbide (Fe2.2C) rather than Hägg carbide (χ-Fe2.5C) and cementite (θ-Fe3C) in iron nano-particles after pre treatment in nanostructured iron catalysts. The olefin/paraffin ratios are decreased from conventional to nano-catalysts because of higher H_2 and lower CO concentration on nano-catalyst surface than conventional catalyst.

One more nanosized iron oxide for, the magnetic Fe_3O_4 has also shown successful yield in catalysis of the alcohol-carbonyl compound (mostly aldehyde) conversion (Sadri et al. 2014). The oxidation of 19 benzylic and aliphatic alcohols by peroxide in presence of magnetic Fe_3O_4 nanoparticles was studied. It has been shown that yields of aldehyde by this method are over 80%, with somewhat longer oxidation times for aliphatic alcohols. This is commercially attractive method since alcohols are convenient raw materials for production of aldehydes, ketones and carboxylic acids. However traditional methods use reagents such as chromium, permanganate or N-chlorosuccinimide, which are all toxic or endangering the environment in some way. Method using the peroxide oxidation enhanced with magnetic Fe_3O_4 nanoparticles has water as the only by-product and very importantly the catalyst nanoparticles show recyclability. Sadri et al. show very particular test results for this. They have performed the catalyst separation from the product by external magnet, washed the powder by acetone and water and then studied their reuse (Fig. 4). Study demonstrated excellent activity and recyclability of the catalyst which adds to the value of this process as the "green" method.

Next, the catalysis by iron oxide nanomaterials to remove toxic ions and organic pollutants from water is discussed. As nicely reviewed by Savage in 2005, the development of future nanosorbents, nanocatalysts, bioactive nanoparticles, catalytic membranes with nanostructures could even more contribute to the solutions of the increasing water pollution problems around the world because nano materials continue to demonstrate larger removal or toxic molecules degradation capabilities than bulk materials. The flower-like structures synthesized by ethylene glycol-mediated self-assembly process (Zhong et al. 2006), followed by calcinations at elevated temperatures, have shown significant removal capacity for As and Cr (Fig. 5). These ions belong to primary pollutants of the drinking water in some countries. In the same work, the adsorption removal capacity of the organic azo-dye, Orange II was studied. It is particularly significant that after the dye

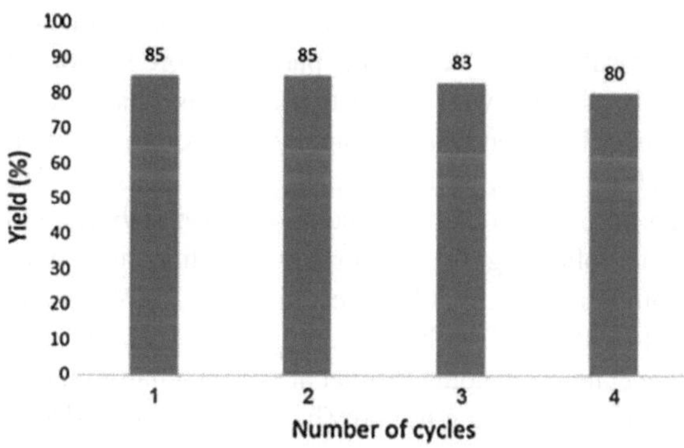

Fig. 4 Recycling of the catalyst for oxidation of 4-bromobenzyl alcohol (Reprinted with permission from Sadri et al. 2014)

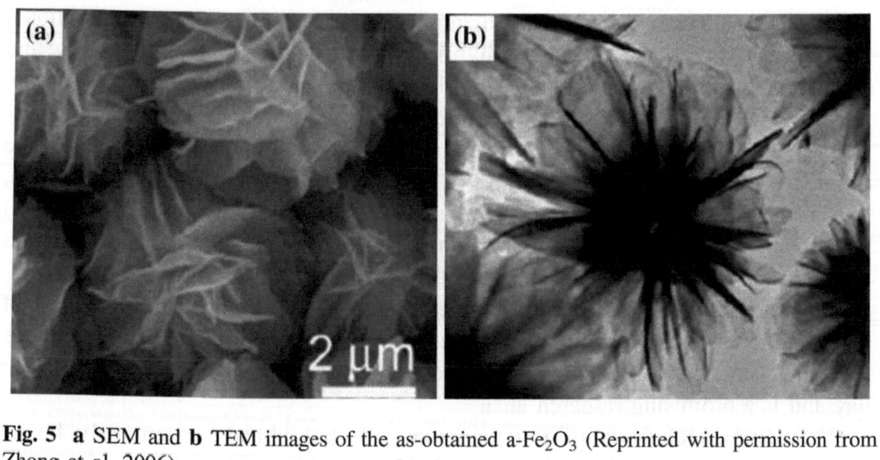

Fig. 5 **a** SEM and **b** TEM images of the as-obtained a-Fe_2O_3 (Reprinted with permission from Zhong et al. 2006)

adsorption, the regeneration by catalytic combustion was possible, with almost completely same adsorption capacity for the same removal. The iron-oxide form of these particles, whose adsorption and catalytic capacity is probably significantly improved by their high surface area, was the α-Fe_2O_3.

Before these new materials arrived, zero-valent iron has been used in the remediation of contaminated aquifers for over two decades. It acts as a reducing agent to transform contaminants, such as chlorinated solvents. In the process, iron is consumed and does not act as a catalyst itself (Comba et al. 2011). However, after the increased potential of nanoparticles comparing to traditionally used millimetre particles have been recognized, numerous other forms of modified nanosized zero-valent iron (nZVI) have been reported. Modifications can be done in several ways: (1) Bimetallic nanoparticles (BNPs) are particles of nZVI that have been coated with a catalyst, such as platinum, gold, nickel, or palladium, and these are the coatings which enhance their performance in reduction chemical processes; (2) Emulsified zero-valent iron (eZVI) are nZVI particles that have been coated with a membrane made from biodegradable oil and water to facilitate the treatment of chlorinated hydrocarbons by making the particles more hydrophobic. This allows the particles to mix directly with dense non-aqueous phase liquids (DNAPL) to increase mass transfer between DNAPL and nZVI through the emulsion membrane (Berge and Ramsburg 2009), and (3) finally, nZVI particles can be covered in different polymers in order to increase the suspension stability and therefore, particle mobility. The following list of surface modifiers is reported as effective: xanthan gum (Comba and Sethi 2009; Vecchia et al. 2009), polyacrylate (Schrick et al. 2004; Kanel et al. 2007), triblock copolymers (Saleh et al. 2008), polyvinyl alcohol-co-vinyl acetate-co-itaconic (Sun et al. 2007), guar gum (Tiraferri and Sethi 2009), carboxymethyl cellulose (He and Zhao 2007), starch (He and Zhao 2005), and poly (4-styrenesulfonate) (Hydutsky et al. 2007).

Particularly attractive in catalysis is the possibility to coat iron nanoparticles, which are obviously much cheaper than Pd, Pt and other proven catalytic materials, which are sometimes few hundred times more expensive than the iron-based ones on the market. However, not only mentioned coatings, but also some additional coatings can give such iron/metal hybrid particles even more applicability and effectiveness. For example, these hybrid nanoparticles, enhanced with certain polyelectrolytes have proven to be more mobile than bare nZVI. The polyelectrolytes slow the deposition of particles by reducing the magnetic attraction of the nZVI to each other and soil grains. nZVI coated with certain polyelectrolytes (e.g., polyasarate-PAP, carboxymethyl cellulose-CMC and polystyrene sulfonate-PSS) can remain mobile for at least eight months after the original injection depending on the hydrochemistry and geochemistry found at a site (Kim et al. 2009) (Fig. 6).

Obviously, the iron oxides with much stable nature should be a material of choice. Therefore the development of their functionality by changing the porous structure, particle size, magnetic properties is one of the challenges of the present day and is a promising research area.

The commercialization of iron nanomaterials has already extensively begun recent years. A wide variety of the particles is available on the world market. The

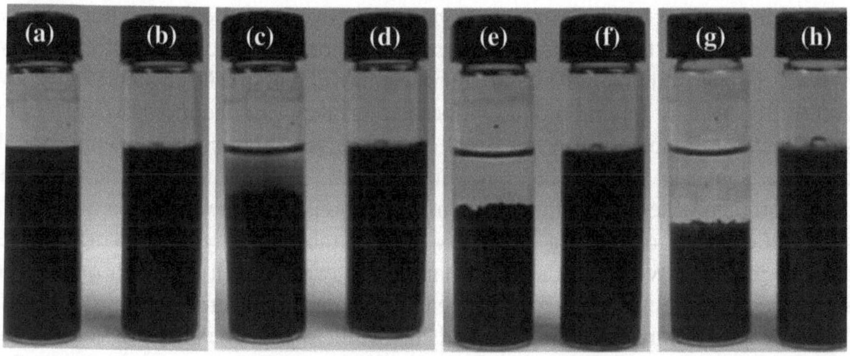

Fig. 6 Vials containing nano-zero-valent iron (nZVI) and stabilized nano-zero-valent iron (SnZVI) at different times: **a**, **b**-1 min; **c**, **d**-10min; **e**, **f**-2 h; **g**-nZVI 2 days and **h**-SnZVI 60 days (Reprinted with permission from Kanel 2008)

products most often include crystalline forms of magnetite, maghemite, hematite and pure zero-valent iron. Particle size and purity is determining the prices. Since it is still not possible to synthesize the particles of perfectly uniform size, the products are arranged by average particle size in classes such as: 20–40, 40–60, and 60–90 nm when the size distribution is wide, or sometimes purely one value for average particle size such as 92 nm. Prices range from 0.38 €/g for a maghemite of 99% purity, up to 2.5 €/g for a 99.5% hematite nanoparticles. It is the narrow size distribution (uniform sizes) that increases the price for the same mineral, not the size itself. It is still a big task as with most of the nanomaterials, to achieve the high levels of repeatability and reproducibility in their characterization process.

With industrial catalysis these tasks are focused on physic-chemical characteristics of the particles which lead to their specific functionalities. On the other hand, with application of iron nanomaterials in biomedical area, the focus is, beside these characteristics, also on the toxicology.

4 Biomedical Applications of Iron Nanoparticles

To be biomedically applicable, the iron nanoparticles have to be non-toxic, and to have specific properties, such as shape and surface, average size, crystallinity, magnetic properties, dispersibility, etc. Nanoparticles must also be made of a non-immunogenic material, with particle size small enough to pass through the capillary systems without causing an embolism (Akbarzadeh 2012). They must have a high magnetization in order to be easily monitored with magnetic field and immobilized to the targeted tissues. The most attractive iron nanoparticles in medicine are superparamagnetic iron oxide nanoparticles (SPIONs), which are considered as novel approach in the area of nanomedicine. SPIONs core is usually based on water insoluble iron oxide crystals, usually magnetite (Fe_3O_4) or maghemite (γ-Fe_2O_3), with 50–100 nm in diameter size. SPIONs are characterized with superparamagnetism, a type of magnetism in which magnetic properties of material can be influenced by external magnetic field, either by physical rotation (the Brownian effect) or by moving the magnetic moment without particle movement (the Néel effect) (Smirnov 2009). SPIONs are having unique properties: (a) they provide the most change in signal (albeit hypointensity) per unit of metal, (b) their core is composed of biodegradable iron, which is biocompatible and thus can be reused/recycled by cells using normal biochemical pathways for iron metabolism, (c) their surface coating can be modified with organic (e.g. polymers like dextran, polyethylene glycol) or inorganic (e.g. gold, silver, silica, apatite) materials or even growth factors, hormones or antibodies, (d) they can be easily detected by light and electron microscopy, and (e) they can be controlled by magnet and change their magnetic properties according to size. To date numerous studies on SPIONs have been conducted, leading to the development of many novel applications and improvements on existing technology using SPIONs, particularly in the field of nanomedicine. SPIONs are successfully used in magnetic resonance imaging (MRI), cellular labelling, hyperthermia and drug delivery (Huber 2005; Gupta and Gupta 2005; Okoli et al. 2011). SPIONs can be either fabricated or commercially procured at present.

4.1 Magnetic Resonance Imaging (MRI)

MRI is non-invasive, non-ionizing imaging technique that became a standard procedure for the detection, diagnosis and advanced interpretation of various types of diseases and medical conditions. Along with the development of modern imaging technologies, a major issue being studied in MRI is the development of the optimal contrast agents for both, clinical applications and scientific research (Na et al. 2009). To date, the most commonly used MRI contrast agents were gadolinium-containing MRI contrasts, which contain a rare metal gadolinium. Since gadolinium is highly toxic and gadolinium-containing MRI contrasts suffer from an inherent high threshold of detectability, the new MRI contrasts have been sought. The introduction of SPIONs opened up a whole new world of magnetic particle imaging (MPI) as a novel tomographic imaging modality, where SPIONs are used as tracer material (Lindemann et al. 2014). The advantages of MPI over current imaging techniques are high spatial resolution and sensitivity, and the potential for high quality real-time imaging. Furthermore, a good signal-to-noise ratio and excellent contrast confirm that MRI is promising imaging modality able to increase the accuracy and the specificity of imaging (Neumaier et al. 2008; Materia et al. 2015).

4.2 Cellular Labelling

Cell labelling by SPIONs has emerged as a potentially powerful tool to monitor trafficking of a large number of cells (e.g. progenitor and mesenchymal stem cells, dendritic cells, natural killer cells, etc.) in a broad spectrum of applications. Since cells labelled with SPIONs possess magnetic and contrast properties, their migration can be non-invasively detected and monitored with a MRI. Cell labelling is basically performed by intravenous injection of long-circulating SPION based MRI contrast agents, which can extravasate and/or undergo a cellular uptake in an amount sufficient to allow an MRI visualization of areas of interest (e.g. inflamed regions or tumours) (Laurent et al. 2009). Available literature suggest that labelled cells are of normal cell surface phenotype and with preserved differentiation potential, suggesting that SPION labelling do not produce detrimental effects on the cells. However, in order for cell labelling to have a widespread use clinically, further studies are necessary. Theoretically, information gained from cell-labelling tracking could be optimized and specifically tailored for treatments of each and every patient.

4.3 Magnetic Hyperthermia (MHT)

In 1957, Gilchrist et al. first used magnetic particles to heat tissues with a magnetic field and opened up the new era of magnetic hyperthermia (MHT). MHT is therapeutic procedure where SPIONs transform the electromagnetic energy into heat during the exposure to an alternating magnetic field (AMF) (Dutz and Hergt 2014; Hilger 2013). Further, heat allows the temperature increase (above 42 °C) in the specific regions in the body where the temperature sensitive tumour cells and nanoparticles are located. So far, this basic principle was mainly used in ex vivo experiments in animal models and some clinical trials treating a prostate cancer and various gliomas and demonstrating some benefit in controlling tumour growth. Recently, MHT has been proposed as a strategy to improve killing of the cytotoxic T lymphocytes of HIV-infected cells and for targeting the HIV latent reservoirs (Williams et al. 2013). The efficiency of the MHT technique is extremely dependent on the variable properties of the particles, such as size, magnetization, amount and the value of the Curie temperature (the maximum temperature reached by a magnetic particle). Further the different parameters of the magnetic field, such as intensity and frequency that will achieve the desired temperature are of great importance. However, to date, the most challenging key factors determining the therapeutic outcome are the amounts of magnetic material and their intratumoral distribution patterns (Dutz and Hergt 2014; Hilger 2013). An improvement of above mentioned parameters is necessary in order to optimize commercially available formulations and consequently MHT. Only optimized MHT can be successfully incorporated in everyday therapy for cancer, either as alone or adjunctive therapy approach.

4.4 Drug Delivery

SPIONs as highly controllable carriers able to transport and release very precise amount of drugs are one of the main challenge in nanomedicine. An efficiency of drug delivery systems is strongly related to neutrality to immune system, ability to carry the drug to the targeted place and to regulate the optimum amount of the drug to be released. One example is SPIONs incorporated into thermoactive polymers or hydrogels, which can reversibly shrink and expand as a result of the temperature increase induced via AMF.

Overall, in order for SPIONs to have a successful clinical application several issues need to be overcome (e.g. low drug loading efficiency, failure of SPIONs to enter the tumour cells from the blood, lack of the magnetization/magnet cannot come into close proximity, toxicity and biodegradation, financial issues concerning production). Therefore, the improvement of established SPIONs with the improved characteristics regarding diffusion through tissues and biocompatibility or the search for the new synthetic routes is necessary. The prices of, so far, commercially available iron nanoparticles for medical applications are reasonably high. As one can expect the guaranty for medical safety, most importantly, confirmed absence of the toxic metals content, will increase the price. Further, the price of the particles will depend on their size; consequently, the price of the smaller particles will be higher.

5 Conclusions

The cost effectiveness of any iron or iron oxide nanoparticles preparation procedure would depend on the quality of the final product and its potential application. Since the nanomedicine requires nanoparticles of the highest quality, comparatively higher cost of production will be tolerable. On the other hand, when the iron or iron oxide nanoparticles as products are to be utilized in industrial synthesis as a catalyst or in catalytic degradation for elimination of toxic substances from waste waters, it is essential to conduct their production with as reasonable prices as possible. Also, use of low cost chemicals during synthesis is essential for making it economical. Therefore, the ideas such as using the iron deposits in the form of mining waste sludge sound promising and sustainable.

As shown, this family of the nanoparticles carries great potential as economical and environmentally-friendly materials with diverse applications. Future nanotechnologies are going to count on iron more and more.

References

Aisen P, Listowsky I (1980) Iron transport and storage proteins. Annu Rev Biochem 49(1): 357–393

Akbarzadeh A, Samiei M, Davaran S (2012) Magnetic nanoparticles: preparation, physical properties, and applications in biomedicine. Nanoscale Res Lett 7(1):144

Aldridge S (1996) The thread of life: the story of genes and genetic engineering. Cambridge University Press, Cambridge

Bang JH, Suslick KS (2010) Applications of ultrasound to the synthesis of nanostructured materials. Adv Mater 22(10):1039–1059

Berge ND, Ramsburg CA (2009) Oil-in-water emulsions for encapsulated delivery of reactive iron particles. Environ Sci Technol 43(13):5060–5066

Bernhoft RA (2013) Cadmium toxicity and treatment. Sci World J. doi:10.1155/2013/394652

Comba S, Sethi R (2009) Stabilization of highly concentrated suspensions of iron nanoparticles using shear-thinning gels of xanthan gum. Water Res 43(15):3717–3726

Comba S, Di Molfetta A, Sethi R (2011) A comparison between field applications of nano-, micro-, and millimetric zero-valent iron for the remediation of contaminated aquifers. Water Air Soil Pollut 215(1–4):595–607

Dutz S, Hergt R (2014) Magnetic particle hyperthermia—a promising tumour therapy? Nanotechnology 25(45):452001

Ebin B, Gürmen S (2011) Aerosol synthesis of nano-crystalline iron particles from iron (II) chloride solution. Metall 65(4):151–154

Emsley J (2011) Nature's building blocks: an AZ guide to the elements. Oxford University Press, Oxford

Gotovac Atlagić S, Malina J, Mionić Ebersold M (2014) From mud to bud-recovering Bosnian forgotten iron. In: Proceedings from the 8th European waste water management conference and exhibition, Aqua Enviro, Manchester, 7–8 Oct 2014

Greener M (2013) Coping with liver disease. Sheldon Press, London

Gupta AK, Gupta M (2005) Synthesis and surface engineering of iron oxide nanoparticles for biomedical applications. Biomaterials 26(18):3995–4021

Gurmen S, Guven A, Ebin B et al (2009) Synthesis of nano-crystalline spherical cobalt–iron (Co–Fe) alloy particles by ultrasonic spray pyrolysis and hydrogen reduction. J Alloy Compd 481:600–604

Hausinger RP (2013) Biochemistry of nickel. Springer Science & Business Media, New York

He F, Zhao DY (2005) Preparation and characterization of a new class of starch-stabilized bimetallic nanoparticles for degradation of chlorinated hydrocarbons in water. Environ Sci Technol 39:3314–3320

He F, Zhao DY (2007) Manipulating the size and dispersibility of zerovalent iron nanoparticles by use of carboxymethyl cellulose stabilizers. Environ Sci Technol 41:6216–6221

Hilger I (2013) In vivo applications of magnetic nanoparticle hypertermia. Int J Hyperthermia 29 (8):828–834

Huber DL (2005) Synthesis, properties, and applications of iron nanoparticles. Small 1(5):482–501

Hydro-Engineering Institute Sarajevo (2014) The project for master plan for remediation of hotspots in Bosnia and Herzegovina. Project report, JICA Sarajevo

Hydutsky BW, Mack EJ, Beckerman BB et al (2007) Optimization of nano- and microiron transport through sand columns using polyelectrolyte mixtures. Environ Sci Technol 41: 6418–6424

Kanel SR, Goswami RR, Clement TP et al (2008) Two dimensional transport characteristics of surface stabilized zero-valent iron nanoparticles in porous media. Environ Sci Technol 42 (3):896–900

Kim HJ, Phenrat T, Tilton RD et al (2009) Fe^0 nanoparticles remain mobile in porous media after aging due to slow desorption of polymeric surface modifiers. Environ Sci Technol 43: 3824–3830

Laurent S, Boutry S, Mahieu I et al (2009) Iron oxide based MR contrast agents: from chemistry to cell labeling. Curr Med Chem 16(35):4712–4727

Leigh GJ (2004) The world's greatest fix: a history of nitrogen and agriculture. Oxford University Press, New York

Licht S, Cui B, Wang B et al (2014) Ammonia synthesis. Ammonia synthesis by N_2 and steam electrolysis in molten hydroxide suspensions of nanoscale Fe_2O_3. Science 345(6197):637–640

Lindemann A, Lüdtke-Buzug K, Fräderich BM et al (2014) Biological impact of superparamagnetic iron oxide nanoparticles for magnetic particle imaging of head and neck cancer cells. Int J Nanomed 29(9):5025–5040

Materia ME, Guardia P, Sathya A et al (2015) Mesoscale assemblies of iron oxide nanocubes as heat mediators and image contrast agents. Langmuir 31(2):808–816

Matos RC, Bessa M, Oliveira H et al (2013) Mechanisms of kidney toxicity for chromium- and arsenic-based preservatives: potential involvement of a pro-oxidative pathway. Environ Toxicol Pharmacol 36(3):929–936

Midander K, Cronholm P, Karlsson HL et al (2009) Surface characteristics, copper release, and toxicity of nano- and micrometer-sized copper and copper (II) oxide particles: a cross-disciplinary study. Small 5(3):389–399

Na HB, Song IC, Hyeon T (2009) Inorganic nanoparticles for MRI contrast agents. Adv Mater 21:2133–2148

Neumaier CE, Baio G, Ferrini S et al (2008) MR and iron magnetic nanoparticles. Imaging opportunities in preclinical and translational research. Tumori 94(2):226–233

Niemantsverdriet JW, Van der Kraan AM, Van Dijk WL et al (1980) Behavior of metallic iron catalysts during Fischer-Tropsch synthesis studied with Moessbauer spectroscopy, X-ray diffraction, carbon content determination, and reaction kinetic measurements. J Phys Chem 84 (25):3363–3370

Okoli C, Boutonnet M, Mariey L et al (2011) Application of magnetic iron oxide nanoparticles prepared from microemulsions for protein purification. J Chem Technol Biotechnol 86: 1386–1393

Ozaki M, Kratohvil S, Matijević E (1984) Formation of monodispersed spindle-type hematite particles. J Colloid Interface Sci 102(1):146–151

Pascal C, Pascal JL, Favier F et al (1999) Electrochemical synthesis for the control of γ-Fe2O3 nanoparticle size. Morphology, microstructure, and magnetic behavior. Chem Mater 11 (1):141–147

Plietker B (ed) (2011) Iron catalysis: fundamentals and applications, vol 33. Springer Science & Business Media, Stuttgart

Pour AN, Housaindokht MR, Tayyari SF et al (2010) Fischer-Tropsch synthesis by nano-structured iron catalyst. J Nat Gas Chem 19(3):284–292

Sadri F, Ramazani A, Massoudi A et al (2014) Green oxidation of alcohols by using hydrogen peroxide in water in the presence of magnetic Fe_3O_4 nanoparticles as recoverable catalyst. Green Chem Lett Rev 7(3):257–264

Saleh N, Kim HJ, Phenrat T et al (2008) Ionic strength and composition affect the mobility of surface-modified FeO nanoparticles in water-saturated sand columns. Environ Sci Technol 42:3349–3355

Savage N, Diallo MS (2005) Nanomaterials and water purification: opportunities and challenges. J Nanopart Res 7(4–5):331–342

Schrick B, Hydutsky BW, Blough JL et al (2004) Delivery vehicles for zerovalent metal nanoparticles in soil and groundwater. Chem Mater 16(11):2187–2193

Schwertmann U, Cornell RM (2008) Iron oxides in the laboratory: preparation and characterization. Wiley, Weinheim

Smirnov P (2009) Cellular magnetic resonance imaging using superparamagnetic anionic iron oxide nanoparticles: applications to in vivo trafficking of lymphocytes and cell-based anticancer therapy. Methods Mol Biol 512:333–353

Smith T (2012) Coping with anaemia. SPCK. Sheldon Press, London

Stević D, Kaneko K, Hattori Y et al. (2014) Precipitation of the highly crystalline iron nanoparticles from the iron mine waste water. In: Proceedings from the international student conference of environmental protection and related sciences applicable environmental protection, University of Novi Sad, Novi Sad, 14–16 Nov 2014

Sun YP, Li XQ, Zhang WX et al (2007) A method for the preparation of stable dispersion of zerovalent iron nanoparticles. Colloids Surf A 308:60–66

Thapa D, Palkar VR, Kurup MB et al (2004) Properties of magnetite nanoparticles synthesized through a novel chemical route. Mater Lett 58(21):2692–2702

Tiraferri A, Sethi R (2009) Enhanced transport of zerovalent iron nanoparticles in saturated porous media by guar gum. J Nanopart Res 11:635–645

Tominaga M, Matsumoto M, Soejima K et al (2006) Size control for two-dimensional iron oxide nanodots derived from biological molecules. J Colloid Interface Sci 299(2):761–765
Twigg MV (1989) Catalyst handbook. Wolf, London
Van der Laan GP, Beenackers AACM (1999) Kinetics and selectivity of the Fischer-Tropsch synthesis: a literature review. Catalysis Rev 41(3–4):255–318
Vecchia ED, Luna M, Sethi R (2009) Transport in porous media of highly concentrated iron micro-and nanoparticles in the presence of xanthan gum. Environ Sci Technol 43(23): 8942–8947
Williams JP, Southern P, Lissina A et al (2013) Application of magnetic field hyperthermia and superparamagnetic iron oxide nanoparticles to HIV-1-specific T-cell cytotoxicity. Int J Nanomed 8:2543–2554
World Health Organization (2011) Guidelines for drinking-water quality, 4th edn. World Health Organization, Geneva
Yang HH, Zhang SQ, Chen XL et al (2004) Magnetite-containing spherical silica nanoparticles for biocatalysis and bioseparations. Anal Chem 76:1316–1321
Zhong LS, Hu JS, Liang HP et al (2006) Self-assembled 3D flowerlike iron oxide nanostructures and their application in water treatment. Adv Mater 18(18):2426–2431

If you have any concerns about our products,
you can contact us on
ProductSafety@springernature.com

In case Publisher is established outside the EU,
the EU authorized representative is:
Springer Nature Customer Service Center GmbH
Europaplatz 3, 69115 Heidelberg, Germany

Printed by Libri Plureos GmbH
in Hamburg, Germany

MIX
Papier aus verantwortungsvollen Quellen
Paper from responsible sources
FSC® C105338

If you have any concerns about our products,
you can contact us on
ProductSafety@springernature.com

In case Publisher is established outside the EU,
the EU authorized representative is:
Springer Nature Customer Service Center GmbH
Europaplatz 3, 69115 Heidelberg, Germany

Printed by Libri Plureos GmbH
in Hamburg, Germany